Protein Geometry, Classification, Topology and Symmetry

A computational analysis of structure

Series in Biophysics

Series Editors: **W Bialek**, Princeton University, USA
C Lowe, University of Cambridge, UK
E Sackmann, Technical University of Munich, Germany

The Series in Biophysics contains monographs and graduate textbooks in all areas of biophysics including bioenergetics, computational biophysics and bioinformatics, modelling, cell biophysics, molecular biophysics including nucleic acids, proteins and supramolecular assemblies, electrophysiology, neurophysiology, membranes, tissue properties, soft condensed matter, photobiophysics, spectroscopy and imaging techniques.

Forthcoming titles in the series

Series in Biophysics

Protein Geometry, Classification, Topology and Symmetry

A computational analysis of structure

William R Taylor and András Aszódi*

Division of Mathematical Biology,
National Institute for Medical Research, London, UK

* *Present address: Novartis Forschungsinstitut GmbH, Vienna, Austria*

Taylor & Francis
Taylor & Francis Group
New York London

Taylor & Francis is an imprint of the
Taylor & Francis Group, an informa business

Published in 2005 by
Taylor & Francis Group
270 Madison Avenue
New York, NY 10016

Published in Great Britain by
Taylor & Francis Group
2 Park Square
Milton Park, Abingdon
Oxon OX14 4RN

Printed in the United States of America on acid-free paper
10 9 8 7 6 5 4 3 2

International Standard Book Number-10: 0-7503-0985-7 (Hardcover)
International Standard Book Number-13: 978-0-7503-0985-1 (Hardcover)

Library of Congress Cataloging-in-Publication Data

Catalog record is available from the Library of Congress

informa

Taylor & Francis Group
is the Academic Division of Informa plc.

Visit the Taylor & Francis Web site at
http://www.taylorandfrancis.com

Contents

Preface **xi**

1 Introduction **1**

1.1 Prologue 1

1.1.1 Scope and aims 1

1.1.2 Why proteins? 2

1.1.3 Outline of the work 4

1.2 Basic principles of protein structure 5

1.2.1 The shapes and sizes of proteins 6

1.2.2 The hydrophobic core 7

1.2.3 Secondary structure 8

1.2.4 Packed layers 9

1.2.5 Barrel structures and β-helices 11

1.2.6 Protein topology 13

1.2.7 Domain structure 15

PART 1
GEOMETRY **17**

2 Ellipsoids and embedding **19**

2.1 Geometric representations of structure 19

2.1.1 Atomic coordinates 19

2.1.2 Torsion angles 22

2.1.3 Distances 26

2.1.4 Inertial axes 27

2.1.5 The shapes of proteins 32

2.2 Distance geometry 34

2.2.1 Out of hyperspace 35

2.2.2 Interpretation of the eigenvalues 40

2.2.3 Hierarchical inertial embedding 41

2.2.4 Gradual projection 45

2.2.5 Practical method specification 47

3 Sticks to strings **49**
3.1 Secondary structure geometries 49
3.1.1 Secondary structure line segments 50
3.1.2 Secondary structure definition 55
3.1.3 Comparison to standard definitions 57
3.1.4 Applications and further developments 61
3.2 Simplified architectures 63
3.2.1 Stick packing 63
3.2.2 Calibrating segment packing 65
3.2.3 Layer architectures 68
3.2.4 Layer-based stick models 70
3.2.5 Polyhedra-based stick models 73
3.2.6 Packing nomenclatures 82
3.2.7 From 3D to 2D 85
3.2.8 From 2D to 1D 87
3.2.9 Uniqueness of string descriptors 87
3.2.10 Predicting helix contacts 89

4 Sheets and barrels **93**
4.1 β-sheet geometry 93
4.1.1 Sheet chirality 93
4.1.2 Geometric models for a twisted sheet 94
4.1.3 The surface of a twisted sheet 99
4.1.4 Sheet bend and curl 111
4.1.5 β-sandwiches 113
4.2 β-barrels 118
4.2.1 Hyperbolic surfaces 118
4.2.2 Shear and stagger 120
4.2.3 Cylindrical β-barrels 121
4.2.4 Hyperbolic β-barrels 125
4.2.5 Optimized barrels 129

PART 2
CLASSIFICATION **131**

5 Networks and domains **133**
5.1 Hydrogen bond networks 133
5.1.1 α-carbon-based β-sheet definition 134
5.1.2 β-sheet dimensions 136
5.1.3 TBC angle distribution 136
5.1.4 β-barrel identification 138
5.1.5 β-sheet classification 140
5.2 Protein domain definitions 143
5.2.1 Physical methods 143

5.2.2 An Ising model 144
5.2.3 Model evolution and domain extraction 146
5.2.4 Conforming to expectation 147
5.2.5 Setting the granularity level 149
5.2.6 Performance and examples 151
5.2.7 Simultaneous definition on multiple structures 155
5.2.8 Probabilistic definitions 156
5.2.9 Reparsing domains 157

6 Protein structure comparison **159**
6.1 Overview of comparison methods 159
6.1.1 Structure representations and degrees of difficulty 160
6.1.2 The `DALI` method 161
6.1.3 Geometric hashing approach 162
6.1.4 Using structural superposition 164
6.1.5 The `SSAP` program 165
6.1.6 Iterated double dynamic programming 167
6.1.7 Secondary structure graph matching 169
6.1.8 Stick-figure comparisons 170
6.2 Assessment of significance 172
6.2.1 Score distributions from known structures 173
6.2.2 Random structural models 174
6.2.3 Randomized alignment models 175
6.2.4 Scoring and biological significance 176
6.2.5 Examples 177

7 Classification and fold spaces **180**
7.1 Protein structure classification 180
7.1.1 Practical approaches to classification 180
7.1.2 Organization of the classifications 182
7.1.3 Analysis of the classifications 183
7.2 Protein fold spaces 185
7.2.1 Distance geometry projection 185
7.2.2 Simplified fold space 187
7.3 A 'periodic table' for protein structures 187
7.3.1 Classification using ideal stick forms 187
7.3.2 Structure layers become valance shells 188
7.3.3 Matching against all stick forms 191
7.3.4 Reintroducing topology 195
7.3.5 Expanding the classification tables 197
7.4 'Evolutionary' steps in fold space 198
7.4.1 Matching ideal forms 200
7.4.2 Largest common fold 200
7.4.3 Trees of structures 203
7.4.4 Links and islands in fold space 207

PART 3
TOPOLOGY **209**

8 Folds, tangles and knots **211**
8.1 Topology and knots 211
8.1.1 Introduction 211
8.1.2 Chemical topology 212
8.1.3 Polymer topology 212
8.1.4 True topology of proteins 214
8.1.5 Pseudo-topology of proteins 216
8.1.6 Topology of weak links in proteins 223
8.1.7 Generalized protein knots 225
8.1.8 Knots in random chains 227
8.2 Random walks in fold space 227
8.2.1 Random walks 229
8.2.2 Secondary structure based fake proteins 230
8.2.3 Off-lattice fold combinatorics 232
8.2.4 Classifying topology in fake proteins 236
8.2.5 Local *versus* global folding 237
8.3 Protein fold complexity 238
8.3.1 Topological indices 238
8.3.2 Local and non-local packing 243
8.3.3 Smoothing folds away 248

9 Structure prediction and modelling **249**
9.1 Random folds from distance geometry 249
9.1.1 Outline of the projection strategy 249
9.1.2 Model specification 250
9.1.3 Hydrogen bonds 252
9.1.4 Chirality 255
9.1.5 Scoring 255
9.1.6 Simple models 255
9.2 Modelling with distance geometry 258
9.2.1 Generic preferences 258
9.2.2 Specific interactions 259
9.2.3 Sources of real data 260
9.3 Protein tertiary structure prediction 261
9.3.1 *Ab initio* prediction 262
9.3.2 Empirical methods 263
9.3.3 Model evaluation 265
9.3.4 Genetic algorithm approach 267

PART 4
SYMMETRY 269

10 Structural symmetry 271
10.1 Symmetry 271
10.1.1 Symmetry from domain duplication 272
10.1.2 Symmetries from secondary structure 277
10.2 A Fourier analysis of symmetry 280
10.2.1 Internal protein structure comparison 281
10.2.2 Smoothing the score matrix 282
10.2.3 Fourier transform 283
10.2.4 Visualizing repeats 285
10.2.5 Analysis of total power 286
10.2.6 Removing expected symmetries 288
10.2.7 Assessment of the Fourier approach 289
10.2.8 Origin of structural symmetry 290

11 Evolution and origins 293
11.1 Evolution of structure and function 293
11.1.1 Gene duplication and fusion 294
11.1.2 Introns and exons 295
11.1.3 Models of structure evolution 297
11.1.4 Evolution of function 300
11.1.5 Selection on random folds 301
11.2 The origins of proteins 303
11.2.1 The emergence of proteins in an RNA world 303
11.2.2 Functions for protoproteins 304
11.3 The secret of life 306

References 308

Contents 325

Preface

This book developed out of a review we wrote for the journal *Reports of Progress in Physics* (Taylor *et al* 2001). That work was aimed at non-(protein) specialists and in particular towards those with a more physical background. This is an aspect that is retained in the current work but we hope the work is, nonetheless, still easily accessible to readers without any (or much) training in physics or mathematics. Some large parts of the review remain unaltered in the current work, specifically, the introductory sections on basic protein structure (just because it is tedious to rewrite the same material). Other parts have reduced in size, largely as the result of comments from less specialized readers who thought they were a bit boring (and were brave enough to tell us!). The main change, however, is the inclusion of much new material which I hope should appeal to those who take a physical view of the world. These include a new section on distance geometry, an 'Ising' model for protein domain definition, an expanded section describing a 'periodic table' of protein structure, and a Fourier analysis of symmetry in structure plus lots more on topology and knots. The quotes on some of these terms indicate that they should be taken to indicate more a similarity to the physical theory rather than an exact application. This is a common feature in the methods (some would say pseudomethods) described later and is true also of many of the descriptions of protein topology. Rather than irritate, I hope that these loose approaches may help inspire someone to re-examine the problem themselves and maybe succeed in dragging the theory of protein structure a little closer to an exact science.

The work does not pretend to be a comprehensive source for the analysis of proteins and there are many methods, even some we have worked on ourselves, that are not covered. In particular, there is virtually no mention of protein sequence analysis, even when this relates to structure (such as in structure prediction and threading). There is also no mention of statistical methods, such as artificial neural nets or associative memory Hamiltonians, even though these (and other statistical methods) have been widely applied in the analysis and prediction of structure. Our decision in this has been to concentrate on simple geometric methods that are deterministic (as distinct from probabilistic). Even with this restriction, there is a large body of work on the mechanical properties of proteins that we have not ventured into. These include analytical methods, such as normal-mode analysis and the calculation of atomic interactions using molecular dynamics. Although probably of considerable interest to those who work in physics, this is a large field in itself that has been covered widely in many books. In contrast to molecular dynamics, our aim in this work has been to concentrate on the more abstract simplifications of protein structure that allow a better understanding of the overall fold of the structure. This view also corresponds with another neglected topic: that analyses the properties of simple force fields on simplified proteins that are represented on a regular lattice. There is some mention of these studies when we deal with protein folds on a secondary structure lattice, but nothing to justify the large body of interesting work on these models.

On the original review we had a few co-authors and are grateful to them for letting us use what remains of their contributions in the current work. They were: Nigel Brown (structure comparison) and Alex May (structure classification). In addition, we were aided in the original Fourier work by Tom Flores, Jaap Heringa and Franck Baud. Linus Johannassen worked with me on the 'fold-trees' while David Jones and Kjell Petersen are thanked for providing some figures. Much of the other material is derived from our own papers in *Nature* ('periodic table' and protein knots), the *Journal of Molecular Biology* (stick models) and *Protein Engineering* (symmetry and domains) plus various sources (including those mentioned) for the distance geometry material. (Specific acknowledgements can be found throughout the text.) We are grateful to those journals for allowing us to reuse some figures and text (with most of the boring bits left out). This material has been heavily edited so any errors in the current work are almost certainly the result of our own recent activity.

Most of the methods described in this book have corresponding computer programs and these can be found (as C source code) at the ftp site of the Division of Mathematical Biology at the National Institute for Medical Research: `http://mathbio.nimr.mrc.ac.uk/ftp/wtaylor`. (Other activities in the Division can be seen by omitting the part following 'uk'). Whilst happy to hear of any 'bugs' in these programs, we cannot provide any help in getting them installed or in running them. The more commonly used geometric routines have been collected in the Web site associated with the book, along with some explanation (which is sadly lacking in most of my other programs). These constitute a little library of utility functions that are widely used in our programs (on the Web site) and an understanding of them may help anyone who is brave enough to venture into these larger bodies of code.

In the course of putting this work together, we have sometimes wondered what sort of a book it is. It is not a textbook, although it does contain some pedagogic material, neither is it a research work even though it does contain some new ideas and results. It sometimes comes close to a review but as such it is particularly biased towards our own work to the extent that, if we were older, it might be mistaken for a retrospective. In the end, we like to think of it as a collection of methods and ideas. Often the two are linked with the method implementing the idea but some of the methods have given rise to new unexplored ideas and some of the ideas need new methods to be developed. Rather than a weakness, we hope this open-ended aspect is what our readers will value most.

Willie Taylor and András Aszódi
1 May 2004

Chapter 1

Introduction

The ultimate rationale behind all purposeful structures and behaviour of living beings is embodied in the sequence of residues of nascent polypeptide chains—the precursors of the folded proteins which in biology play the role of Maxwell's demons. In a very real sense it is at this level of organization that the secret of life (if there is one) is to be found. If we could not only determine these sequences but also pronounce the law by which they fold, then the secret of life would be found—the ultimate rationale discovered!

Jaques Monod (1970)
from *Chance and Necessity*
loosely translated from the French (and Latin).

1.1 Prologue

1.1.1 Scope and aims

Proteins are the main essential active agents in biochemistry: without them almost none of the metabolic processes that we associate with life would take place. Consequently, most reviews of proteins concentrate on these catalytic abilities: on their chemical kinetics, interactions and the detailed stereochemical arrangement of the catalytic groups that allow catalysis (or binding) of substrate and other macromolecules. From this biochemical viewpoint, the overall structure of the protein (which is much larger than the active-site) is viewed as a relatively uninteresting supporting scaffold for the chemistry. In this review, however, proteins will be viewed from a different angle—indeed, their biology and chemistry will be completely ignored. Instead, their overall structure will form the central topic and within this, an emphasis will be placed on abstracting an overview rather than concentrating on chemical or structural details. The underlying theme of the work is: 'why do proteins adopt the forms that we see?'

leading to the supplementary question: ‘do the proteins we know represent a fraction or a full sample of the possible forms?’. The answers to these questions are not only of interest from a structural/biochemical viewpoint but also have implications for our ideas of molecular evolution and the origin of life.

The text of this work will be aimed at readers from the physical and mathematical sciences and, as such, will not rely on any significant biochemical knowledge on the part of the reader. Each topic will be fully explained from first principles with an emphasis on basic concepts rather than applications or occurrences. Much of the text will also concentrate on computational methods, again focusing on the basic algorithms rather than their application or implementation. As such, while essentially a review, little attempt has been made to provide an exhaustive coverage of the specialized literature. Rather, effort has been directed towards communicating ideas and methods that might have some resonance for those with a more physical background.

Many of the aspects of proteins that will be explored have been investigated by molecular biologists (such as ourselves) who have been enticed into more abstract areas. Along the way we have usually taken a pragmatic approach to each investigation, sometimes inventing new methods (which often turn out to be re-inventions) or ‘borrowing’ methods and approaches from other fields (especially physics). It is our hope in writing the current work, that some more specialized readers, perhaps having seen a frightening misapplication of their favourite method, might take up the challenge and ‘do it properly’. There are also some problems discussed for which we, at least, see no way forward (or more generally, no satisfactory way forward). We hope that these topics might inspire consideration from a fresh (ideally, orthogonal) viewpoint and allow some new directions to be identified.

1.1.2 Why proteins?

1.1.2.1 Catching a demon

There are many large biological molecules, including: nucleic acids, carbohydrates, lipids and proteins. While each play a vital (and interesting) part in life, there is something special about proteins. From a physicist’s point of view, the essence of this uniqueness might be captured by saying that, mechanically (if not thermodynamically), proteins are about as close as we can come to capturing a real-life Maxwell’s demon (figure 1.1).

Of the components that make up life, almost all but proteins are relatively inert and are, generally, the substrates that are chopped and changed by the action of proteins. In doing this, proteins do not act using some abstract bulk property (as do lipids and carbohydrates) but are individual agents (rather like demons) that latch onto their ‘victims’ (substrates) and cut and change them (sometimes even using the chemistry of sulfur). Indeed, when located across a lipid membrane, they are also quite good at opening and shutting trapdoors!

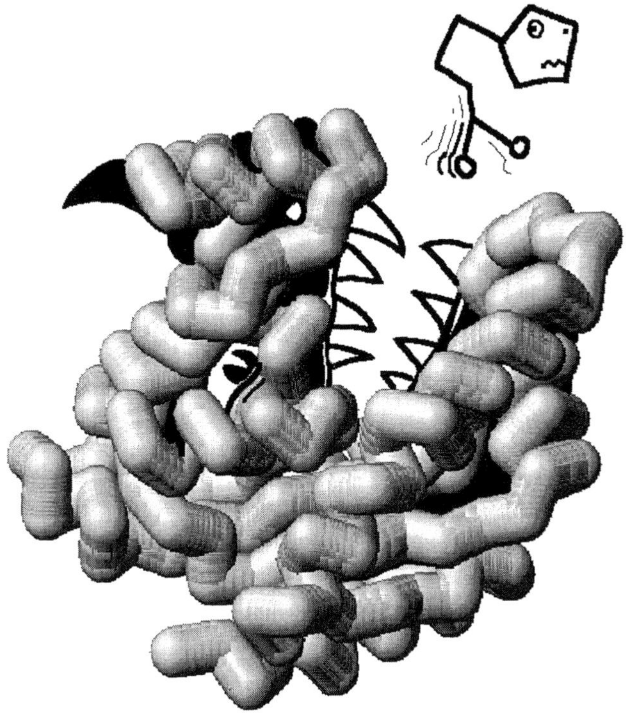

Figure 1.1. A small enzyme approaches its substrate. Against all thermodynamic reason, some people have likened proteins (such as this adenylate kinase molecule) to Maxwell's demons. The active (or catalytic) site of the molecule is indicated by elongated triangles.

To a large extent, understanding the action of proteins is the key to understanding the spark of life itself and this has been stated quite explicitly in the quotation by Jacques Monod (one of the 'founding fathers' of molecular biology) that opens this section. As indicated by Monod in the same quotation, proteins also occupy a unique position in the hierarchy of physical organization: lying in a grey region between chemistry and biology. For a chemist, proteins are large complicated molecules that even polymer chemists would have difficulty in modelling. From the biological side, although any individual protein would not be considered to be alive, it does not take many of them (plus a bit of nucleic acid) before life-like behaviour begins to emerge. For example; some of the smallest viruses, such as HIV, which might be considered to be on the borderline of life, operates with only 10 different types of protein.

1.1.3 Outline of the work

Hopefully, the preceding thoughts and speculations have proved to be sufficiently intriguing to persuade the less biologically-oriented reader that proteins are a fascinating topic and certainly one central to the understanding of life. In the following chapters, these themes are elaborated with various digressions into the details of methods and their applications. The overall progression is one of scale—starting from an analysis of basic secondary structure units, through their assembly into domains to their distribution throughout the known protein structures. From this apex, there is then something of a gradual decent into the unknown as we consider how this variety might have arisen, ending with one of the great unknowns: the origin of life itself.

1.1.3.1 Basic structure and geometry (chapters 1 and 2)

In the first chapter, we will leave these broader considerations and lay down some basic groundwork on protein structure so that all readers, irrespective of background, will have a common foundation on which some of the later more technical chapters can build. As promised, we will try to avoid the standard 'biochemistry textbook' approach to the topic.

The second chapter provides much the same function for some geometric methods that are employed widely in various contexts throughout the following chapters. This chapter begins simply and progresses from the basic properties of angles and distances, through the inertial properties of point-sets to the method of distance geometry. These methods are seldom explained fully anywhere (at a level that is easy to understand) and we have taken the trouble to go through them in detail without any annoying phrases like 'it is then obvious that...' (followed by an equation that seems to bear no relationship to the previous).

1.1.3.2 Secondary structures (chapters 3, 4 and 5)

The following chapter immediately applies these methods to reduce proteins to simple 'stick' figures that allow secondary structure packing to be easily analysed. This also 'prepares the ground' for the later development of a 'periodic table' of proteins.

The structure of β-sheets is not ideally represented by stick models and chapter 4 develops more elaborate models for these based on the recursive tiling of a simple twisted surface. This analysis culminates in an analysis of the geometric constraints necessary for the formation of β-barrels.

The analysis of β-sheets continues in chapter 5 but with the more messy problem of defining what a β-sheet is when observed in the known structures. The Ising-like method employed to dissect sheets is then extended to the definition of autonomous compact units of structure called domains. This leads to the more general problem of the best way to 'parse' or dissect proteins into their component parts (domains) which is a prerequisite for any classification system.

1.1.3.3 *Comparison and classification (chapters 6 and 7)*

To a large extent, domains are defined by their recurrence in different proteins so a necessary extension of their analysis involves the comparison of structures. Some of the many methods for proteins structure comparison are reviewed in chapter 6, concentrating on those that are suitable for the comparison of distant similarities and the rough (α-carbon) models that are considered through this work, including those based on the 'stick' models mentioned earlier.

Chapter 7 continues the analysis of structures from the viewpoint of finding a suitable way to systematically classify them. This reviews the more common hierarchical clustering methods and progresses to the alternative 'pigeonhole' approach of allocating protein structures to a position in a 'periodic table' constructed from idealized stick-figures. The chapter concludes with some speculations on the meaning of pathways through such a table.

1.1.3.4 *Topology and modelling (chapters 8 and 9)*

Structure comparison raised the problem of what is meant by a protein fold which leads to the more abstract consideration of protein topology. Chapter 8 starts from the rigorous end in which topology is the problem of knots in strings and gradually softens this approach to develop a method that gives more interesting results for proteins.

Using this method, the occurrence of knots in random chains is considered. These chains are then progressively made more protein-like and in the end, the test for knots and tangles becomes just a filter to discard reject structures. With increasing constraints, this leads through *ab initio* structure prediction to structure modelling using distance geometry.

1.1.3.5 *Symmetry and evolution (chapters 10 and 11)*

In chapter 10, the regularities observed in protein structures are analysed as those arising through direct duplication to those that may just have arisen from structural constraints, with the latter being quantified by a Fourier analysis.

This leads, in chapter 11 to the probable origins and evolution of proteins through successive duplications of structure. Finally, these threads are gathered together in a concluding section on a more philosophical consideration of where proteins have come from (whether they will be found on Mars) and, of course, the secret of life.

1.2 Basic principles of protein structure

In this section the basic principles that determine protein structure will be reviewed. Although many aspects of these topics will be returned to in greater detail in the following sections, it is better at first to gain an overview of these together in one place rather than encounter important definitions scattered throughout the text. Further information on many of these topics, including greater biological background, can be found in Brändén and Tooze (1991) or Chothia (1984) for a review concentrating more on packing.

1.2.1 The shapes and sizes of proteins

From a chemical viewpoint, proteins are linear hetropolymers. However, unlike most synthetic polymers, which are condensed from one or a few monomer units, proteins can draw on a mix of 20 different monomers. A further distinction is found in their organization: while polymers are generally very large extended molecules forming a matrix (typically crosslinked as a gel), the majority of proteins fold as relatively small self-contained structures. These factors balance: although small (for a polymer), the variety of monomers gives an almost unlimited scope for the construction of different protein molecules. Perhaps the most remarkable feature of proteins, however, is the observation that each protein found in nature has a specific three-dimensional structure and that this structure is determined (effectively) only by the sequence of the monomers themselves. To give names to these parts: the monomer units are amino acids which condense with the formation of a peptide bond linking them: hence, the resulting chain is often referred to as a polypeptide. The linked amino acids are then referred to as residues: an odd name deriving from the stuff at the bottom of test tubes when proteins were sequenced by chemical means in the early days of protein chemistry.

There is great variety in the structure of the 20 different (natural) amino acids but despite this, the variation (with one exception) is all confined to the side groups leaving a constant unit that polymerizes into a regular backbone chain. (See Taylor (1986a), Taylor (1999a) for some further discussion of amino acid properties.) Furthermore, even though amino acids contain a chiral centre (on their α-carbon), only one enantiomer is used to make proteins. As we shall see later, this regularity in the polypeptide chain allows the formation of semi-regular substructures that are the building blocks of proteins. The polypeptide chain is also very flexible: although the peptide bond is not free to rotate, the two flanking bonds are, giving two reasonably free rotations for each residue.

1.2.1.1 Fibrous proteins

There is no (reasonable) physical limit to the length of a polypeptide chain but those occurring naturally tend to be less than 1000 residues. This may represent a constraint derived from the fidelity of translation in the synthesis of the protein (or a historical relic from the days when fidelity was poorer) or it may simply be a consequence of the time needed to synthesize the protein. There are, of course, many exceptions and the largest known protein has about 100 000 residues (Higgins *et al* 1994). Clearly, to fold such a protein into a unique structure would be a formidable task and proteins of this size are composed of repeated units: either of like or mixed type. When the repetition is regular, involving a single (or few) type(s) then the resulting structure takes the form of a general helix—providing there is good interaction between the repeats. Otherwise, if the repeats form independent units, the structure has the form of a flexible string of beads. These proteins are referred to as fibrous and tend to play a more inert structural role in the cellular functions.

1.2.1.2 Globular proteins

Of greater interest are the proteins that have a unique structure derived from a non-repetitive sequence. These tend to fold in to fairly compact units and are, correspondingly, referred to as globular proteins. This class is composed predominantly of proteins in the size range of a hundred to several hundred residues. They include the majority of proteins that catalyse metabolic processes (enzymes) and those that regulate replication and expression of the genetic material. Clearly this covers most of the interesting functions of life and this richness is reflected in a corresponding richness of structure. Fortunately, this class is also that about which most is known structurally. This is a consequence of the ability of many globular proteins to crystallize and hence have their structure determined by x-ray crystallography. For the smaller members of the family, the technique of nuclear magnetic resonance (NMR) is also yielding an increasing number of structures.

1.2.1.3 Membrane proteins

A third class of proteins is restricted to the unique environment of the phospholipid bilayer membrane that surrounds all cells and many sub-cellular organelles. These proteins cover a range from globular proteins that happen to have a small tail that anchors them to the membrane through proteins that are half-in/half-out of the membrane, to proteins that are fully embedded in the membrane. In function, they cover the transport of material across the enclosing cell membrane, ranging from simple ions to the import of nutrients and the export of products that can influence the surrounding environment. For multicellular organisms, one aspect of the latter function is to influence the state or behaviour of neighbouring cells. This can be effected through the secretion of chemicals that others detect (and, again, the detection involves membrane bound proteins called receptors), or through direct physical contact between receptors.

1.2.2 The hydrophobic core

Globular proteins generally exist in the aqueous ('soup'-like) environment of the cellular cytoplasm. The basic organizing principle of their structure is to get the amino acid side chains that are not soluble in water (referred to as hydrophobic)

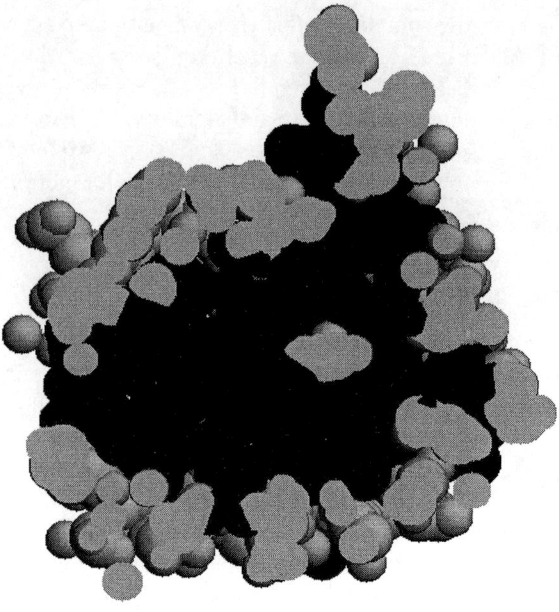

Figure 1.2. The hydrophobic core. A section (slab) has been taken through the core of a small protein (PDB code: 3chy) and displayed (using RASMOL) to show the van der Waal's surface of all the (non-hydrogen) atoms. These are coloured as grey for polar amino acids and black for hydrophobic amino acids. The black hydrophobic core can be clearly seen but (as with all 'rules' concerning protein structure) there are some exceptions and a (grey) hydrophilic residue can be seen in the core and a (black) hydrophobic residue on the surface. The former is probably hydrogen bonded to another hydrophilic side chain or to main chain polar groups, while the latter may make contact with another protein.

together in a core and surround them with a shell of water-soluble amino acid side chains (referred to as hydrophilic or polar) which provide an interface to the solvent (figure 1.2). This arrangement generally results in a protein that is itself soluble in water and prevents unspecific protein–protein aggregation as might occur if the 'sticky' hydrophobic residues were exposed.

1.2.3 Secondary structure

One complication of this simple scheme, however, is that all residues also have polar atoms in their main chain and this includes the hydrophobic residues which we would otherwise like to see buried in the core. Burying these residues will now necessarily entail the burial of a polar amide (N-H) and carbonyl (C=O) group with each residue (each of which carry a partial charge).

A solution to this problem is to form a hydrogen bond between these unlike charges using groups from different parts of the main chain. When mutually satisfied in this way, the bonded pair can then be 'safely' buried away from solvent. One might imagine that such a pairing could be achieved in an *ad hoc* manner (simply matching-up whatever pairs came nearby)—but possibly as a consequence of the complexity of connecting such a network, the hydrogen bonded networks found in proteins are remarkably regular.

Hydrogen bonded pairings are dominated by the shortest local connection along the chain that can be made without significant distortion of the bond geometry—bonding the carbonyl group of residue i to the amide group of residue $i + 4$. When repeated along the chain, this arrangement is a helical structure of period 3.6 residues, known as the α-helix. The second, and almost only other solution of structural importance in proteins (known as β structure), is formed by two remote parts of the chain lining-up to form a 'ladder' of hydrogen bonds between them. This 'ladder' of bonds can be formed either when the juxtaposed chains run parallel or antiparallel. Each β-strand can contribute to two ladders, allowing the hydrogen bonded network to extend indefinitely in either direction, resulting in a general sheet structure, referred to as a β-sheet.

Together the α-helix and β-sheet structures are referred to as secondary structure, being intermediate in a structural hierarchy in which the polypeptide chain is primary and the folded chain is tertiary. However, there is a wide variety of other commonly occurring substructures that cannot be ignored in a more detailed analysis (Efimov 1993) including recurring combinations of secondary structures commonly referred to as super-secondary structure (Efimov 1991a, b, 1987).

1.2.4 Packed layers

With the main chain atoms tied-up in secondary structure, a core can be constructed using any mixture of α or β as building blocks. The simplicity of having effectively only two secondary structures is that there are only three (pairwise) combinations of them that can be used to construct proteins; so giving the three major structural classes: (1) α with α, (2) α with β and (3) β with β (Levitt and Chothia 1976). For a more detailed analysis of each class, see: (1) Chothia *et al* (1981), Lesk and Chothia (1980); (2) Cohen *et al* (1981), Chothia and Janin (1981), Chothia and Janin (1982); and (3) Cohen *et al* (1982).

Incorporation of a β-sheet, however, imposes a long-range constraint across the structure. The β-sheet has free hydrogen bonds on its two edges, which consequently prevents the sheet from terminating in the hydrophobic core. This divides the core into two and, if considered more generally, imposes a layered structure onto the further arrangement of secondary structures in the protein. (See figure 1.3 for examples and both Chothia and Finkelstein (1990) and Finkelstein and Ptitsyn (1987) for further consideration of protein structure along these lines.)

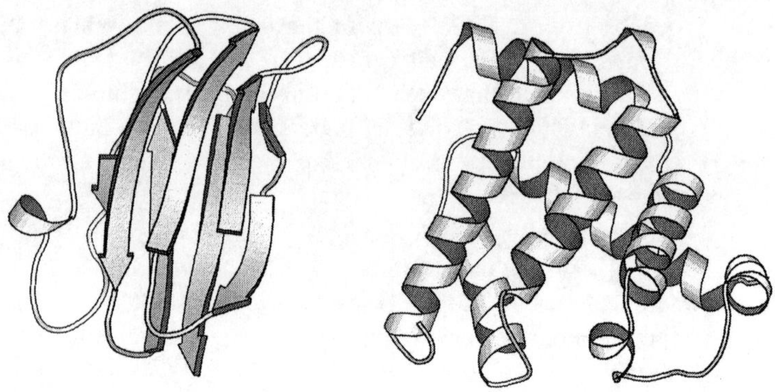

(*a*) All-β protein (*b*) All-α protein

Figure 1.3. Protein structures with one secondary structure type. (*a*) An all-β protein (immunoglobulin) with two packed β-sheets. (*b*) An all-α protein (globin) showing packed α-helices.

1.2.4.1 All-α proteins

The all-α protein class is dominated by small folds, many of which form a simple bundle with helices running up then down (figure 1.3(*b*)). The interactions between helices are not discrete (in the way that hydrogen bonds in a β-sheet are either there or not) which makes their classification more difficult (Lesk and Chothia 1980). Set against this, however, the size of the α-helix (which is generally larger than a β-strand) gives more interatomic contacts with its neighbours (relative to the a β-strand) allowing interactions to be more clearly defined (figure 1.3(*b*)).

1.2.4.2 All-β proteins

The all-β proteins are often characterized by the number of β-sheets in the structure and the number and direction of β-strands in the sheet. This leads to a fairly rigid classification scheme (Richardson 1977) which can be sensitive to the exact definition of hydrogen bonds and β-strands. Being less rigid than an α-helix, the β-sheets can be relatively distorted—often with differing degrees of twist and fragmented or extra strands on the edges of the sheet (figure 1.3(*a*)). Various patterns can be identified in the arrangement of the β-strands, often giving rise to the identification of recurring motifs (Hutchinson and Thornton 1993).

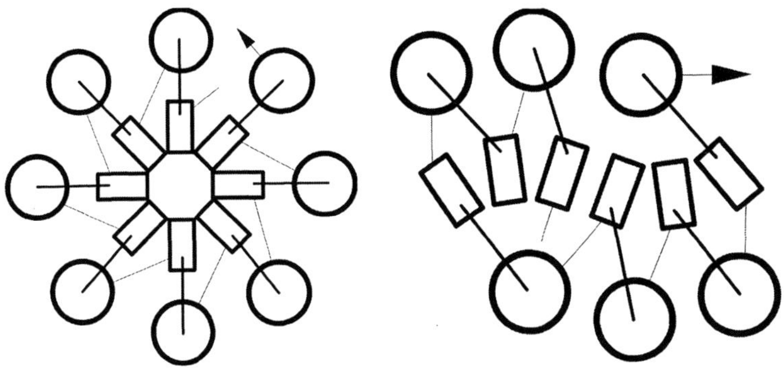

(*a*) Eight $\beta\alpha$-units (*b*) Six $\beta\alpha$-units

Figure 1.4. Folding options for tandem β-α units. β-strands are represented by rectangles and α-helices are represented as bold circles. All strands run parallel and progress towards the viewer. In reality, the strands are both curved and twisted which is suggested by their nonlinear alignment and α-helices are about twice as broad as a β-strand. The direction of the chain is indicated by a terminal arrow-head. The structural implications of the size difference between the two secondary structure types (combined with chirality constraints on their connection) are shown for a concatenation of both eight and six β-α units. (*a*) With eight units the sheet can form a barrel and the different radii of this circular form at the β and α level accommodate their different size. The barrel structure is found in many (unrelated) enzymes, typified by triosephosphate isomerase (TIM) and is referred to as a *TIM barrel* (see figure 1.5). (*b*) Six units cannot form a barrel forcing an inversion in one half of the sheet to allow helices to be placed both above and below. The resulting arrangement has two-fold symmetry and occurs widely among di-nucleotide binding proteins. It is typified by the dehydrogenases where it is referred to as a *Rossmann fold*.

1.2.4.3 $\beta\alpha$ proteins

The α-β protein class can be subdivided roughly into proteins that exhibit a mainly alternating arrangement of α-helices and β-strands along the sequence and those that have more segregated secondary structures. The former class includes structures in which the secondary structures are arranged in layers and those that form a circular of barrel-like arrangement (figure 1.4). Recurring folds can also be identified in the latter type (Orengo and Thornton 1993).

1.2.5 Barrel structures and β-helices

Structural solutions can be found to tie-up the 'loose' hydrogen bonds on the edge of a β-sheet. One commonly encountered, is to twist the sheet so that

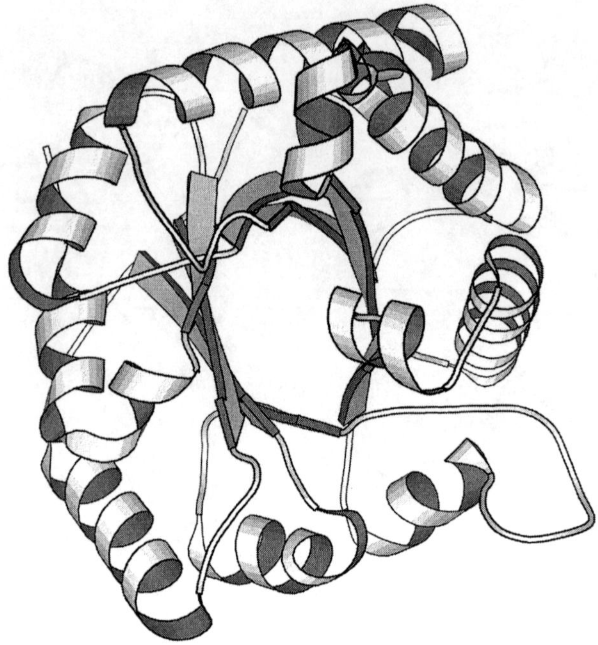

Figure 1.5. Eight-fold alternating β/α barrel protein. The protein chain spirals (as a toroid) while alternating between β and α secondary structure type, giving rise to a closed ring or barrel β-sheet in the centre surrounded by a larger ring of α-helices on the outside. The structure, first seen in the enzyme triosephosphate-isomerase (after which it is often named as the TIM barrel) has been seen many times in unrelated proteins.

the two edges meet and can hydrogen bond to each other—forming a closed barrel-like network of hydrogen bonds (Chou *et al* 1990). This cannot easily be accomplished with less than six strands and if only β-structure is used, then the barrel must incorporate antiparallel pairings. However, in combination with α-helices it is possible to link one (open) end of the barrel to the other and allow the formation of a, predominantly, or pure parallel sheet. A particularly striking example of this arrangement is seen in the eight-fold β-α-barrel $(\beta\alpha)_8$ which was found originally in the enzyme triosephosphate isomerase and is often referred to as the TIM barrel (figure 1.5). (See Murzin *et al* (1994a, b) for a full analysis.)

A barrel can also be formed with the β-strands running in the orthogonal direction (leaving free hydrogen bonds on the open ends of the barrel). This structure, however, completely dictates the course of the protein chain (as a simple helix) giving little scope for evolutionary exploitation of the fold for different functions. (See Chothia and Murzin (1993) for some examples). This type of structure is associated more with structural (fibrous) proteins. Some examples are considered in chapter 10.

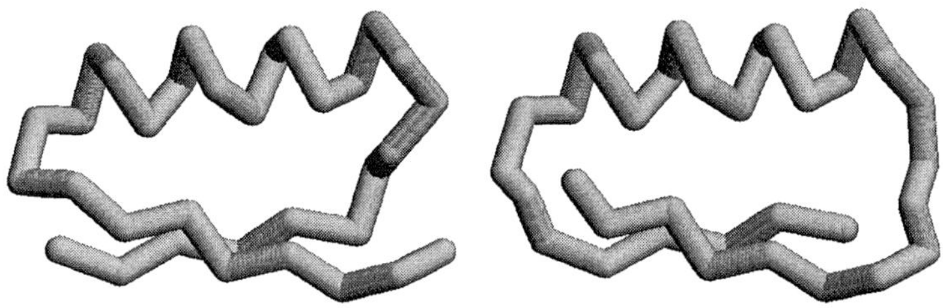

(*a*) Right-handed unit (*b*) Left-handed unit

Figure 1.6. Handedness in secondary structure connections. An α-helix linking two β-strands (hydrogen bonded in a sheet) is shown as a backbone (alpha-carbon) trace in: (*a*) the common right-handed configuration, and (*b*) with the rare left handed connection. The different chiralities can be appreciated if the whole chain is viewed as a super-helix: in the R-hand form clockwise rotation would drive it into the page (like a screw or corkscrew) while the same rotation would extract the L-hand form.

1.2.6 Protein topology

The path of the chain through the various layers of packed secondary structures described earlier (sometimes referred to as frameworks or architectures) is referred to as the fold of the chain. The names of folds are arbitrarily derived, sometimes from the (fertile) imagination of those working in the field—resulting in names such as 'jelly roll' or 'Greek key' or after the crystallographers who determined the first example of the fold, such as the 'Rossmann fold'. More commonly, the name comes from a particular function with which the fold was once exclusively associated but has later been found in more diverse proteins: as has been seen with the TIM barrel, the immunoglobulin (Ig) fold and the oligosaccharide-binding (OB) fold. While these names provide useful mnemonics for different aspects of the structure or function, an automated analysis requires a more general and systematic scheme even if this results in a less 'colourful' nomenclature.

As most descriptions of protein folds entails various degrees of crosslinking through hydrogen bonds, it is also possible to, loosely, view them from a topological perspective. This topic will be returned to in detail in Part 3 while, here, a few basic aspects will be considered that are relevant to the later discussions. (See Ptitsyn and Finkelstein (1980) for a general review.) The course of the chain through the secondary structure frameworks is largely unrestricted. Two constraints, however, are well observed. The strongest is that two loops cannot cross on the same face between layers (Ptitsyn and Finkelstein 1980)[1]. The source of this constraint is a simple consequence of the bulk of the

[1] An exception has been found in the protein with PDB code `2csmA`.

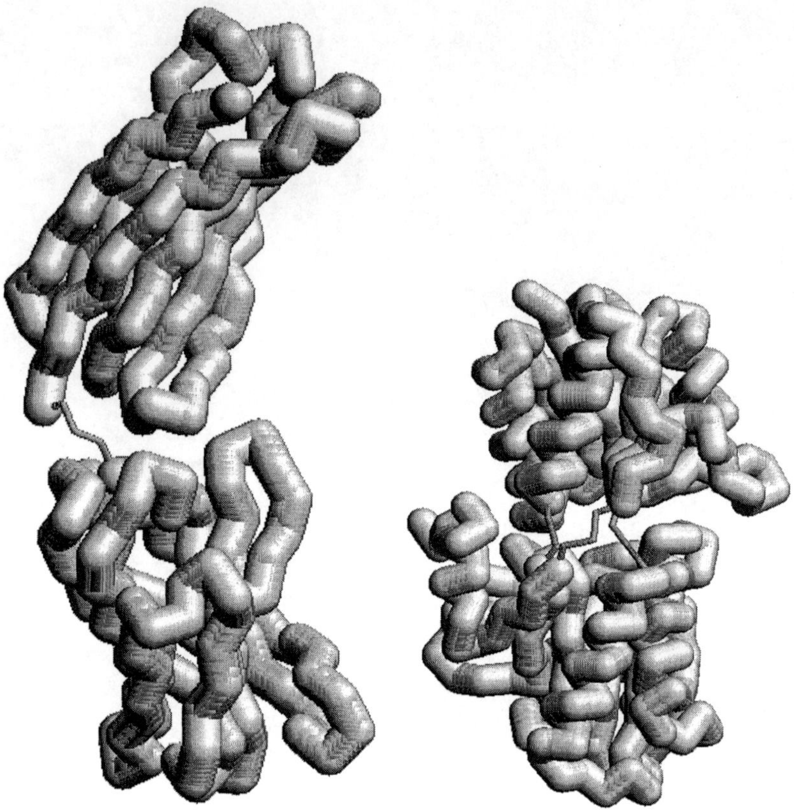

(*a*) Single link (*b*) Multiple links

Figure 1.7. Simple and complex domain connections. (*a*) Two immunoglobulin domains linked by a single connection. (*b*) Two more closely packed domains (arabinose-binding protein) between which the chain passes three times. (The linkers have been drawn thinner for clarity.)

polypeptide chain: if two loops cross, one will be buried by the other which will be energetically unfavourable unless the buried loop can satisfy its main chain hydrogen bonds. Having done this however, the loop is now probably a secondary structure and so the rule that loops do not cross is preserved.

The second strong constraint derives from the chiral nature of the central (α) carbon in each residue. This favours a particular (right)-handedness for the α-helix and a corresponding twist to the β-sheet which is left-handed when viewed along the chain direction. Together, these local chiralities result in a strong preference for connections between strands in the same sheet to be right-handed (even when there is no α-helix involved). The few exceptions to this rule are seen when the chain meanders to a remote part of the structure (another domain) and the 'context' of the local constraint is lost (Sternberg and Thornton 1977b) (figure 1.6). Some chiral effects are also detected in the $\beta\beta\alpha$ and $\alpha\beta\beta$ arrangements (Kajva 1992) and in the packing of four α-helices (Weber and Salemme 1980, Presnell and Cohen 1989).

1.2.7 Domain structure

Large hydrophobic cores are not found in globular proteins, probably because of limitations in the folding kinetics and stability. Single compact units of more than 500 residues are rare with the typical size lying more around half this size (200–300 residues). As a consequence, large proteins are organized into units of this size referred to as domains (Rose 1979, Richardson 1981, Janin and Chothia 1985) (figure 1.7).

The definition of a domain is problematic—one suggestion is that, if the chain were to be cut, then the two parts would remain stable (with each having its own hydrophobic core). With well segregated domains (like beads on a string) this is undoubtedly true but with more closely interacting domains (and in particular, those in which the chain crosses between the domains more than once), such an experiment cannot be carried out without exposing surfaces that are not optimally evolved for solvation.

Various working definitions of domains have been derived (Holm and Sander 1994a, Swindells 1995a, Siddiqui and Barton 1995, Islam *et al* 1995, Sowdhamini and Blundell 1995) but in the more difficult examples, these seldom agree. The problem with all these methods is that they try to imitate (human) expert definitions and it is clear that these definitions entail the synthesis of many abstract ideas such as biological function, recurrence and symmetry, all of which are difficult to capture in an automatic method. A recent approach to this problem has been based (loosely) on an Ising model, in which structural domains evolve in competition with each other for residues in the protein (Taylor 1999c). (This approach is more fully explained in section 5.2.)

PART 1

GEOMETRY

Atomics is a very intricate theorem and can be worked out with algebra but you would want to take it by degrees because you might spend the whole night proving a bit of it with rulers and cosines and similar other instruments and then, at the wind-up, not believe what you had proved at all.

Flann O'Brien (from the Third Policeman)

Chapter 2

Ellipsoids and embedding

2.1 Geometric representations of structure

2.1.1 Atomic coordinates

Through the previous descriptions of structure, proteins have been represented in a variety of ways using different levels of detail. Although little of it has been seen hitherto, the full representation of proteins has all atom coordinates specified, including hydrogens. For most x-ray analyses of structure, however, the hydrogen positions are not normally visible and the standard representation is generally to use all heavy (non-H) atoms (figure 2.1(a)). While this level of representation is required for detailed analysis of substrate binding, packing and catalysis, in the present work, we have concentrated more on the overall fold of the protein and for this a representation of the protein backbone path is usually sufficient. This can be shown in many ways: some of which incorporate features derived from the more detailed levels, such as secondary structure. The simplest representation is to connect a central atom in each residue (and for this the α-carbon is the obvious choice) resulting in a trace that shows the overall fold of the protein clearly and in which secondary structure (if present) can also be seen (figure 2.1(b)).

Additional levels of information can be represented along the backbone trace and this can be done either with or without explicit definition of the secondary structures. The orientation of the (flat) peptide plane ($> N - C <$) can be used to guide the surface of a ribbon representation (Carson 1991) (figure 2.1(c)) or with explicit secondary structures resulting in a similar representation but now the β-strand components have been 'labelled' with an arrow-head (Sklenar *et al* 1989, Thomas 1994) (figure 2.1(d)).

These different representations are shown together in figure 2.1 for comparison. Each image was generated from the program `RASMOL` (Sayle and Milner-White 1995) which is principally intended to display these representations interactively. Finer quality but static representations can be generated from other programs such as `Molscript` (Kraulis 1991), some examples of which can be seen in figure 1.3 and figure 1.5 in section 1.2. Given the quality of the latter representations, it is now rare for the cartoon figures to be drawn by hand as they once were (Richardson 1985).

While it is a 'simple' computational problem to generate a smooth curve for a protein backbone chain, it is less simple to define on this where the secondary structure elements begin and end. Where the smooth chains described above have been been 'labelled' with an arrow-head (Thomas 1994) these definitions have been generated by an 'expert' (usually the scientist who determined the structure) or by an automatic algorithm that has explicitly considered H-bonding networks such as the `DSSP` program (Kabsch and Sander 1983). This problem will be returned to later.

2.1.1.1 *Smoothed traces*

The α-carbon trace can be smoothed to different degrees to simplify 'unimportant' details in surface loops (Feldman 1976). The simplest method of smoothing the α-carbon trace is to average the coordinates over a moving window and the smallest of any practical use is three residues. Indeed it is not necessary to use a larger window, as the triplet averaging can be repeatedly applied to the smooth coordinates. If the coordinates are represented by the vector set $\{x_1 \ldots x_N\}$ then the smoothed coordinate of residue i (x_i') is: $(x_{i-1} + x_i + x_{i+1})/3$. Note that care should be taken to ensure that position i does not take its new value until after the average is taken of the subsequent triple.

A few rounds of smoothing in this way is sufficient to reduce elements of secondary structure to almost straight lines and by five iterations an acceptably smooth curve is obtained which provides a very immediate visual aid for assessing the overall fold of the protein figure 2.2. Depending on the analysis, this procedure can have some disadvantages. Most obviously, as smoothing progresses, the size of the protein shrinks. This can be countered by continually inflating the chain to maintain a constant radius-of-gyration (but better alternatives can be found). Where smoothing is carried to an extreme degree, then it is also possible that the topology of the protein might have changed (since the chain is free to pass through itself). An application where this is prevented will be described later and leads to a practical algorithm for finding knots and tangles in proteins. It should also be noted that, using the algorithm described earlier, the two end residues will not move so the exact result of the smoothing will be sensitive to their positions.

A more exact way to represent the smooth chain is to fit a curve to the points. Both splines (Sklenar *et al* 1989, Carson 1991) and Chebychev polynomials (Thomas 1994) have been fitted to proteins. Having an analytic curve has the advantage that continuous measures of curl (the amount of bending) and writhe (the amount of twisting) can be calculated to characterize the curve. This approach has been more extensively applied to the analysis of nucleic acid chains (Cozzarelli and Wang 1990).

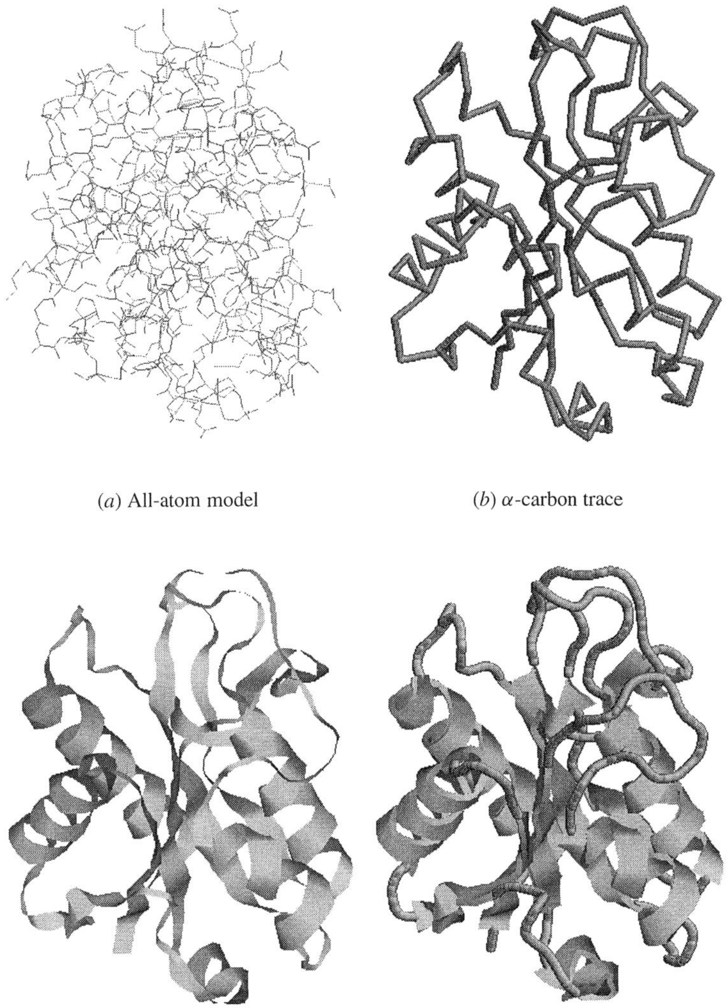

(*a*) All-atom model

(*b*) α-carbon trace

(*c*) Ribbon trace

(*d*) Cartoon representation

Figure 2.1. Protein structures representations A small β/α protein (flavodoxin) is shown in four representations. (*a*) Showing bonds between all non-hydrogen atoms, (*b*) with lines connecting sequential α-carbon atoms, (*c*) as a flat trace (ribbon), drawn to follow the orientation of the peptide planes, and (*d*) with explicit secondary structure definitions represented by 'cartoon' objects. The figures were produced by the program RASMOL.

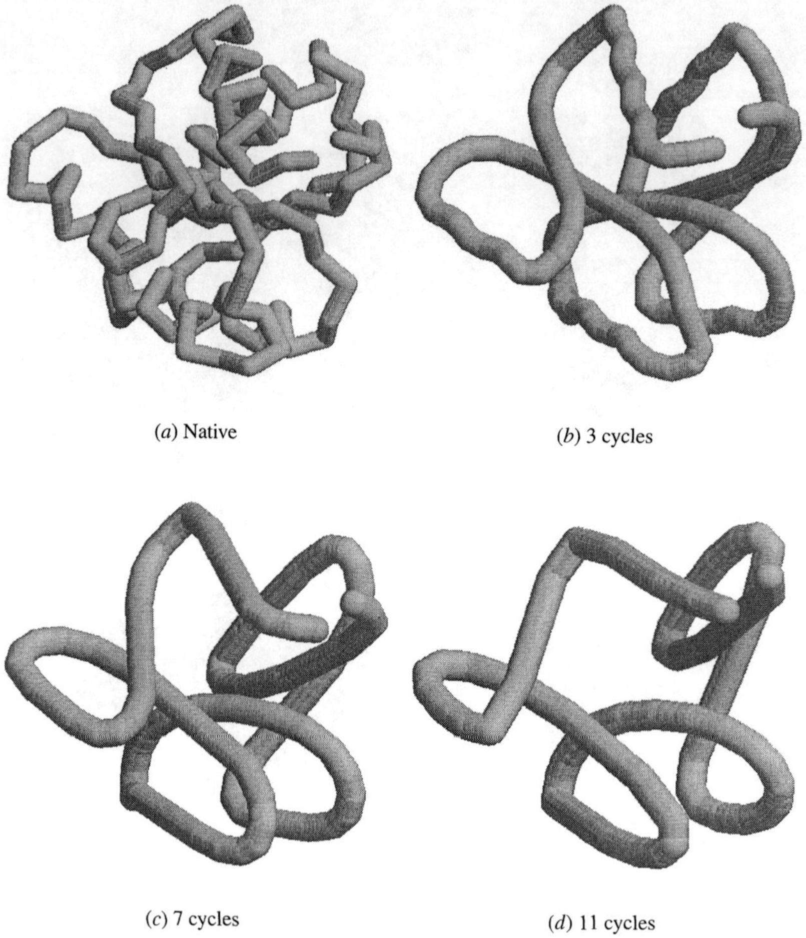

(*a*) Native (*b*) 3 cycles

(*c*) 7 cycles (*d*) 11 cycles

Figure 2.2. Smoothed protein structures A small β/α protein (flavodoxin-like) is shown as a backbone (α-carbon) trace as: (*a*) the native structure, (*b*) with three cycles of backbone smoothing, (*c*) seven smoothing cycles, and (*d*) 11 cycles. In each cycle the central atom in a triple of α-carbon atoms is replaced by their average position. The figures were produced by the program `RASMOL`.

2.1.2 Torsion angles

2.1.2.1 Phi-psi angles

The peptide bond has a partial double bond character and so cannot rotate freely. It was the realization of this that gave Linus Pauling the 'edge' in predicting the secondary structure of proteins. This implies that there are only two bonds on the protein backbone that can rotate relatively freely: N–α-carbon (ϕ rotation) and α-carbon–C (ψ rotation) (see figure 2.3). Much analysis has been carried out in the parameter space of these two angles, referred to as the Ramachandran plot (figure 2.4). In particular, exploiting their close correlation with secondary structure.

Due to steric collisions between side chain and main chain atoms, the dihedral angles ϕ and ψ of each residue, are naturally restricted. This knowledge, specific to each amino acid type, can be utilized directly in the calculation of molecular conformations to reduce the search space. An advantage of this representation is that it is intuitive and compact. It is linear in nature which is convenient for some optimization algorithms such as genetic algorithms (Petersen and Taylor 2003). It is also straightforward to represent ideal secondary structure elements by their repeated dihedral angles.

The main disadvantage of this representation is that it is very context sensitive in the way that residue i is defined on the basis of residue $i - 1$. This is sufficient for local interaction such as in secondary structure elements, but global interactions in the protein structure, are only indirectly represented through a chain of local interactions. When one dihedral angle is changed, it usually has a large effect on global conformation and adjusting neighbouring dihedral angles to counteract the global change is not trivial.

2.1.2.2 Virtual torsion angles

A simplified approach based on the C_α chain itself can be used, giving one torsion angle per residue where ϕ and ψ angles results in two variables per residue. However, both a virtual bond angle θ_V as well as the virtual torsion angle ω_V are required for an accurate representation (see figure 2.5).

As with the true torsion angles (ϕ and ψ) the α-carbon virtual torsion angle can also be used to define secondary structure (Levitt 1983b). By scanning the chain for consecutive lengths of three or longer stretches of residues with strand or helix conforming values for θ_V and ω_V, secondary structure elements can be detected. When dealing with (sometimes rough) protein models, these parameters have been found to be less sensitive to the variance in secondary structure coordinates and are essential when dealing with models that contain only α-carbons.

As well as being used to define secondary structure, the relationships used by Levitt (1983b) can also be reversed to produce a reasonable regeneration of the full main chain atom positions. The full code for this procedure (called `ca2main`) is given in the Web site associated with the book. The more conventional approach to this problem (and the more accurate) is to fit short fragments of structure from the PDB onto the α-carbon backbone (Purisima and Scheraga 1984, Claessens *et al* 1989, Holm and Sander 1991). These methods would be recommended in situations where an accurate model is required but

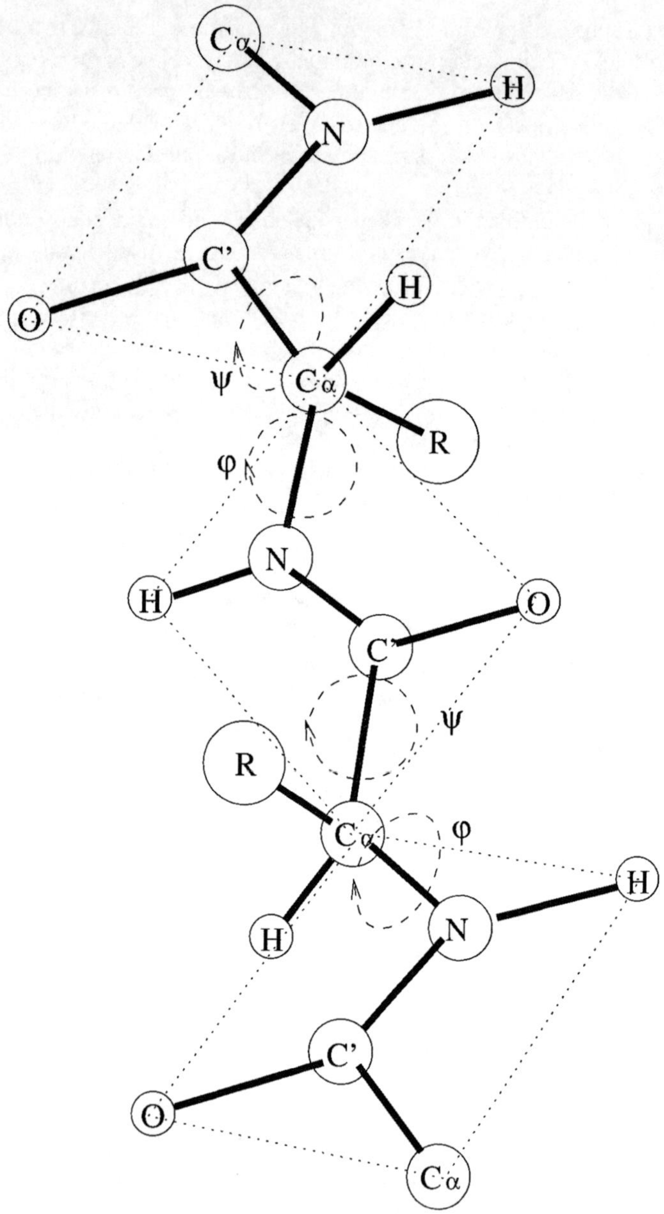

Figure 2.3. Schematic representation of the backbone with ideal bonding distances and angles.

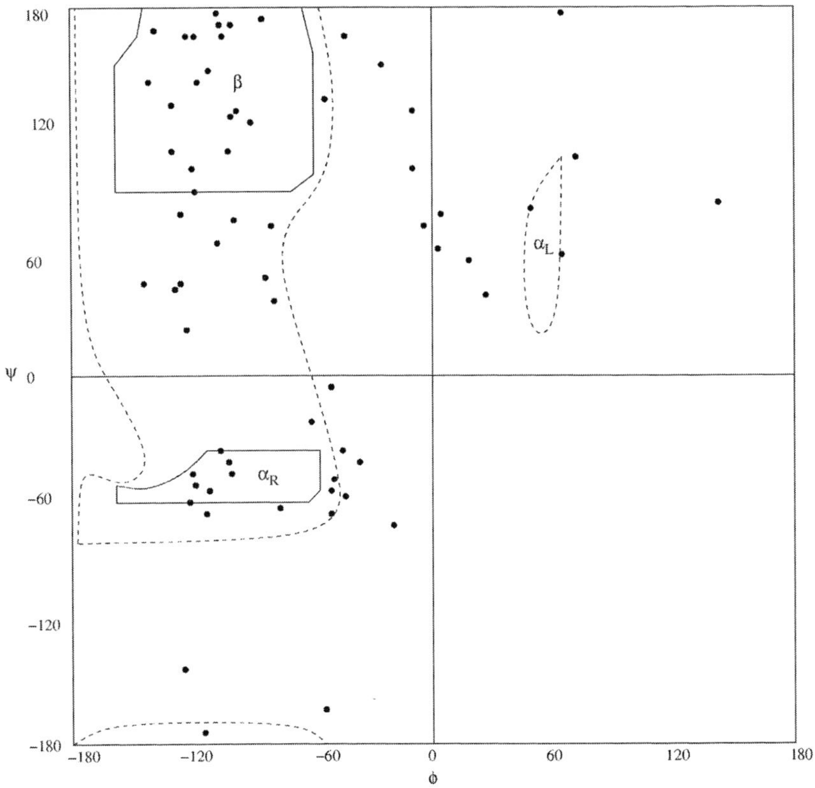

Figure 2.4. Ramachandran diagram indicating the sterically allowed ϕ, ψ values and their associated secondary structure. The subscripted 'R' and 'L' on the α conformation designate right- and left-handed helices. The data points are taken from the native structure of a small protein.

with many of the rough models considered in this work, the simple reverse-Levitt method is sufficient.

2.1.2.3 Angle/coordinate interconversion

While the representation of a protein backbone as a set of torsion angles appears to provide a more compact representation (ϕ, ψ *versus* x, y, z), the peptide bond is not uniformly flat and the torsion along the peptide (ω) must also be known to allow a reconstruction of the chain by 'dialling-up' successive torsion values. Even with this information, the resulting chain conformation is so sensitive to small errors in the angle values that it is not a practical transformation.

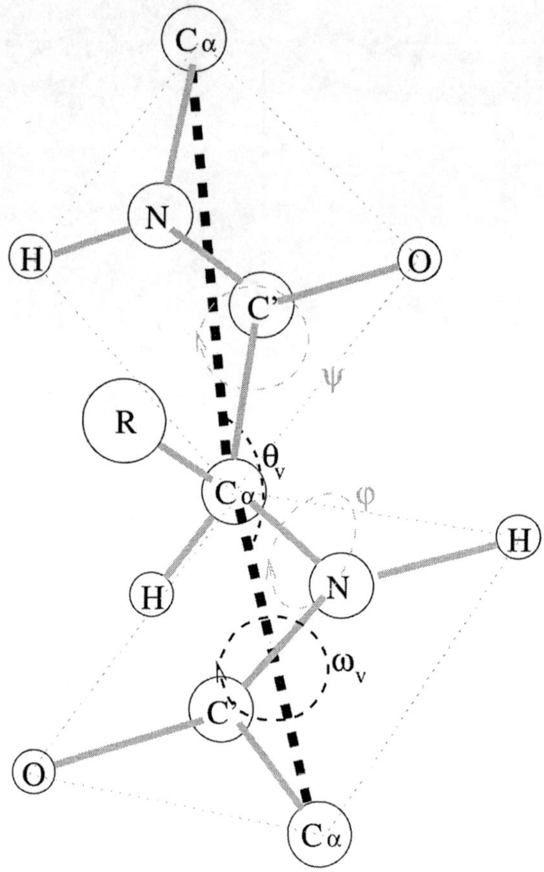

Figure 2.5. Virtual variables for C_α chain.

2.1.3 Distances

Instead of using a Cartesian coordinate system, a point set can be described as a set of interpoint distances (sometimes called internal coordinates). This is clearly not done for reasons of storage efficency since, for N points, a full interpoint set of distances requires $(N^2 - N)/2$ values to be held (a top-triangular matrix excluding the diagonal) which, for $N > 7$, is always larger than the equivalent $3N$ coordinates (and much more for big N).

However, a distance matrix has some advantages, both for visualization and computation, that make it a useful representation for protein structure. This is especially so when the data from experiments (such as NMR) are available as interatomic distances.

2.1.3.1 Distance plot

The original and simplest use of a distance matrix, originally pioneered by David Phillips, is to look at it. If the elements are plotted and shaded by size (or plotted just below a specified distance cut-off), then a very immediate impression is gained of the atomic interactions within the structure (Phillips 1966). This insight is gained, however, at the expense of a clear image of the fold of the protein.

The visual attractiveness of the distance plot has led many to use it as a basis for comparing protein structures ranging from the qualitative (Kuntz *et al* 1976) to quantitative (Holm and Sander 1993b) but these applications will be described later in chapter 6. An example plot is shown in figure 2.6 for a small flavodoxin-like protein (chemotaxis Y protein, PDB code: 3chy). This protein has alternating β and α secondary structures and the $\beta\beta$ packing can be seen as short dark off-diagonal bars running parallel to the diagonal. The interactions of the larger α-helices are more diffuse and have a characteristic 'tartan' pattern as they spiral closer and farther away from each other on their helical course.

The plot is, of course, strictly symmetric about the main diagonal but in this particular protein, a weaker symmetry can be seen roughly around the opposing diagonal, indicating a degree of two-fold symmetry in the structure. The extraction of this information using a Fourier method will be described in section 10.2 (chapter 10).

2.1.4 Inertial axes[1]

We have seen previously that it is useful to look at the smoothed backbone of the protein and that several cycles of smoothing can reduce convoluted loops to smooth curves and secondary structures to (almost) straight lines. However, because of the distortion introduced by smoothing, it is more desirable to have a direct method to reduce secondary structures to lines and, in theory, the repetitive structure of the secondary structure elements should allow a line to be found that corresponds to their helical axis.

From a physical viewpoint, the simplest solution to this task is to determine the inertial axes of the points that constitute the secondary structure element (SSE). Alternatively, an ideal helix could be fitted but this would require the adjustment of its helical parameters for each type of (if not each individual) secondary structure. The inertial axes have the advantage that they can be applied 'blindly' to all types of secondary structure and, of course, any other non-helical point set.

The derivation of the inertial axes will be described here in some detail as it is a technique that recurs in many of the applications described in this work and also has links with the method of distance geometry that will be described later in this chapter.

[1] Parts of this section are based on Taylor *et al* (1983) with the kind permission of Oxford University Press.

Figure 2.6. Distance plot. The α-carbon–α-carbon distances in a small β/α protein (figure 2.2(a)) are plotted as a matrix. Each cell represents the distance between two residues and is shaded from black (zero distance) to white for long distances. The exact mapping is the Gaussian function: $z = 1 - \exp(-5d_{ij}^2)$, where d_{ij} is the interatomic distance between residues i and j in the structure.

2.1.4.1 Moments and products of inertia

If, 'by chance' the inertial axes happened to be coincident with the coordinate frame, then the moments of inertia about each axis could be simply found as the sum of the squared distances of the points from each axis[2]. If the coordinates are

[2] In this section (and all subsequent sections), a matrix will be represented by a single bold uppercase letter ($\boldsymbol{M}$) with its component (row) vectors in bold (single-subscripted) lowercase ($\boldsymbol{m}_i$) and its individual components in italic (double-subscripted) type (m_{ij}). The same applies to vectors but with one less subscript (e.g. $\boldsymbol{w}$ and w_i). The dot (or scalar) product of vectors will be represented as a big dot (e.g. $\boldsymbol{u} \bullet \boldsymbol{v}$). while the vector (or cross) product will be represented by a circled 'times' symbol

held in a set of vectors $\boldsymbol{A} = \{\boldsymbol{a}_1, \boldsymbol{a}_2 \ldots \boldsymbol{a}_N\}$, each with components a_{ix}, a_{iy} and a_{iz} (for a position i), then about the X axis the moment of inertia (I_X) is:

$$I_X = \sum_{i=1}^{N} a_{iy}^2 + a_{iz}^2 \tag{2.1}$$

and similarly for the Y and Z axes (giving I_Y and I_Z).

Generally, an inertial axis will not conveniently correspond with a coordinate axis and it is necessary to calculate the moment of inertia about an arbitrary direction from the origin (specified by a unit vector $\boldsymbol{w}$). Considering (for the moment) just a single point ($\boldsymbol{a}$), its perpendicular distance, d, to $\boldsymbol{w}$ is $|\boldsymbol{a}| \sin\theta$, where θ is the angle between $\boldsymbol{w}$ and the line from the origin to the point. Since sin θ can be expressed as the cross-product of the vectors ($\sin\theta = |\boldsymbol{a} \otimes \boldsymbol{w}|/|\boldsymbol{a}|$), the distance, or its square, is simply found by Pythagoras' theorem on the components of the cross-product:

$$d^2 = (\boldsymbol{a} \otimes \boldsymbol{w})^2 \tag{2.2}$$

which expands to ...

$$\begin{aligned} &= (a_y w_z - a_z w_y)^2 \\ &\quad + (a_z w_x - a_x w_z)^2 \\ &\quad + (a_x w_y - a_y w_x)^2 \end{aligned} \tag{2.3}$$

becomes very messy,

but 'simplifies' to ...

$$\begin{aligned} &= A w_x^2 + B w_y^2 + C w_z^2 \\ &\quad - 2(F w_y w_z + G w_z w_x + H w_x w_y) \end{aligned} \tag{2.4}$$

where $A = a_y^2 + a_z^2$, $B = a_x^2 + a_z^2$, $C = a_x^2 + a_y^2$ and $F = a_y a_z$, $G = a_x a_z$, $H = a_x a_y$. By comparison with equation (2.1), A, B and C can be recognized as the moments of inertia about each coordinate frame axis while F, G and H are called the products of inertia and are equivalent to covariance in statistics.

2.1.4.2 *Rotation into the inertial frame*

The problem should now be clear: we need to rotate the coordinate frame until the products of inertia are zero. This can be found by differentiating equation (2.4) with respect to each of the components of $\boldsymbol{w}$, which gives a vector $\boldsymbol{u}$ with components:

$$u_x = \delta I/\delta w_x = 2A w_x - 2G w_z - 2H w_y \tag{2.5}$$

$$u_y = \delta I/\delta w_y = 2B w_y - 2F w_z - 2H w_x \tag{2.6}$$

$$u_z = \delta I/\delta w_z = 2C w_z - 2F w_y - 2G w_x. \tag{2.7}$$

(e.g. $\boldsymbol{u} \otimes \boldsymbol{v}$). This distinguishes these vector operations from the simple dot or 'times' symbol used for scalar multiplication. The (scalar) length of a vector is denoted as: $|\boldsymbol{w}|$ (or $|\boldsymbol{m}_i|$) and the transpose of a matrix (or vector) is indicated by a superscripted 'T' (e.g. $\boldsymbol{M}^{\mathrm{T}}$).

This can be written neatly as the multiplication of the vector $\boldsymbol{w}$ with a symmetric matrix formed from the moments and products of inertia:

$$\boldsymbol{u} = 2\begin{pmatrix} A & -H & -G \\ -H & B & -F \\ -G & -F & C \end{pmatrix}\begin{pmatrix} w_x \\ w_y \\ w_z \end{pmatrix} \tag{2.8}$$

or more compactly as:

$$\boldsymbol{u} = \boldsymbol{T}\boldsymbol{w} \tag{2.9}$$

where $\boldsymbol{T}$ is the matrix of moments and products of inertia (sometimes called the inertial tensor). As we will only be interested in the direction of $\boldsymbol{u}$, the factor of two has been dropped.

So far, this analysis has referred to one point ($\boldsymbol{a}$) relative to one line ($\boldsymbol{w}$). For many points, the A, B, C and F, G, H terms simply become sums over the set of points $\boldsymbol{A}$ (as in equation (2.1)) while the unit vector $\boldsymbol{w}$ becomes a set of three orthogonal vectors forming a rotation matrix $\boldsymbol{W}$. Similarly, the vector of partial derivatives becomes the (3×3) matrix $\boldsymbol{U}$ and equation (2.9) becomes:

$$\boldsymbol{U} = \boldsymbol{T}\boldsymbol{W}. \tag{2.10}$$

The matrix $\boldsymbol{W}$ will rotate the original coordinate frame axes into the new reference frame and we need to find what it is when the matrix $\boldsymbol{U}$ is made diagonal[3]:

$$\boldsymbol{U} = \boldsymbol{W}\boldsymbol{\Lambda} \tag{2.11}$$

where $\boldsymbol{\Lambda}$ is a diagonal matrix. Substituting for $\boldsymbol{U}$(from equation (2.10)) gives:

$$\boldsymbol{T}\boldsymbol{W} = \boldsymbol{W}\boldsymbol{\Lambda}. \tag{2.12}$$

Post-multiplying each side by $\boldsymbol{W}^{\mathrm{T}}$ (and remembering that the transpose of a rotation matrix is its inverse) we have,

$$\boldsymbol{T} = \boldsymbol{W}\boldsymbol{\Lambda}\boldsymbol{W}^{\mathrm{T}}. \tag{2.13}$$

This equation has the form of a 'classic' eigenvalue problem: in other words, to find the reference frame in which the products of inertia are zero, we need to diagonalize the inertial tensor ($\boldsymbol{T}$) and the resulting eigenvectors (in $\boldsymbol{W}$) specify the axes of inertia, while the eigenvalues (the diagonal elements of $\boldsymbol{\Lambda}$) will give the moments of inertia. Explicitly, $A = \boldsymbol{\lambda}_{11}$, $B = \boldsymbol{\lambda}_{22}$, $C = \boldsymbol{\lambda}_{33}$ (while the products, $F = G = H = 0$).

[3] The reason for this condition comes from the physical interpretation of the vector $\boldsymbol{u}$ as the surface normal to the general ellipsoid specified by equation (2.4). When the coordinate frame and $\boldsymbol{U}$ coincide, then $\boldsymbol{U}$ will be a diagonal matrix. At this point, the off-diagonal products of inertia terms will be zero.

2.1.4.3 *Momental and equivalent ellipsoids*

The effect of the rotation by $\boldsymbol{W}$ has been to eliminate the product of inertia terms from equation (2.4), reducing it to a generalized form of equation (2.1) as:

$$I = Ax^2 + By^2 + Cz^2. \tag{2.14}$$

This is equivalent in form to the equation for an ellipsoid:

$$x^2/a^2 + y^2/b^2 + z^2/c^2 = 1 \tag{2.15}$$

where a, b and c are the semiaxis lengths.

Combining equations (2.14) and (2.15) gives the semiaxis lengths of the ellipsoid describing the moments of inertia as:

$$a = 1/\sqrt{A} \qquad b = 1/\sqrt{B} \qquad c = 1/\sqrt{C} \tag{2.16}$$

or in terms of the eigenvalues, as:

$$a = 1/\sqrt{\lambda_1} \qquad b = 1/\sqrt{\lambda_2} \qquad c = 1/\sqrt{\lambda_3} \tag{2.17}$$

where the eigenvalues are now singly subscripted to indicate that they have been sorted by size. In general, the distance from the centre of the ellipsoid to its surface (in any direction) gives the moment of inertia about that line as $1/\sqrt{I}$ and the ellipsoid is named the momental ellipsoid.

We return (at last) to the original problem of finding the best axis for a secondary structure element. This will be the inertial axis about which most of the points lie closest and will have the smallest corresponding moment of inertia and hence, (by equation (2.16)) the largest semiaxis length. In other words, it will be the eigenvector corresponding to the smallest eigenvalue of $\boldsymbol{T}$. While this has solved the original problem, it is not sufficient if we want to construct a simple ellipsoidal model for a point-set (as might be required for visualization or packing analysis). To construct such a model we require the shape of a solid object that would have the same inertial properties as the point-set. This is similar, but not identical with, the momental ellipsoid.

The principal moments of a homogeneous solid ellipsoidal body in terms of its semiaxes (a, b, c) are:

$$\begin{aligned} A &= (b^2 + c^2)/5 \\ B &= (c^2 + a^2)/5 \\ C &= (a^2 + b^2)/5. \end{aligned} \tag{2.18}$$

However, we require the semiaxes in terms of the moments and their squares are found to be:

$$\begin{aligned} a^2 &= k(-A + B + C) \\ b^2 &= k(A - B + C) \\ c^2 &= k(A + B - C) \end{aligned} \tag{2.19}$$

where k is a constant. Substituting for the moments (A, B, C) in terms of the coordinates brings about a dramatic simplification to:

$$\begin{aligned} a^2 &= kx^2 \\ b^2 &= ky^2 \\ c^2 &= kz^2. \end{aligned} \tag{2.20}$$

The extraction of the principal axes for this ellipsoid, called the equivalent ellipsoid (or Chaucy's ellipsoid) follows as for $\boldsymbol{T}$ but using instead the second moment matrix $\boldsymbol{S}$ which has simply x^2, y^2 and z^2 terms along the diagonal (and hence corresponds exactly to the variance/covariance matrix used in statistics).

$\boldsymbol{S}$ is related to $\boldsymbol{T}$ by the radius of gyration, r, $(r^2 = x^2 + y^2 + z^2)$ as can be clearly seen when $\boldsymbol{S}$ is written in terms of the moments (A, B, C) and products (F, G, H) of inertia:

$$\boldsymbol{S} = \begin{pmatrix} r^2 - A & -H & -G \\ -H & r^2 - B & -F \\ -G & -F & r^2 - C \end{pmatrix}. \tag{2.21}$$

The diagonalization of $\boldsymbol{S}$ can progress as with $\boldsymbol{T}$ (above) but now the semiaxes are found directly as the square-root of the eigenvalues:

$$a = \sqrt{\lambda_1} \qquad b = \sqrt{\lambda_2} \qquad c = \sqrt{\lambda_3} \tag{2.22}$$

where the eigenvalues have been sorted by size (as in equation (2.17)) but note that the longest axis corresponds now to the largest eigenvalue.

This construct is properly known as the equivalent ellipsoid but is sometimes wrongly referred to as the momental ellipsoid or loosely as the inertial ellipsoid. Finally, it is worth noting that $\boldsymbol{S}$ can be neatly expressed in terms of the matrix product of the coordinate set with itself transposed:

$$\boldsymbol{S} = \boldsymbol{A}\boldsymbol{A}^{\mathrm{T}}. \tag{2.23}$$

2.1.5 The shapes of proteins

Some applications of the inertial ellipsoid will be considered in detail in chapter 3, while the more direct applications of simply characterizing (and visualizing) proteins as ellipsoids will be described in this section.

2.1.5.1 Rugby balls and flying saucers

Although most soluble proteins are fairly globular, there are considerable deviations of the molecular shape from the perfect sphere. One way to accomplish this is to fit an ellipsoid

$$\frac{x^2}{A^2} + \frac{y^2}{B^2} + \frac{z^2}{C^2} = 1 \tag{2.24}$$

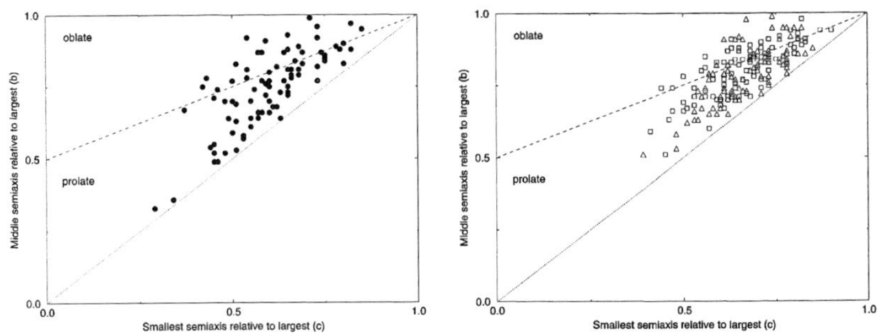

(*a*) Real Proteins (*b*) Model proteins

Figure 2.7. Eccentricity plots of protein shape. (*a*) The ellipsoidal eccentricity values (*b*) and (*c*) are plotted for a selection of native protein structures. (*b*) 'Random' protein models constructed by the program DRAGON are plotted with and without a hydrogen bonding constraint (squares and triangles, respectively). (See section 9.2 for modelling details.)

with semiaxes $A \geq B \geq C > 0$ to the molecule using the methods already described. All ellipsoids can then be represented as points on the plot of the normalized semiaxes $b = B/A$ versus $c = C/A$: the perfect sphere is at the ($b = 1, c = 1$) point, oblate ellipsoids ('flying saucers') are in the region $b > (1+c)/2$, whereas prolate ellipsoids ('rugby balls') are below the $b = (1+c)/2$ line (Taylor *et al* 1983). This plot is referred to as the *eccentricity* plot (figure 2.7).

On the eccentricity plot (figure 2.7(*a*)), the majority of proteins are not spherical (none appear at the spherical point on the top right of the plot) and there is a trend towards being prolate. The plot can also be used to analyse model proteins and a selection constructed by the distance geometry based program DRAGON (section 9.2) are plotted using two different sets of constraints (figure 2.7(*b*)). The calculated distributions match the observed distribution very well and if the spread of values seen in the models is taken as a 'random' background distribution then it can be inferred (with a few exceptions) that most native proteins have an unexceptional globular shape.

The eccentricity values can also be used as a filter to select good models (or more commonly, to reject unfolded structures) but often the isotropic radius-of-gyration value provides an adequate test for this.

2.1.5.2 Measuring protein density

Ellipsoids can be used to estimate the density of protein models. The α-carbon chain coordinates of a selection of 84 protein molecules were approximated by inertial ellipsoids with the length of the semiaxes (A, B and C) proportional to the three moments of inertia (the square root of the eigenvalues of the moment matrix). Each semiaxis was scaled by a factor $f_{0.9}$ so that the ellipsoid contained 90% of the point-set. This empirical scaling was necessary to compensate for the concave crevices present on most protein surfaces. The density of an N-residue polypeptide chain was then calculated as:

$$\rho = \frac{3N}{4\pi ABC} = \frac{3N}{4\pi f_{0.9}^3 (\lambda_1 \lambda_2 \lambda_3)^{1/2}} \tag{2.25}$$

where $\lambda_1, \lambda_2, \lambda_3$ were the eigenvalues of the moment matrix. The average density of the proteins was found to be $\rho = (6.3 \pm 1.3) \times 10^{-3}$ residues/Å^3, in good agreement with other studies (Taylor 1993b, Gregoret and Cohen 1991).

The density of subselections of coordinates, measured in a similar way using ellipsoids, was used by Rose (1979) to define compact domains in proteins. (The problem of domain definition will be considered more fully in section 5.2.)

2.1.5.3 *Domain movements and juxtapositions*

A more general use of ellipsoids is to use the coordinate frame specified by the inertial ellipsoid to quantify the juxtaposition between two proteins or two instances of the same protein in different positions. The latter is ideal when a part or domain of a protein has shifted in position, typically, on activation or binding another protein. The rotation matrix between the 'before' and 'after' positions of the inertial frame can be used to quantify the shift.

2.1.5.4 *Protrusions and antigenic epitopes*

By gradually shrinking an equivalent ellipsoid around a protein, a measure can be obtained for the parts that are protruding most above the bulk (or surface) of the protein. This can be a useful measure in defining the regions of sequence that may be most susceptible to mutation or most likely to be bound by other proteins, in particular antigens (where the bound fragment is called an epitope) (Rothbard *et al* 1986, Thornton *et al* 1986).

2.2 Distance geometry

Three ways of representing a protein chain have been described: (1) Cartesian coordinates, (2) torsion angles and (3) internal distances. Coordinates can be used to calculate both angles and distances and we have seen above that angles can be used to generate coordinates—but at the risk of loosing accuracy. As Sergeant Pluck advises in the opening quote to this part: the careful application of ruler and cosines (and other such instruments) can convert angles to distances but in the calculation of torsion angles from distances, the chirality of the angle is lost. This is obvious from the observation that the distances viewed in a mirror are unaltered.

In this section, the missing link in the triangle: the direct extraction of coordinates from distances is described. The method is called *distance geometry* (DG) and makes use of the full matrix of interatomic distances. As such, it is extremely robust and can tolerate quite large deviations from the correct distances which makes it very attractive for dealing with uncertain predicted distances or experimentally determined distances with significant errors. There is only one weakness to DG which is that the overall chirality of the structure is lost and either the true structure or its mirror image is arbitrarily obtained. However, this is still much better than the indirect route of local distances $\rightarrow$ angles $\rightarrow$ coordinates, where the handedness of every successive torsion angle would be arbitrary.

While it is straightforward to calculate the interpoint distances of a point set represented by Cartesian coordinates, the inverse problem, once called the 'fundamental problem in distance geometry' (Crippen and T F Havel 1988) is more difficult. This mathematical problem consists in finding an arrangement of points in space such that the interpoint distances correspond to prescribed values. In macromolecular modelling applications, the goal is to generate three-dimensional conformations which satisfy a set of distance restraints obtained from experiments and/or from theoretical considerations. A variety of algorithms have been discovered over the years which solve the fundamental problem, e.g. the method of alternating projections (MAP) (Glunt *et al* 1990) or the method of spectral gradients (Glunt *et al* 1992) but the method followed here is closest to the technique known as multidimensional scaling in the multivariate statistics literature. For an introduction to this, see the books by Torgerson (1958) or Krzanowski (1988) (and references therein) or for molecular embedding in computational chemistry, see: Crippen and T F Havel (1988), Kuntz *et al* (1989) and Havel (1991).

2.2.1 Out of hyperspace

As with the method for finding the inertial axes of a set of points, the central operation in DG is the diagonalization of a matrix. In the calculation of inertial axes it was possible to have a physical 'feel' for what was happening (rotating the coordinate frame until the products of inertia vanished), however, it is not so simple to grasp what happens in DG: except in the vague terms of 'a multidimensional rotation into an orthonormal basis set'. To many, this is no more obvious than the process by which the starship 'Enterprise' emerges from hyperspace. Because of this conceptual difficulty, we have kept the following description of the method at quite a symbolic level and while most descriptions start with distances and work towards coordinates, we will begin with the answer and instead just show why, if not how, (or maybe how if not why) the method works.

2.2.1.1 Back to eigenvalues

Consider the classic eigenvalue problem:

$$\boldsymbol{M}\boldsymbol{W} = \boldsymbol{W}\boldsymbol{\Lambda} \tag{2.26}$$

where $\boldsymbol{M}$ is a symmetric matrix, $\boldsymbol{\Lambda}$ is a diagonal matrix and $\boldsymbol{W}$ is a rotation matrix (all of which have rank $N \times N$). Although it is not really relevant, in words, this equation states that when $\boldsymbol{M}$ is rotated 'forward' by $\boldsymbol{W}$ and $\boldsymbol{\Lambda}$ is rotated 'back' by $\boldsymbol{W}$, they become equal.

Equation (2.26) can be rearranged as:

$$\boldsymbol{M} = \boldsymbol{W}\boldsymbol{\Lambda}\boldsymbol{W}^{\mathrm{T}}. \tag{2.27}$$

Each side has simply been post multiplied by $\boldsymbol{W}^{\mathrm{T}}$ and since the transpose of a rotation matrix is its own inverse, the product $\boldsymbol{W}\boldsymbol{W}^{\mathrm{T}}$ on the left hand-side is the identity matrix (and vanishes). The diagonal elements of $\boldsymbol{\Lambda}$ are the eigenvalues while the rows of $\boldsymbol{W}$ are the eigenvectors, each of which are orthogonal unit vectors (as required for them to constitute a rotation matrix).

For reasons that will become apparent later, let us now assume that the matrix $\boldsymbol{M}$ consists of the dot products of the coordinate vectors of a set of N points: $\boldsymbol{A} = \{\boldsymbol{a}_1, \boldsymbol{a}_2 \ldots \boldsymbol{a}_N\}$, each with components a_{xi}, a_{yi} and a_{zi} (for a position i). Specifically, the $\{j, k\}$ element of the matrix $\boldsymbol{M}$ is:

$$m_{jk} = \boldsymbol{a}_j \bullet \boldsymbol{a}_k \tag{2.28}$$

where $\bullet$ designates the dot (or scalar) product. Conveniently, the matrix of dot products can be expressed as the multiplication of the coordinate set with its transpose:

$$\boldsymbol{M} = \boldsymbol{A}^{\mathrm{T}}\boldsymbol{A}. \tag{2.29}$$

Combining equations (2.27) and (2.29) forges a link between the coordinates ($\boldsymbol{A}$) and a diagonal matrix $\boldsymbol{\Lambda}$. There may not seem to be much point to this yet, but if we set:

$$\boldsymbol{A} = \boldsymbol{\Lambda}^{1/2}\boldsymbol{W}^{\mathrm{T}} \tag{2.30}$$

and substitute this for $\boldsymbol{A}$ in equation (2.29), then,

$$\begin{aligned}\boldsymbol{M} &= (\boldsymbol{\Lambda}^{1/2}\boldsymbol{W}^{\mathrm{T}})^{\mathrm{T}}(\boldsymbol{\Lambda}^{1/2}\boldsymbol{W}^{\mathrm{T}})\\ &= (\boldsymbol{W}\boldsymbol{\Lambda}^{1/2})(\boldsymbol{\Lambda}^{1/2}\boldsymbol{W}^{\mathrm{T}})\\ &= \boldsymbol{W}\boldsymbol{\Lambda}\boldsymbol{W}^{\mathrm{T}}.\end{aligned} \tag{2.31}$$

(Remember that the transpose of a matrix product is the product of the transposed matrices in reverse order and, more obviously, the transpose of a diagonal matrix does nothing.)

2.2.1.2 Lagrange to the rescue

This result in equation (2.31) shows that we can take a set of coordinates ($\boldsymbol{A}$), make a matrix of dot products ($\boldsymbol{M}$), diagonalize this matrix and get the coordinates back again from the eigenvalues ($\boldsymbol{\Lambda}$) and eigenvectors ($\boldsymbol{W}$). In summary: $\boldsymbol{A} \to \boldsymbol{M} \to (\boldsymbol{\Lambda} \text{ and } \boldsymbol{W}) \to \boldsymbol{A}$. However, since we knew the coordinates in the first place, this might seem to be a rather sterile exercise designed simply to test the consistency of geometric transformations. This would be so but for a method, found many years ago, which allows us to break into this cycle by calculating the metric matrix ($\boldsymbol{M}$) directly from a set of distances ($\boldsymbol{D}$) (Lagrange 1870). This method, named the Lagrange theorem, means that the progression $\boldsymbol{D} \to \boldsymbol{M} \to (\boldsymbol{\Lambda} \text{ and } \boldsymbol{W}) \to \boldsymbol{A}$, becomes possible: which is what was originally desired.

More explicitly, given the $N \times N$ matrix of squared interpoint distances $\boldsymbol{D} = [d_{ij}^2]$, we want to find the dot product between each pair of points from their centroid and collect them in the $N \times N$ metric matrix[4] $\boldsymbol{M} = [m_{ij}]$. If we also knew the distances of each point from the centroid, this could be done using the cosine rule:

$$m_{ij} = \boldsymbol{a}_i \bullet \boldsymbol{a}_j = d_{i0} d_{j0} \cos\alpha = \tfrac{1}{2}(d_{i0}^2 + d_{j0}^2 - d_{ij}^2) \tag{2.32}$$

where d_{i0} and d_{j0} are the centroid distances for the points i and j, respectively. It is these distances that the Lagrange theorem allows us to find as follows:

$$d_{i0}^2 = \frac{1}{N}\sum_{k=1}^{N} d_{ik}^2 - \frac{1}{N^2}\sum_{j<k}^{N} d_{jk}^2. \tag{2.33}$$

2.2.1.3 Lagrange theorem

The Lagrange theorem is the key step that makes DG possible. For this reason and as it is difficult to get the original book, the derivation of the theorem is given in this section. Slightly more accessible sources can be found in Flory (1969) or Levitt (1983a) but often these derivations are rather condensed with little explanation of how one step leads to another. In the following explanation, each step is explained and little more is required than knowledge of the cosine rule.

In the previous sections we dealt with a set on N points $\{\boldsymbol{a}_1, \boldsymbol{a}_2, \ldots \boldsymbol{a}_N\}$. In this section we need to refer to the origin which will be included as the additional point $\boldsymbol{a}_0$. It is also necessary to refer to interpoint vectors which require an additional subscript: thus $\boldsymbol{a}_{ij}$ is the vector from point i to point j, while $\boldsymbol{a}_{0j}$ is the vector from the origin to j (previously just $\boldsymbol{a}_j$). If we assume that the set of

[4] Similarly to the distance matrix, the metric matrix is also invariant under transformations in the Euclidean group, and can be regarded as an internal coordinate representation (Young and Householder 1938).

points has a centroid at the origin, then (by definition),

$$\sum_{i=1}^{N} \boldsymbol{a}_{0j} = 0. \tag{2.34}$$

We want to find the centroid distances in terms of the interpoint distances. If we ignore distances for the moment and consider vectors, this can be easily obtained since the vector from one point i to any other point j can always be expressed in terms of two vectors passing through the origin:

$$\boldsymbol{a}_{ji} = \boldsymbol{a}_{0i} - \boldsymbol{a}_{0j} \tag{2.35}$$

or rearranging,

$$\boldsymbol{a}_{0i} = \boldsymbol{a}_{ji} + \boldsymbol{a}_{0j}. \tag{2.36}$$

Then summing over all i points gives:

$$N\boldsymbol{a}_{0i} = \sum_{i=1}^{N} \boldsymbol{a}_{ji} + \sum_{i=1}^{N} \boldsymbol{a}_{0j}. \tag{2.37}$$

By definition (equation (2.34)), $\sum_{i=1}^{N} \boldsymbol{a}_{0j} = 0$, giving:

$$\boldsymbol{a}_{0i} = \frac{1}{N} \sum_{i=1}^{N} \boldsymbol{a}_{ji}. \tag{2.38}$$

This useful result specifies a single centroid vector in terms of the interpoint vectors and if an equivalent of this can be found in terms of distances, then we have reached our goal.

The left-hand side of equation (2.38) can be transformed into a distance expression since the squared length of the centroid vector ($\boldsymbol{a}_{0i}$) is given by its dot product with itself:

$$d_{0i}^2 = \boldsymbol{a}_{0i} \bullet \boldsymbol{a}_{0i} = |\boldsymbol{a}_{0i}| \cdot |\boldsymbol{a}_{0i}| \cdot \cos\theta \tag{2.39}$$

(since $\cos\theta = 1$). Substituting for $\boldsymbol{a}_{0i}$ from equation (2.38) gives:

$$d_{0i}^2 = \left(\frac{1}{N} \sum_{i=1}^{N} \boldsymbol{a}_{ji} \right) \bullet \left(\frac{1}{N} \sum_{i=1}^{N} \boldsymbol{a}_{ji} \right). \tag{2.40}$$

Like normal (scalar) multiplication, the dot produce is distributive[5] which means that the dot product of two sums can be rewritten as a double sum of a dot product:

$$d_{0i}^2 = \frac{1}{N^2} \sum_{j=1}^{N} \sum_{k=1}^{N} \boldsymbol{a}_{ij} \bullet \boldsymbol{a}_{ik}. \tag{2.41}$$

For the right-hand side of equation (2.41), the dot product $\boldsymbol{a}_{ij} \bullet \boldsymbol{a}_{ik}$ can also be re-expressed in terms of distances by applying the cosine rule:

$$\boldsymbol{a}_{ij} \bullet \boldsymbol{a}_{ik} = d_{ij} \cdot d_{ik} \cdot \cos\theta = (d_{ij}^2 + d_{ik}^2 - d_{jk}^2)/2. \tag{2.42}$$

Replacing the dot product in equation (2.41) with this distance expression gives:

$$d_{0i}^2 = \frac{1}{2N^2} \sum_{j=1}^{N} \sum_{k=1}^{N} d_{ij}^2 + d_{ik}^2 - d_{jk}^2 \tag{2.43}$$

which now has a centroid distance expressed entirely in terms of interpoint distances.

It remains only to slightly tidy up equation (2.43) which can be done by splitting it into individual summations:

$$d_{0i}^2 = \frac{1}{2N^2} \left(\sum_{j=1}^{N} \sum_{k=1}^{N} d_{ij}^2 + \sum_{j=1}^{N} \sum_{k=1}^{N} d_{ik}^2 - \sum_{j=1}^{N} \sum_{k=1}^{N} d_{jk}^2 \right). \tag{2.44}$$

Since the first two distances are summed only over one index, the 'redundant' sum simply counts them N times, allowing N to be extracted as a factor leaving a single summation. In addition, as both sums are equal, a factor of 2 can also be extracted, leaving:

$$d_{0i}^2 = \frac{1}{N} \sum_{k=1}^{N} d_{ik}^2 - \frac{1}{2N^2} \sum_{j=1}^{N} \sum_{k=1}^{N} d_{jk}^2. \tag{2.45}$$

This result is now equivalent to equation (2.33) except that in equation (2.33) the double sum is only over half the distance matrix (top-triangle), so a factor of two does not appear in the denominator.

[5] That is: $a(b+c) = ab + ac$, or in this case,

$$\begin{aligned}
&(\boldsymbol{a}_1 + \boldsymbol{a}_2 + \boldsymbol{a}_3 + \cdots + \boldsymbol{a}_N) \bullet (\boldsymbol{a}_1 + \boldsymbol{a}_2 + \boldsymbol{a}_3 + \cdots + \boldsymbol{a}_N) \\
&= (\boldsymbol{a}_1 \bullet \boldsymbol{a}_1 + \boldsymbol{a}_1 \bullet \boldsymbol{a}_2 + \boldsymbol{a}_1 \bullet \boldsymbol{a}_3 + \cdots + \boldsymbol{a}_1 \bullet \boldsymbol{a}_N) \\
&+ (\boldsymbol{a}_2 \bullet \boldsymbol{a}_1 + \boldsymbol{a}_2 \bullet \boldsymbol{a}_2 + \boldsymbol{a}_2 \bullet \boldsymbol{a}_3 + \cdots + \boldsymbol{a}_2 \bullet \boldsymbol{a}_N) \\
&+ (\boldsymbol{a}_3 \bullet \boldsymbol{a}_1 + \boldsymbol{a}_3 \bullet \boldsymbol{a}_2 + \boldsymbol{a}_3 \bullet \boldsymbol{a}_3 + \cdots + \boldsymbol{a}_3 \bullet \boldsymbol{a}_N) + \cdots.
\end{aligned}$$

Which can be written as the double summation in equation (2.41). (Note also that any indices can be used in the summation as it is over all pairs of points).

The double sum in equation (2.33) (and equation (2.46)) is the mean-squared interpoint distance which is sometimes written as: $\langle D^2 \rangle$, giving the more succinct formulation:

$$d_{0i}^2 = \frac{1}{N} \sum_{k=1}^{N} d_{ik}^2 - \langle D^2 \rangle. \tag{2.46}$$

2.2.2 Interpretation of the eigenvalues

The substitution used to demonstrate that the extraction of the coordinates could be cast as an eigenvalue problem (equation (2.30)), gives each individual coordinate as:

$$a_{ki} = \lambda_k^{1/2} w_{ik} \tag{2.47}$$

where λ_k is the kth eigenvalue and w_{ik} is the ith coordinate of the corresponding kth eigenvector. However, the matrix $\boldsymbol{W}$ has rank $N \times N$ which means that the coordinates are N-dimensional. If the distances were measured on a three-dimensional (3D) point-set (without error) then this is not a problem since there will only be three (non-zero) eigenvalues. Generally, the number of non-zero eigenvalues specify the dimension of the Euclidean space into which the points can be embedded without error. This can be a useful measure if one is searching for a metric to minimize the dimensionality of a relationship between objects (Taylor and Jones 1993).

If errors are made in the distance measurements of a 3D point-set then there will be more than three (non-zero) eigenvalues. Assuming that we want a three-dimensional object (and with protein structures more than three dimensions becomes difficult to visualize), then the best course is to base the coordinate set on the three largest eigenvalues/vectors and discard the rest. Although no proof will be offered, this provides the optimal (least-squares) fit between the 'observed' distances and the distances between the calculated points. However, if the distance matrix contained incompatible (non-metric) entries that do not obey the triangle (or higher-order) inequalities then some of the eigenvalues will be negative. This presents a problem since in equation (2.47) it is necessary to take the square-root of the eigenvalue and so these components must be discarded. When the distance data is of very poor quality the number and size of the negative eigenvalues can become a serious problem and in this situation it is necessary to pre-filter the data to try and 'patch-up' both triangle inequalities (Aszódi and Taylor 1994a) and even tetrangle inequalities (Havel 1991).

If the distances are perfectly embedded in an M-dimensional space (where M is at least one less than the number of points N), it might still be preferable to embed the points in less than M dimensions and, typically, in three for interactive visualization or two so they can be easily plotted. This useful option, called subspace embedding or projection, is used widely to visualize data and, in the following applications, to make simplified representations of proteins. If the $(M - 3)$ dimensions that must be discarded have near-zero eigenvalues, then the projection will provide a good representation of the distances but if the summed value of the discarded eigenvalues constitutes a significant fraction of the total, then the resulting projection will be shrunken (in proportion to the lost eigenvalues) and distorted. In the prediction of protein structure, this is the usual situation and the projection of a poor set of distances directly into 3D results in an structure that is unrecognizable as a protein. Methods to correct for this will be considered in the following sections.

2.2.3 Hierarchical inertial embedding

Most of the computational burden in the distance geometry (DG) method is associated with the diagonalization of $\boldsymbol{M}$. Even the most efficient diagonalization algorithm, the 'QL algorithm with implicit shifts' (Press *et al* 1992) is an $\mathcal{O}(N^3)$ process (i.e. the CPU time requirement grows with the third power of the size of the number of points). For large matrices this can become a serious problem and a way to reduce this was found through exploiting the relationship between the inertial axes and embedding described in the preceding section. This was achieved by breaking the problem into smaller subproblems by carrying out separate embeddings on subsets of the original point-set. The relative orientation of the subsets were then determined by an additional embedding and the final coordinates of the full point set were obtained by rigid-body translations and rotations.

This new approach, called hierarchical inertial embedding (HIP), hinges on a relationship between distance geometry and inertial axes that will be described in the next subsections.

2.2.3.1 Correspondence of DG and inertial axes

Some readers cannot have failed to notice the remarkable similarity in the origin of the central matrices used in calculating inertial axes and DG. If the coordinates are in the $N \times 3$ matrix $\boldsymbol{A}$ then the (3×3) inertial tensor is $\boldsymbol{A}\boldsymbol{A}^{\mathrm{T}}$ while the $(N \times N)$ metric matrix of DG is $\boldsymbol{A}^{\mathrm{T}}\boldsymbol{A}$. Furthermore, the largest three eigenvectors of the metric matrix also correspond with the inertial axes. As is shown below, this is not a coincidence! (but anyone who is happy to take this statement 'on trust' can skip to the next section).

In the calculation of the axes of inertia, the (k, l)th element of the inertial tensor $\boldsymbol{T} = [t_{kl}]$ is defined as:

$$t_{kl} = \sum_{i=1}^{N} a_{ki} a_{li} \tag{2.48}$$

and by substituting the expression for the point coordinates in DG (equation (2.30)), we obtain:

$$\begin{aligned} t_{kl} &= \sum_{i=1}^{N} (\lambda_k^{1/2} w_{ik})(\lambda_l^{1/2} w_{il}) \\ &= \sum_{i=1}^{N} \lambda_k^{1/2} \lambda_l^{1/2} w_{ik} w_{il} \\ &= (\lambda_k \lambda_l)^{1/2} \sum_{i=1}^{N} w_{ik} w_{il} \\ &= (\lambda_k \lambda_l)^{1/2} \cdot (\boldsymbol{w}_k \bullet \boldsymbol{w}_l). \end{aligned} \tag{2.49}$$

Since $\boldsymbol{w}_k$ and $\boldsymbol{w}_l$ are orthogonal unit vectors (they come from the rotation matrix $\boldsymbol{W}$), their dot product will have the value 1 when $k = l$ and are otherwise 0. It then follows that

$$\begin{aligned} t_{kk} &= \lambda_k \\ t_{kl} &= 0 \qquad k \neq l \end{aligned} \tag{2.50}$$

which means that, in this coordinate frame, the inertial tensor $\boldsymbol{T}$ is diagonal. Since its diagonal elements are the principal moments of inertia, these must then correspond to the eigenvalues of the metric matrix.

2.2.3.2 *Embedding ellipsoids*

The relationship described previously can be used to decompose a large DG calculation into a series of smaller problems. The distances between subsets of points can be selected from a large distance matrix as block-diagonal sub-matrices and independently projected into Euclidean space. The inertial ellipsoids of these subsets can then be treated as 'solid' objects themselves and embedded as independent units into a Euclidean space.

In more detail, the procedure consisted of the following steps (outlined in figure 2.8):

(i) Divide the point-set into a number of smaller sets called *clusters* in which every point in the original set should belong to one and only one cluster.
(ii) Construct the *local embedding* for each cluster, using the traditional embedding approach outlined above (section 2.2). The cluster points are represented in a local coordinate system centred on the centroid of the cluster, with axes aligned to the axes of inertia of the cluster.
(iii) Represent each cluster with its centroid and a number of dummy *inertial points*. The pth inertial point is on the pth local axis of inertia, $\sqrt{\sigma_p}$ distance away from the centroid, where σ_p is the pth principal moment of inertia of

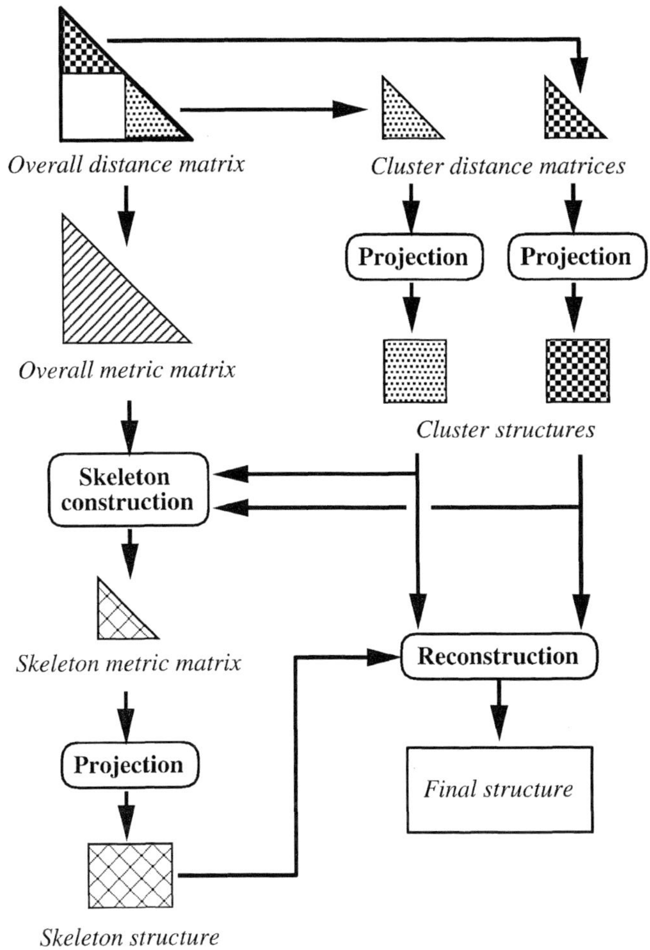

Figure 2.8. Flowchart of the hierarchic inertial projection algorithm. The initial point set is divided into clusters, and the distance matrices of the clusters are embedded to yield the local structures. Their mutual orientation is obtained by the embedding of the skeleton (a point set derived from the centroids and inertial points of the clusters). Finally, the whole structure is reconstructed from the skeleton and the local coordinates. See text for details.

the cluster, known from the local embedding performed in the previous step. A cluster, which was successfully embedded in D_{clu} dimensions, can thus be represented by $D_{\text{clu}} + 1$ points.

(iv) Deduce the mutual orientation of the clusters by generating an embedding of the *skeleton*, a point set composed of the inertial points and centroids of all clusters (figure 2.9). This can be accomplished by constructing the metric

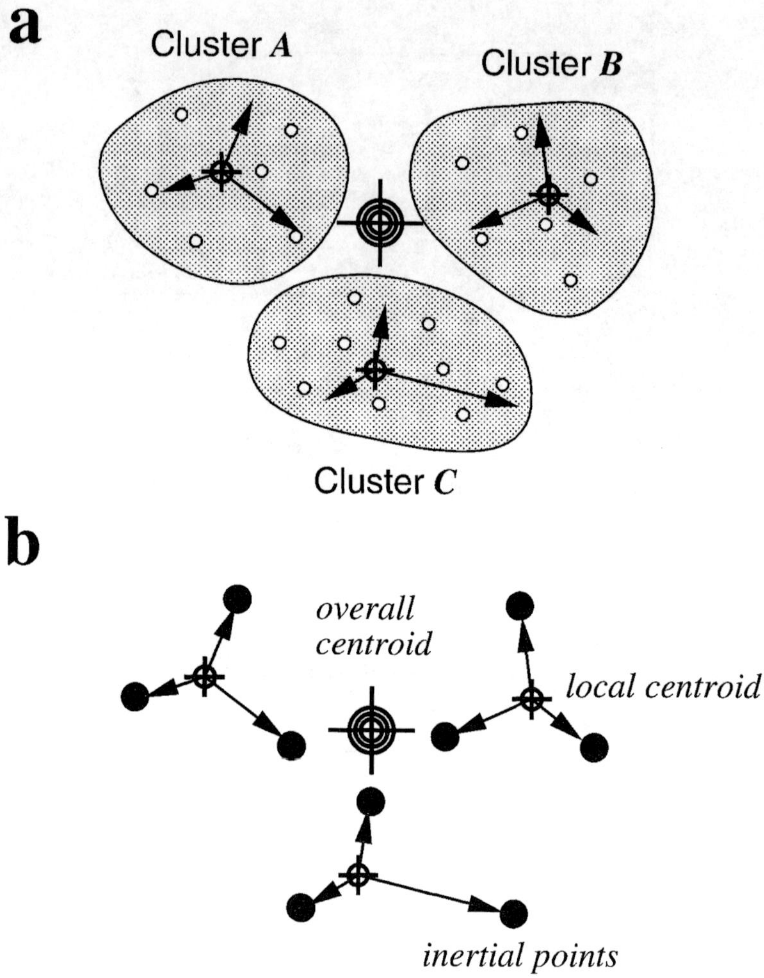

Figure 2.9. Geometric relationships in HIP between the local centroids (a) and inertial points (b) in the skeleton shown for three clusters, A, B and C (shaded).

matrix of the skeleton from the original metric matrix of the full point-set and the local embeddings.

(v) Reconstruct the full embedding of the point-set in the global coordinate system centred on the overall centroid by suitably translating and rotating the local embeddings so that the local centroids and inertial points match the corresponding points in the skeleton.

2.2.3.3 Applications

The HIP algorithm was found to be considerably faster than the traditional method and has a number of applications including the fast generation of model conformations from a set of distance restraints and macromolecular docking simulations.

This algorithm can replace traditional embedding in all distance geometry applications, especially when the number of points to be embedded is large. Although the algorithm described above uses only one level of hierarchy, it is possible to extend it so that the cluster embeddings themselves would be performed by HIP embeddings of subclusters. The projections of very large point-sets, structured as a tree of hierarchic clusters, could thus become computationally tractable. In addition, as each subprojection is independent, they can be calculated in parallel.

Another powerful application of the algorithm is the embedding of rigid bodies. If the Euclidean coordinates of the points in the clusters are known, the cluster embeddings are not necessary, only the cluster inertial axes and moments have to be precomputed once. Afterwards, their mutual orientation is determined by a single skeleton embedding, for which the metric matrix elements are obtained by a procedure similar to the one described in section 2.2.3, using the intercluster distances as input data. This modification provides an elegant way to assemble models of large proteins from smaller fragments of known structure (such as α-helices and β-sheets) and to perform macromolecular docking simulations.

2.2.4 Gradual projection

2.2.4.1 Choice of method

Protein molecular modelling and structure prediction are problems of constraint satisfaction. Both contain very exact local constraints (bondlengths and angles and steric repulsion) that must be satisfied (otherwise the model quickly ceases to be recognizable as a protein). Progressing to a higher level of organization, both contain elements of secondary structure that can impose slightly longer range constrains (although with less certainty where these have been predicted). Above this, a modelling problem is distinguished by having some global constraints that can, at least restrict, or more typically define a unique chain fold. In contrast, the prediction problem lacks specific global constraints and instead has constraints (or more accurately, restraints) of a generic nature, such as the general property that hydrophobic residues tend to pack together.

These various constraints are most commonly expressed as relationships between pairs of objects (typically atoms, residues or secondary structures) and if the relation is not already a simple distance it can often easily be expressed as such. This leads to distance geometry as the most general method to formulate the problem. However, the term is often used to cover two quite distinct techniques.

As described in the previous sections, a set of distances (which may be incompatible in three dimensions) can be embedded into three dimensions such that the sum of squares of the errors from their ideal lengths is minimal. This is the method of subspace projection (or multidimensional scaling) referred to previously as distance geometry but will be referred to in this section simply as the *projection* approach.

Alternatively, a starting configuration of points can be specified in three dimensions which is then refined, through incremental steps, towards an optimal resolution of the distance constraints. This approach is a standard minimization problem and can be tackled using a very wide variety of algorithms and will be referred to here as the *refinement* approach (Brünger and Nilges 1993).

2.2.4.2 *Projection* versus *refinement*

The two approaches introduced here have differing advantages and disadvantages which has often resulted in their application to qualitatively different problems.

Perhaps the most fundamental difference in the approaches is that the projection approach does not require a starting model from which to begin its refinement. This is a great advantage since with a 3D space refinement approach the effect of the starting configuration must be evaluated and if there is no *a priori* guide to this, the best that can be done is to repeat the refinement with a reasonable number of random starting configurations.

If the distance estimates are very uncertain it can also be desirable to use random starting models as a tool to explore the conformational space and test the stability of solutions. In the projective approach, however, this ability is not lost as the initial distance matrix can be perturbed by a random value or using the protocol outlined later, started with a random set of distances (Aszódi and Taylor 1994a).

2.2.4.3 *Distance weighting*

In the refinement approach, individual distances can be given different weights reflecting the degree of certainty with which they are expected to obtain their ideal value. Properly, each weight should be proportional to the reciprocal of the variance of the observed length, but in practice values are often chosen to reflect an intuitive feel for the relative importance of the property. Thus, for example, bondlengths can be given a relatively high weight (reflecting their small variance) to maintain their ideal lengths in the final model. By contrast, in the projective approach, individual distances cannot be weighted and although a mass can be assigned to each point, this affects all its associated distances. Clearly, where we desire a final model with undistorted bondlengths, angles and steric exclusion, this behaviour is very undesirable and is probably the main reason that DG methods have not been more widely adopted in optimization problems.

2.2.4.4 *Chirality*

A further disadvantage of the projective approach is a lack of control over chirality. Any point-set and its mirror image generate the same set of distances so in the inverse transformation the enantiomer that emerges from the projection cannot be predetermined. With such obviously chiral structures as proteins, it is important to have the correct hand. With a good (metric) set of distances this is not a problem since the mirror image can be recognized at atomic resolution by the hand of the asymmetric α-carbon, or at the residue level, by the hand of the α-helix. However, at the residue level if there are no helices, or there is an equal mix of right- and left-handed forms, then we have a problem. This problem, however, is not restricted to the projection approach as the refinement approach also initially lacks this chiral information. The difference lies in the mode of solution: the projection approach calculates the final conformation directly giving no scope for interference whereas the gradual refinement approach allows local chiral structures to be imposed which (it is hoped) propagate their asymmetry to favour the formation of the correct enantiomer. If no asymmetric structures can be imposed, however, then the approaches are equivalent.

2.2.4.5 *Kinetic traps*

To achieve a final fold the stepwise refinement approach must be able to get there along an energetically favourable path and it is possible that low energy folds exist that have no favourable approach path. These would typically involve structures that have some non-local cooperative interaction (such as knots). This problem can be overcome by random sampling (Monte Carlo minimization) but this is a very inefficient method that is unlikely to randomly jump to conformations that involve multiple dependencies. Efficency can be improved using a more 'intelligent' method such as a genetic algorithm (Dandekar and Argos 1992, Petersen and Taylor 2003). In contrast, with its direct materialisation from hyperspace, the projective approach is independent of any kinetic (pathway) bias. This property can be viewed either as an advantage or disadvantage. It can be argued that it allows the projective approach to sample undesired conformations (such as knots), alternatively, it can be argued that the limited simulation time of the refinement approach (relative to real folding times) cannot allow sufficient exploration of the conformational space, resulting in a bias towards structures with sequentially local interactions.

2.2.5 Practical method specification

2.2.5.1 *Combined approach*

For the purpose of *ab initio* structure prediction it is attractive to have a method that is independent of the starting configuration (since there is no preferred initial state) and also free of any kinetic dead-ends (since we want to avoid prolonged calculation times). The problems associated with this approach (weighting and chirality) can be dealt with by a modification to the projection protocol (described later) and a subsequent phase of (real-space) refinement to correct bondlengths and steric violations and impose correct chirality. Viewed from a different point it might be seen that we are simply using the projective approach to provide a starting configuration for refinement—so avoiding (or minimizing) the need to perform repeated refinements from random starting configurations.

2.2.5.2 Dealing with highly non-metric data

The projection of highly non-metric data typically results in a jumbled mass of points that bears little resemblance to a protein structure. A major contribution to this unrecognizable state is the violation of bond-length and steric volume constraints. Ideally, these should be given higher weights in the projection. To avert this problem, and the general problem of individual distance weighting, the distances can be projected not directly into three dimensions, but into a higher dimensional space. In this space the distances between points (represented as multicomponent vectors) can be refined towards their ideal values. The resulting new positions can then used to generate another distance matrix that can subsequently be re-projected. Furthermore, in this matrix, any distance values that were not refined in real-space can be reset to their desired values (so maintaining a 'soft' bias towards the desired packing). Generally this is best done with a degree of strictness reflecting the importance of the effect. This process can then be repeated and the dimensionality of the projection reduced in each subsequent cycle until three dimensional space was reached (Aszódi and Taylor 1994a).

This introduction of intermediate cycles of projection allowed weighting to be introduced but it does not, unfortunately, allow chirality to be refined as the handedness we anticipate in three dimensions is ambiguous in higher dimensions—in the same way that left and right are ambiguous in three dimensions where the direction of view can be altered. Chirality refinement therefore, by necessity, must be carried out entirely in three-dimensional space. The problems that this unavoidable aspect poses for the construction of (highly chiral) protein models will be returned to in a later chapter where we describe the application of DG methods to molecular modelling. In the meantime, the basic method of DG will find many applications in the visualization of complex datasets of pairwise interactions, not all of which need to preserve chirality.

Chapter 3

Sticks to strings

3.1 Secondary structure geometries[1]

The analysis of protein three-dimensional structure often involves the representation of structures in a simplified form. This not only allows the structure to be appreciated more easily by visual inspection but can also lead to considerable savings in computation when many structures are analysed. The degree of simplification must be made with care: too much and important details can be omitted or too little and efficiencies will not be gained. The type and degree of simplification also depends on the aspect of structure being studied. In this work, the main aspects addressed are the overall fold of the protein and the packing of secondary structures. Both of these are central to the comparison of protein structures and to their classification (which will be considered in chapter 6).

The analysis of protein structure using secondary structure line segments has been widely used in many structure analysis and prediction methods over the past 20 years. Its use in methods that compare protein structures at this level of representation is becoming more important as an increasing number of protein structures become determined through structural genomic programmes. The standard method used to define line segments is to fit an axis through each secondary structure element. This approach has difficulties however, both with inconsistent definitions of secondary structure and the problem of fitting a single straight line to a bent structure. In this section we describe a method developed by Taylor (2001) to avoid these problems by finding a set of line segments independent of any external secondary structure definition. This allows the segments to be used as a basis for secondary structure definition by taking the average rise/residue along each axis to characterize the segment. This practice has the advantage that secondary structures are described by a single (continuous) value that is not restricted to the conventional classes of α-helix, 3_{10} and β-strand. This latter property allows structures without 'classic' secondary structures to be

[1] This section is reproduced in part from Taylor (2001) with the kind permission of Elsevier.

encoded as line segments that can be used in comparison algorithms. The method was encoded as a computer program called `STICK`.

3.1.1 Secondary structure line segments

Early simplifications of protein structure often represented secondary structures as line segments (Chothia *et al* 1981, Cohen *et al* 1981) and although some sophisticated alternative schemes have been devised (Sklenar *et al* 1989), the most common method is to fit a least-squares line to the α-carbons of each secondary structure element (Taylor *et al* 1983). This introduces a great saving in structural description without a significant loss of detail. The information that is discarded is the phase of the helix or strand relative to the rest of the protein along with any minor distortions (kinks, bulges or bends). The line segments can then be connected with loops represented with different degrees of detail that can range from the full α-carbon trace, through increasing degrees of smoothing to the situation in which the link between secondary structures is represented only by an abstract line or curve. Alternatively, the connection between secondary structures can be completely neglected, leaving a set of unconnected sticks (which can either be directed or directionless). These 'bones' of the structure are sometimes referred to as its *architecture*.

The economy of the stick description has resulted in great savings in computational time in various structure comparison methods. In general, the number of points is reduced tenfold and for algorithms that typically require execution times with cubic or quadratic order dependency on the number of points, then savings can be considerable (Eidhammer *et al* 2000). Consequently, it is at this level of representation—at which greatest simplification has been achieved with least loss of structural information—that it is convenient to gain an overview of the full range of protein structure and to devise ways in which it can be systematically represented and compared.

3.1.1.1 Problems with current criteria

One of the problems that bedevils the analysis of protein structure at the level of secondary structures is to find a robust automatic definition of secondary structure. Difficulties arise because trivial differences at the atomic level can propagate upwards to become obvious differences at the higher level of representation. A difference of as little as a fraction of an Ångstrom in the position of a main chain hydrogen bonding group might lead to the failure of an algorithm to recognize a potential hydrogen bond. This might then leave a β-strand (on the edge of the sheet) to be too short to be incorporated into the sheet which could lead to a secondary structure representation with one less element between otherwise identical proteins. (See the opening quote to part 2).

One of the advantages of a manual definition of secondary structure is that experts can 'gloss over' these minor abberations and tend to make a more regular

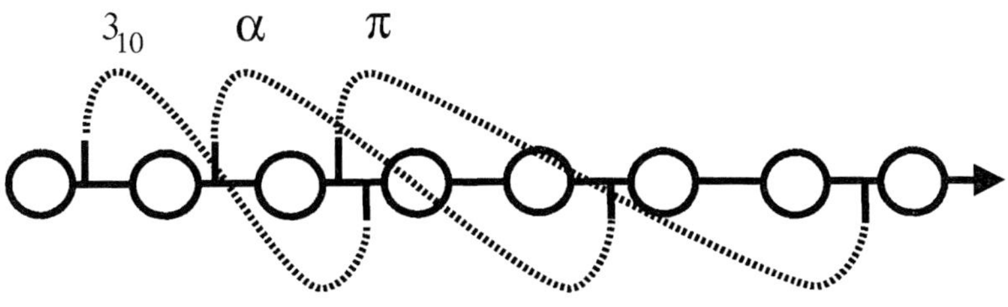

Figure 3.1. A difficult secondary structure assignment. A schematic polypeptide chain is shown with α-carbons represented as circles and peptide bonds as lines from which amide groups project downwards while carbonyl groups point upwards. Three local hydrogen bonds are drawn (broken curves) at three different separations: $O_1 \cdots N_4 = 3_{10}$, $O_2 \cdots N_6 = \alpha$ and $O_3 \cdots N_8 = \pi$. As these ranges overlap, it would be difficult to make a single designation for the segment.

or 'tidy' definition of secondary structure. While good for an overview, if one is analysing disruptions in secondary structure then this is not a very useful approach. To minimize these difficulties, automatic methods tend to have a flexible definition of hydrogen bonding and also tend to base their definition on larger scale structures—such as hydrogen bonded ladders (as in the `DSSP` program of Kabsch and Sander (1983)), so giving some degree of robustness. Differences in methods can also be partially overcome through taking a consensus definition (Colloc'h *et al* 1993) but, if possible, a single robust method is preferable.

A further problem, not well dealt with either by 'eye' or automatically is in deciding what the secondary structure is when there are only a few irregular hydrogen bonds involved. This might seem to be simple since the hydrogen bonds are discrete: progressing through the various helices of 3_{10} , α and π in steps of one residue in the nearest bonded neighbour. However, consider the pathological example in which each of the three helix types follows in progression (figure 3.1). On the fourth residue, the three different helix types overlap each other and while this information can be recorded (as in the `DSSP` program) it cannot be simply encoded in a single string representation.

Finally, relying purely on the secondary structure line segments can result in too great a loss of information for structure comparison. The most extreme situation being when a protein does not contain any secondary structure. In less severe situations, however, large loop regions can be ignored which might well contain characteristic structure that would help in comparison. These can include linear elements (such as the 'pathological' helix in figure 3.1) or fully extended segments that have not been defined as β-structure because they lack hydrogen bonding partners.

3.1.1.2 Line segments from inertial axes

The axis of a secondary structure is typically taken as the line with minimum deviation (least-squares) from the α-carbons and this can be found as the principal axis of the equivalent inertial ellipsoid (Taylor *et al* 1983) (see section 2.1.4). If the size of the three inertial axes are given by A, B and C (in descending order), then for a good linear structure, the ratio (r):

$$r = \frac{A}{B + C} \tag{3.1}$$

will be large. This ratio can be calculated for all segment sizes at all residue positions and the problem is then just to find the optimal combination of segments.

To make the calculation more equivalent over β-strands and α-helices, the protein structure was initially smoothed by averaging successive triples of α-carbons, as described in chapter 8 (Taylor 2000a). Two cycles of smoothing were needed to reduce α-helices and β-strands to roughly linear segments with comparable ratios when calculated using equation (3.1)—resulting in a more 'even-handed' treatment in the further processing of the segments described later. Smoothing also avoids the problem that helices shorter than six residues do not have a unique dominant inertial axis.

3.1.1.3 Dynamic programming solution

As with many problems that incorporate a linear-ordering constraint, the optimal solution (for a given scoring scheme) can be found by the application of the dynamic programming algorithm. This approach has been applied to the current problem, initially by Park and Levitt (1996) (for the definition of secondary structure in rough models) and later by Taylor (2001) to derive a more robust definition of secondary structure in known protein structures. The problem is also similar to the definition of trans-membrane segments (Jones *et al* 1994).

In most applications of dynamic programming, the values (scores) associated with each object are assumed to be additive whereas the current application lies far from this situation. For example; if helical transmembrane propensity is being considered, then the score for concatenating two (identical) helices will be twice the single value. However, the axial ratios defined by equation (3.1) will be quite different depending on whether one helix extends the other or doubles-back (as in a hairpin).

The basic working construct is a matrix of which the dimensions are sequence position against window size and for each of the components in the matrix, the value of the inertial ratio $A/(B + C)$ was calculated. Generally, long thin structures will have a high value but so also will small structures: indeed, for the trivial case of two residues, the value will be infinite. To prevent the unwanted solution of a series of very short segments, a minimum segment size of five was imposed. Trials were then made which indicated that the required normalization for larger segments lay somewhere between a factor of the segment length and its square. However, when tested in the fully developed method (as elaborated later), no simple polynomial normalization could reproduce the observed distribution of segment lengths simultaneously in α and β structure. This problem was overcome by assigning to segments the sum of the values of all their sub-segments. Defining the window at residue m to encompass residues $m - w$ to $m + w$ (that is a window of size $2w + 1$), then:

$$s_{m,w} = \sum_{j=0}^{w-2} \sum_{i=-j}^{j} r_{m+i,w-j} - aw - c \tag{3.2}$$

where s designates the summed scores and r is the raw ratio of the inertial axes for the current window (w) on residue m. The subtraction of the terms aw and c in equation (3.2) prevents the summed score from monotonically increasing with window size. They are somewhat equivalent to the use of the two gap penalties in sequence alignment with c being a constant penalty and a controlling the increase of the penalty with segment size.

It should be noted that the construct of a window with a central residue allows only segments with odd numbers of residues. However, given the resolution of the model, this was not considered a serious problem compared with the complications of asymmetry introduced into the algorithm by considering even numbered segments also. As a bonus, calculation time is halved. Should the full range of secondary structure lengths be required, the method can be simply extended to odd and even lengths by doubling the number of residues in the structure through the addition of 'dummy' atoms midway between the real α-carbon atoms.

The choice of a and c in equation (3.2) controls the typical segment size: if these are zero then one big segment will be obtained, dropping through a series of shorter segments with increasing a and c. This can be seen in the example in figure 3.3 in which a bent helix can be defined as either one or two segments.

3.1.1.4 Algorithmic details

The two segments selected in figure 3.2 (centred on residues 8 and 20) are the obvious choice as each correspond to a local maximum in the summed scores (58 and 341, respectively). While not completely coincidental (as the two segments form a β-α hairpin), it can easily be imagined that the first local maximum could lie closer (say, on residue 9) or require a bigger window—if this were so then the selected segments would overlap. To avoid this unwanted situation, the algorithm 'looks' from one point only to other segments where such a situation would not arise. In figure 3.2(*b*), the allowed choices include only positions lying on or beyond the rising diagonal of values 3,34,58,32,1. More formally, if the current segment (i) is centred on residue position m_i and extends for w_i residues either side, then the preceding segment ($i - 1$) must have $m_{i-1} + w_{i-1} < m_i - w_i$. In

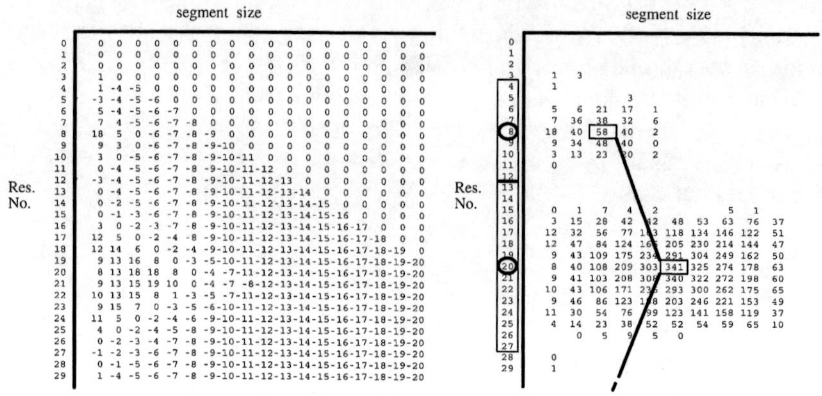

(*a*) Raw values (*b*) Summed values

Figure 3.2. Line segmentation of protein structure. Two matrices are shown at stages of the calculation to segment protein structure using dynamic programming. Each matrix has the protein sequence running downwards and the segment (or window) size increasing towards the right. (*a*) The raw scores: being the inertial ratio $A/(B+C)$ (see text for details) less the penalty $am + c$ with $a = c = 1$ (see equation (3.1)). (*b*) The summed matrix (showing only positive values). The dynamic programming algorithm selects a maximum sum of scores under the constraint that segments do not overlap. In the example, the selected segments are centred on residues 8 and 20 with window sizes (w) of 4 and 7, respectively. (Values are not shown for the trivial columns with $m \leq 1$.)

the current example (figure 3.2(*b*)), $m_2 = 20$ with $w_2 = 7$ while $m_1 = 8$ with $w_1 = 4$ and the inequality: $(8+4) < (20-7)$ is clearly held.

In the application of the method it was considered desirable to have an unassigned position between segments giving the modified inequality:

$$m_{i-1} + w_{i-1} < m_i - w_i - 1. \tag{3.3}$$

Under this condition, the locally optimal score for segment 1 (58 in the current example) cannot now be reached and the next best is to use a smaller window size (scoring 40). However, a better combined score can be obtained by shifting segment 2 to position 21, losing only one unit of score. To consider all such shifts, a matrix of summed scores is compiled that records the best score that can be obtained between the current position, $\{m_i, w_i\}$, and the start of the structure. This can be stated using a recursive relation:

$$t_{m_i,w_i} = s_{m_i,w_i} + \max(t_{m_{i-1},w_{i-1}}) \tag{3.4}$$

where the relationship between segments i and $i+1$ are subject to the condition expressed in equation (3.3) and s is the score defined in equation (3.2). It should

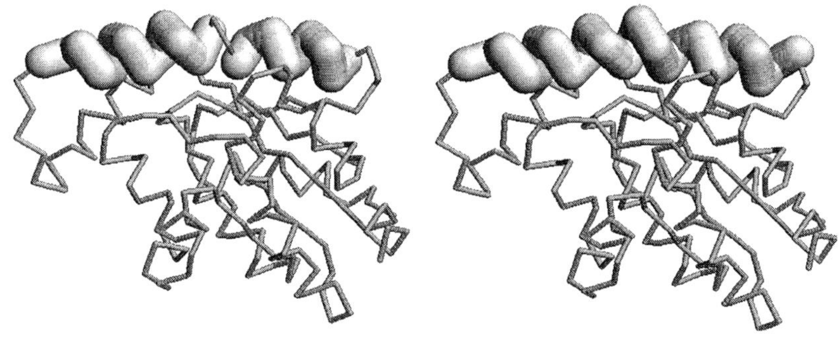

(*a*) default 'gap' penalty (*b*) stricter 'gap' penalty

Figure 3.3. Line segment variations. A small β/α protein (adenylate kinase, `3adk`) segmented under different 'gap' penalties. (a and c in equation (3.2)). In the region of variation, the segment differences are emphasized using a thick curve representation. (*a*) Using the default parameters $a = c = 1$ a long helix is broken into two parts. (*b*) With $a = 0.5$ ($c = 1$) a longer (slightly kinked) helix is selected.

be noted that at this point a conventional gap penalty could be inserted but it was considered that this aspect is better controlled through the use of the 'pseudo' gap penalty incorporated in equation (3.2) which controls the overall size of the segments and not the gap between them. The form of this penalty will be further discussed and refined later.

As in sequence alignment algorithms, a set of pointers is maintained for each $\{m, w\}$ pair, recording the location of the preceding segment from which they 'inherited' their score contribution. These are then traced backwards from the highest scoring position $\{m_N, w_N\}$, so giving the set of N highest scoring segments. Unlike some segmentation algorithms, the current method does not require the number of segments to be specified beforehand.

Also as in sequence alignment, rather than search all possible values allowed under equation (3.3), for each position, a set of maxima can be recorded for each diagonal ('row') of constant $m - w$.

3.1.2 Secondary structure definition

3.1.2.1 'Continuous' secondary structure types

This approach parses the protein structure into lines and each line can be characterized by the residue/length (density) or by the length/residue (rise). The length/residue along the axis is referred to later as the 'rise' (d) and is related to the pitch of the helix as: $d = p/n$, where p is the pitch and n is the number of residues/turn (Dickerson and Geis 1969). The helices important in proteins are: π ($n = 4.4$, $d = 1.0$), α ($n = 3.6$, $d = 1.5$), 3_{10} ($n = 3.0$, $d = 1.9$) and the 2_7 ribbon ($n = 2.0$, $d = 2.6$). Either of these measures is effectively a definition of secondary structure but, unlike the definition of secondary structure based on hydrogen bonds, it is not discrete and it is thus unnecessary to make explicit definitions of secondary structure type—so allowing more freedom for ambiguous structures (loops, 3_{10}-helices or distorted β-strands) to assume different rôles. Indeed, the problem of the pathological structure described in figure 3.1 is resolved, as it becomes identified as a clearly linear segment with a residue rise approximating the α-helix.

3.1.2.2 Variance weighting

Along with the helical rise, the variation of the residue spacings within a segment was also calculated. It was thought that this might provide a useful contribution to the raw score and the following formulation was tested:

$$r' = r \cdot \exp(-\sigma^2 f) \tag{3.5}$$

where r is the raw score defined in equation (3.2) and σ^2 is the variance of the residue spacings along the segment when projected perpendicularly onto their axis. The exponential is a Gaussian function, the 'decay' of which is controlled by the parameter f. Values of $f = 5$, 10, and 20 were tested. Despite the use of variance-based clustering as the principal means of segmentation in other methods (Hawkins and Merriam 1973, Bement and Waterman 1977), this modification was of secondary importance in the current application. However, the approach has the potential to detect disruptions in regular secondary structure that do not result in a change in direction or even, in the extreme, of a colinear α-helix and β-strand.

3.1.2.3 β-sheet definition

The algorithm described to this point will define linear segments and characterize them by their average helical rise. However, it cannot distinguish between an extended segment hydrogen bonded in a β-sheet and one in a loop region. Such a distinction can be made beforehand on the basis of the α-carbon coordinates or afterwards based on the relationship between the line segments. Only the first option will be considered here as it leads to a more direct parametrization of the method by keeping the secondary structure and the segment definitions distinct.

The algorithm of Kabsch and Sander (1983) defines β-structure based on 'ladders' of hydrogen bonds between strands using a simplified hydrogen bond model. Using just α-carbons, it is necessary to consider larger fragments of structure to avoid spurious bonding partners and the current approach is based on that used in the definition of protein domains (Taylor 1999c) (see section 5.1). In the current application the slightly smoothed coordinates are used (as already defined) which averages over random fluctuations in the distance between strands.

A base separation of $d = 4.7$ Å was taken as the ideal separation of adjacent α-carbons hydrogen bonded in a β-sheet. Onto this value, an incremental margin of error e was added to allow for increasingly tenuous relationships. All pairs of residues (i and j) with separations less than $d + e$ were considered and those in which the adjacent pairs in the 'ladder' ($i + 1, j \pm 1$ and $i - 1, j \pm 1$) were both within $d + 2e$ were declared to be bonded. In addition, if three residues are aligned as: j–i–k, then the more relaxed condition ($d + 2e$) was tested for the i, j and i, k pairs and a further relaxation of $d + 3e$ was used for the $i \pm 1$ partners.

For each pair of residues identified by these tests, an element in a (symmetric) score matrix (N) was incremented by 1, resulting in high scores for the links between residue pairs most deeply buried in sheets. This range of scores was exploited below in the tuning of the method to reflect the observed secondary structure composition of proteins.

3.1.3 Comparison to standard definitions

The `DSSP` method (Kabsch and Sander 1983) defines a variety of helices, turns and sheets including both α and 3_{10} helices as well as residues in β-sheets and those with isolated hydrogen bonds (β-bridges). For comparison with `DSSP`, β-bridges were ignored and the remaining β-residues (designated 'E' in `DSSP`) were compared with those involved in the sheet network defined above. A distinction was made between helix type on the basis of axial rise with an ideal α-helix having a rise of 1.5 residue/Å while a β-strand has 3.1 residues/Å and a range between these, from 1.7 to 2.3, was taken as 3_{10} helix. It should be noted that this designation of 3_{10} helix is only a convenient shorthand as the range of rise encompasses a wider variety of structure from distorted α-helix to highly twisted strands.

The segment-based method was applied to a reduced PDB data set of structures and the parameters adjusted to produce definitions that matched the secondary structure composition calculated by the `DSSP` program on the same data set. This was done for different values of the parameter b and with a few cycles of iteration, the composition produced by each parameter set could be matched to the `DSSP` composition to within a fraction of a percent across all the structure types. (See Taylor (2001) for further details.)

3.1.3.1 Analysis of length distributions

The distribution of the lengths of α-helices showed greatest deviation for small helices: with a large number (1400) of single turn helices (3–4 residues) being defined in the `DSSP` records and, of course, none with the current method (as they were disallowed). The number found in the PDB was intermediate. These disparities converged, with the frequency of helices up to ten residues being closer in all three methods. Beyond this, the `DSSP` and the PDB distributions decreased together while the current method continued to a maximum around 11–

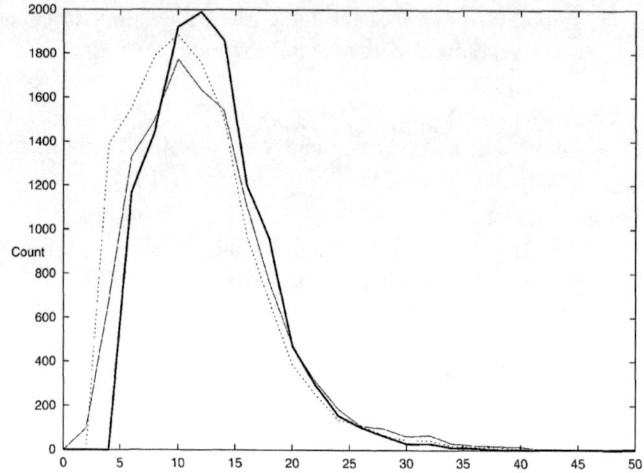

(a) α

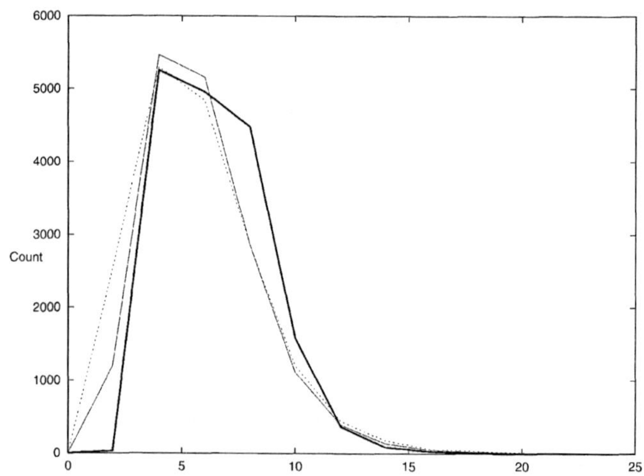

(b) β

Figure 3.4. Secondary structure length distributions. (a) α-helix distribution and (b) the β-strand distribution. On each plot the distributions are shown for the `STICK` program (bold full curve) with the PDB (thin full curve), `DSSP` (broken thin curve). Both distributions were 'binned' in a class interval of two and the lines connect the higher value in each bin. (For example; the point at 10 in the α-helix plot contains the counts for helices of length 9 and 10.)

12 residues before falling-off in a similar manner (figure 3.4(*a*)). This extended peak would be expected in compensation for the loss of very small helices (since the distributions have equal area under the curves). A further difference could be seen in the region of large helices where the PDB and DSSP declared helices as large as 82 and 80 residues (respectively) while the largest found by the current method was 49 residues. These large helices, however, have marked supertwists and cannot be expected to be represented by a single straight line segment. Care must also be taken when dealing with PDB secondary structure definitions as some apparently long structures are simply multiple secondary structures with no 'break' (unassigned residue) between them (e.g. `1ppf`).

The distribution of the lengths of β-strands again showed marked differences in the shortest lengths where the DSSP method had 2500 strands of length less than three compared to none for the current method (where they are forbidden). At these lengths, however, the current method was in closer agreement with the PDB definitions which had less than 600 strands in this length range. All the distributions peaked around five residues and dropped off to almost zero by 15 residues. Beyond this, the PDB and DSSP definitions allowed strands up to 48 and 34 residues (respectively). As with the long helices, it would be unreasonable to expect these twisted structures to be fitted by a single line segment and the longest strand found by the current method was 19 residues. The loss of strands at both the low- and high-length ranges again led to a broader shoulder on the upper edge of the β distribution (figure 3.4(*b*)).

3.1.3.2 Refined definitions

Looking across the full range of variation and rise values (figure 3.5(*a*)), two dominant clusters are apparent around $d = 3$ and $d = 1.5$, corresponding to the main regions of β and α structure, respectively. Between these regions, a more minor cluster can be seen, corresponding to the rise of the 3_{10} helix (1.9 Å/residue). The region 1.7–2.3 taken to span this cluster can be narrowed slightly to 1.75 and 2.20 Å/residue. It should be remembered, however, that the term 'rise' is only a convenience and these substructures are not necessarily internally hydrogen bonded or even helical.

In the β region, distinguishing between the 'B' (bonded) and 'E' (extended) states revealed that the latter was slightly less regular (virtually no segments with $\sigma < 0.1$) with the bulk of its distribution lying below the ideal β-strand rise of 3.1 Å/residue. However, the two distributions have considerable overlap as the 'E' state contains a large number of 'true' β-strand that are only found to be bonded when the full multichain and multimeric state of the protein is considered (figure 3.5(*b*)).

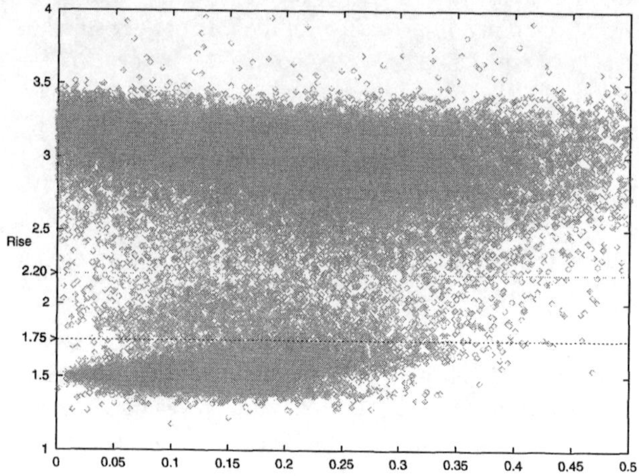

(*a*) full range

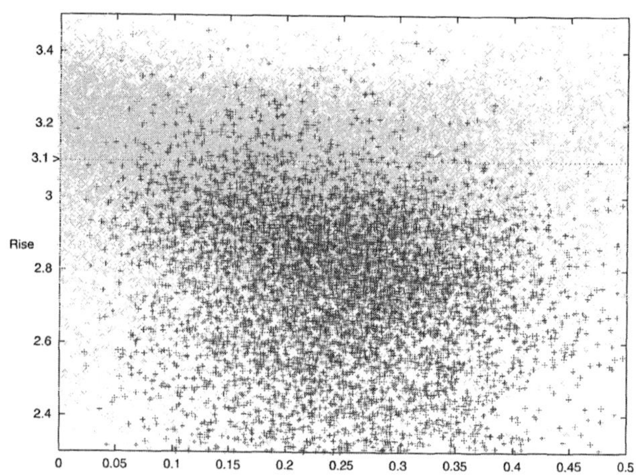

(*b*) β-region

Figure 3.5. Axial rise plotted against variation. The average rise of the residues along the axis of each segment (in Å/residue) is plotted against the standard deviation of the rise within each segment for the 45 000 secondary structures defined in the reduced PDB data set. (*a*) Shows the full range of rise with clear clusters at 1.5 and 3 (α and β structure) and a minor cluster between corresponding largely to 3_{10} structure. The lines at 1.75 and 2.20 mark the refined range of the 3_{10} class. (*b*) Shows an expansion of the β region in which bonded β-strands ('B') are distinguished from isolated strands ('E') as grey diamond and black cross, respectively.

3.1.4 Applications and further developments

The method described in this section has addressed the problem of how to decompose a protein structure automatically into a series of line segments. Where previous approaches have started with a definition of secondary structure (on which line segments are then based), the current method takes the definition of line segments as the principal element of structure and investigates how they relate to secondary structure.

The basic method requires just α-carbons and has only one key parameter (a in equation (3.2)) which controls the average size of the structural segments (similar to a gap-penalty). This was modified by giving a small boost to 'encourage' the selection of shorter β-strands, which were in turn defined from the α-carbons with just one adjustable parameter. Together these three parameters gave sufficient freedom to control the composition and the length distributions of the main secondary structure types. The latter deviate only for the shortest and longest segments with the difference in short segments arising because they were specifically avoided by the current approach while the longer secondary structures were often 'broken' by the `STICK` program into shorter (more linear) subsegments.

3.1.4.1 'Continuous' secondary structure

By considering the rise of the residues along the line segment as an indicator of secondary structure type, the `STICK` program is not constrained to definitively assign each segment as α, β, 3_{10} or extended. This becomes useful later in protein structure comparison (and threading) in which a segment of ambiguous type will retain the ability to be matched against different secondary structure types in a comparison program.

This property can be extended to structures that have few or no conventional secondary structures. For example, the structure of the HIV-1 transactivator protein (`1tiv`) has no secondary structure defined in the PDB or by `DSSP`, yet the `STICK` program finds a variety of linear segments (figure 3.6) all of which would provide a basis for comparison with other structures.

3.1.4.2 Definition on rough and incomplete models

One of the less automatic components of the work described previously was dealing with the many 'unusual' definitions of secondary structure found in the PDB. It would be useful to perform such checks at an earlier stage of the analysis—ideally, when the structures are initially deposited in the PDB. Such a check might also be adequately made using the `DSSP` (or similar) program. However, with low resolution structures or structures determined by NMR, `DSSP` does not always perform well on this quality of data (figure 3.6) and does not perform at all on structures which have only α-carbons. As more structures are determined by high-throughput automatic methods (as proposed for structural

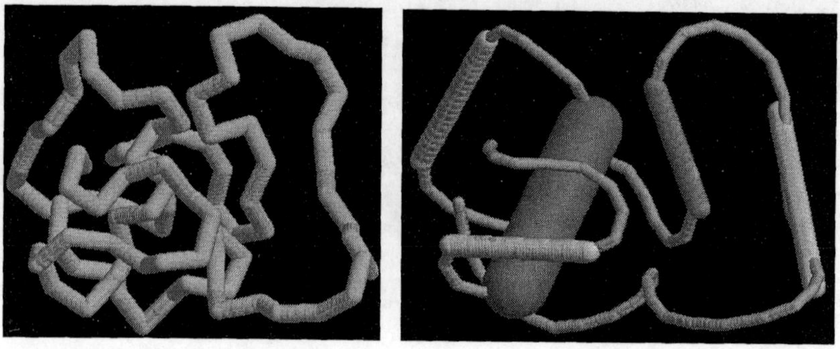

(*a*) α-carbon (*b*) STICK segments

Figure 3.6. Linear segments in 1tiv. The small protein 1tiv (HIV-1 transactivator protein) contains no conventional secondary structures (as found in the PDB or defined by DSSP). Nonetheless, it contains a variety of linear segments that are identified by the STICK program (shaded and drawn with a thickness proportional to their rise/residue value).

genomics programmes) it will become increasingly important to have a robust method for secondary structure assignment.

A problematic aspect of protein structure analysis is that differing definitions of secondary structure can arise depending on the segment of chain that might have been extracted from the full crystal structure. This can make a dramatic difference to the conventional definitions. In this situation, the STICK program is rather more robust than a hydrogen bond based method as it will at least find an extended segment in the position of an isolated β-strand, where, by contrast, DSSP will find no structure.

The ability to work on just the α-carbon data is useful when assessing rough models that might have resulted from molecular modelling or from *ab initio* methods (Aszódi *et al* 1995b). This ability allows such structures to be compared quickly to a collection of native structures represented as 'stick' structures (Taylor 2000c). The use of the STICK method in this context will be returned to in later chapters.

3.1.4.3 Other areas of application

The current method need not be restricted to structural data but could similarly be used to segment (or parse) predicted secondary structure 'probabilities' in such a way that the expected composition and length distributions would result. This would provide a function similar to that played by the second (post-processing) layer of neural-nets in the PHD prediction program (Rost and Sander 1993).

Other areas in the analysis of protein structure that involves segmenting the sequence is in the definition of domains (see section 5.2). Here the linear (sequence) component is less important as the domains can be composed of multiple chain segments (figure 1.7), however, when the domains are constrained to be colinear as when they occur as repeated segments (see chapter 10) then the algorithm could be applied.

3.2 Simplified architectures

As we have seen in the previous section, secondary structures are extended (helical) objects and, because of their linear axis, often pack in a roughly aligned manner as in a bundle of rods. This allows the structure of proteins to be displayed in a very simplified manner, either as idealized stick-figures or by neglecting the extended dimension, portraying only the ends of the 'rods' in a two-dimensional representation.

In this section, some basic framework structures (also referred to as 'architectures' or 'forms') will be described. In these descriptions, we will concentrate mainly on the packing of the secondary structure elements (SSEs). The possible ways in which they can be connected by a protein chain will be deferred to the following chapters on protein topology (Part 3) while methods that allow them to be identified in the stick representations derived from real proteins will be considered in the chapter on protein structure comparison (chapter 6).

3.2.1 Stick packing[2]

The transformation of a stick model of a protein into a more regular form, or its reduction to a simpler 2D representation, requires a measure of how strongly the sticks in the structure interact. The relationship between those with no interaction can then be distorted to preserve the more important representation of those with a strong interaction.

3.2.1.1 Line segment overlap

The degree of interaction between two line segments would be expected to be greatest for long segments that run close together and least for remote end-to-end juxtapositions. A geometric measure that captures these features is the degree of overlap between the segments. This can be quantified by the length of the region over which the two segments can be connected by a series of lines with end-points equidistant from the contact normal[3] (figure 3.7). However, the overlap length

[2] This section is reproduced in part from Taylor (2002b) with the kind permission of the *Journal of Cellular and Molecular Proteomics* (Highwire Press).

[3] The contact normal between two extended lines is the unique line that is perpendicular to both. Occurring within a curve, this is sometimes called the 'doubly critical distance'. An explanation of its calculation (including computer code) can be found in the Web site associated with the book.

must be modified by how closely the two line segments lie together—with close lines attaining a higher interaction score. A simple reciprocal of the approach distance was considered but this was found to give too great a score to close lines and instead a Gaussian function was used. This was further modified by setting a base level below which the score was set to 1 and above which the Gaussian damping applied, as follows:

$$a = \begin{cases} 1 & x \leq b \\ \exp\left(-\frac{(x-b)^2}{d^2}\right) & x > b. \end{cases} \tag{3.6}$$

In this equation, x is the distance between two points that lie equidistant from the end-points of the contact normal between the two lines. The parameters b and d are the distance cutoff (b) beyond which the Gaussian decay is applied with a damping factor determined by d. The values of b (for base) and d (for decay) will be adjusted later.

The interaction of the line segments was measured by summing the inverted distance (a) over a series of lines that have end-points on the two line segments equidistant from the contact normal of the lines (figure 3.7). The set of lines have a separation of 0.1 Å giving typically 100 lines summed for an average interaction. The summed measure will be referred to below as the 'inverse overlap area' or IOA.

3.2.1.2 Solvent accessible surface area changes

The interactions of the segments of protein structure corresponding to the SSEs were measured using the change in the solvent accessible surface area (ASA) observed when each segment was removed from the intact structure. This approach follows earlier studies (Richmond and Richards 1978, Cohen *et al* 1981) but used the `DSSP` program (Kabsch and Sander 1983) to calculate the solvent areas rather than the original program of Lee and Richards (Lee and Richards 1971).

The ASA (summed over each residue) was calculated firstly using the intact protein (length N) giving a set of residue areas $\{C_1 \ldots C_N\}$. The two segments were then removed in turn and the areas recalculated giving two further sets of residue areas $\{A_1 \ldots A_N\}$ and $\{B_1 \ldots B_N\}$ (in which the residue numbering of intact protein is retained). If the first segment runs from $a_n \ldots a_c$ and the second runs from $b_n \ldots b_c$, then the combined effect of removing each segment on the other (summed area change, or $\boldsymbol{SAC}$) can be found as:

$$\boldsymbol{SAC} = \sum_{i=a_n}^{a_c} (A_i - C_i) + \sum_{i=b_n}^{b_c} (B_i - C_i). \tag{3.7}$$

Note that (within the error of the area calculation) the areas of the intact protein (C_i) are always less than those after removal of a segment. In addition, the linear

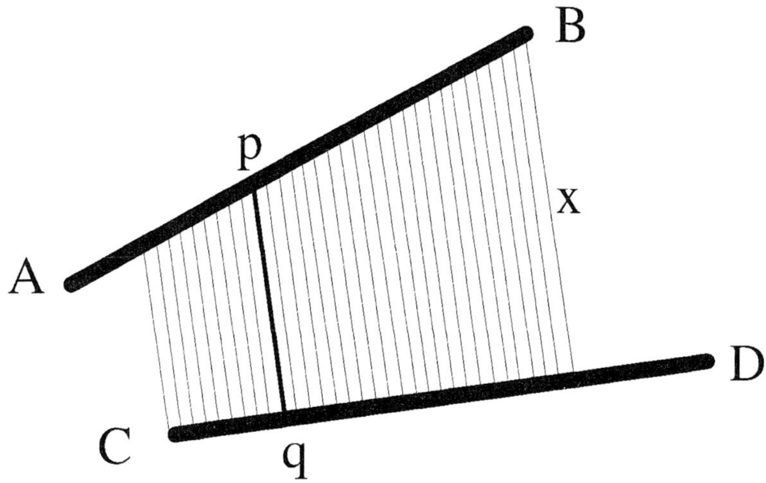

Figure 3.7. Line segment overlap measure. Two line segments corresponding to secondary structure elements are shown (***A*** $\rightarrow$ ***B*** and ***C*** $\rightarrow$ ***D***) as thick lines with their mutually perpendicular connecting line (***p*** and ***q***) shown at medium thickness. (This may lie outside one or both of the line segments.) A series of fine lines cover the span in which the line segments overlap, the end-points of which are equidistant from their corresponding ends of the mutual perpendicular. A measure of interaction is calculated from this as a summation of the lengths (x) of these lines as specified in equation (3.6).

segments were always separated by at least one residue to avoid covalent bonded surfaces being exposed and counted (section 3.1.1.4).

3.2.2 Calibrating segment packing

3.2.2.1 Continuous packing classes

Each segment was characterized only by the length/residue (rise) along the segment axis (Taylor 2001) (see also in the previous chapter). This can be extended to pairs of segments as: $R_{ij} = r_i + r_j$, where r is the rise along the axis of the segment for segments i and j. Plotting the combined rise of both segments against their interaction strength (figure 3.8(a)) shows that the $\alpha\alpha$, $\beta\beta$ and $\beta\alpha$ classes remain sufficiently distinct and that little information is lost by reducing the pair of values to one number.

3.2.2.2 Initial adjustment

The solvent accessible surface area changes (SACs) were calculated for each pair of segments as described in section 3.2.1.2 for a sample of 300 proteins, all of which were free of any errors reported by the `DSSP` program (such as missing atoms or chain breaks). The IOA was also calculated for each segment pair, using an initial estimate of $b = 5$ and $d^2 = 40$ and plotted against the corresponding SAC value. When broken down into the different packing types ($\alpha\alpha$, $\beta\beta$ and $\beta\alpha$), it was clear that the $\alpha\alpha$ class has less SAC/IOA relative to the $\beta\beta$ class and that the latter also has a large number of IOA interactions with little or no SACs. To adjust for this, the IOA parameters b and d were made functions of the segment types: giving the $\alpha\alpha$ interaction both a longer flat region and a slower decay than the $\beta\beta$ type (with the $\beta\alpha$ intermediate). This was done by setting $b = d = p - R_{ij}$, where p becomes the new parameter to be adjusted and R is the joint rise of both segments i and j, as calculated above. The rise along the α-helix is 1.5 and 3.1 along a β-sheet ($R_{\alpha\alpha} = 3.0$ and $R_{\beta\beta} = 6.2$) making a value around 10 a suitable estimate for p. While this reformulation helped with the problems outlined above, it did not completely eliminate the larger average SAC/IOA ratio for $\alpha\alpha$ packing. This was then corrected by applying a small explicit multiplying factor to the IOA values of: $1+1/R_{ij}$. This increases the $\alpha\alpha$ IOAs by 14% relative to the $\beta\beta$ values giving the improved correspondence plotted in figure 3.8(*b*).

3.2.2.3 *Secondary structure packing plot*

For two segments i and j, their combined rise (R_{ij}) can be plotted against their interaction strength as estimated by the IOA measure. This gives a very quick visualization of the type of protein in terms of its secondary structure packing (or architecture). Furthermore, the comparison of two of these plots can give a rough measure of the similarity of the packing between two proteins. Without resolving the specific identity (or sequence order) of the SSEs, it is possible to match-up similar interactions. For example: a small $\beta\alpha$ protein (`5nul`) with two helices above a five stranded sheet and three helices below, will have three $\alpha\alpha$ interactions and four $\beta\beta$ interactions plus various $\beta\alpha$ interactions. When plotted with a structurally equivalent protein (`1fxl`), corresponding interactions are apparent to the 'eye' (figure 3.9). The use of this approach in protein structure comparison will be returned to in chapter 6.

3.2.2.4 *Weights for distance geometry*

As mentioned above, the strength of interaction between two SSEs can be used as a measure of how important it is to maintain their juxtaposition in a simplified representation of structure. These values can be used directly with the method of distance geometry (DG) described in section 2.2. Using the term in its wider sense, the weights can be directly incorporated into a refinement method in which the minimal weighted error is found between the original distances and those in the simplified model. Thus, the relationship between pairs of weakly interacting SSEs can be more distorted compared to the well-packed pairs. The refinement approach may be best for large structures but when dealing with a secondary

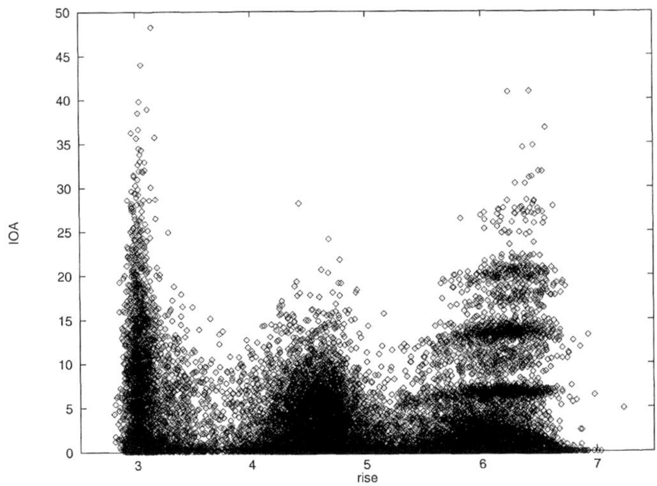

(*a*)

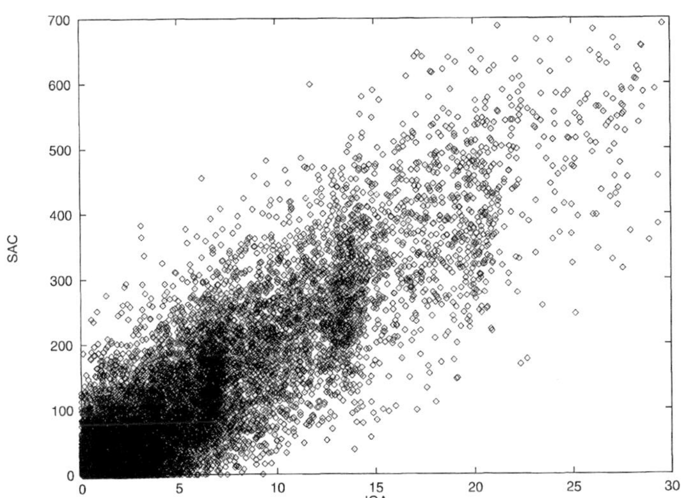

(*b*)

Figure 3.8. (*a*) The interaction of pairs of SSE lines (IOA) is plotted against their combined helical rise ($3.0 = \alpha\alpha$, $4.6 = \beta\alpha$, $6.2 = \beta\beta$). The stratification in the $\beta\beta$ peak results from the addition of pairs of residues to bonded strands and not to separation in the sheet. (*b*) The solvent accessible surface area change (SAC) seen on extracting the SSEs is plotted against the normalized line segment overlap area (IOA).

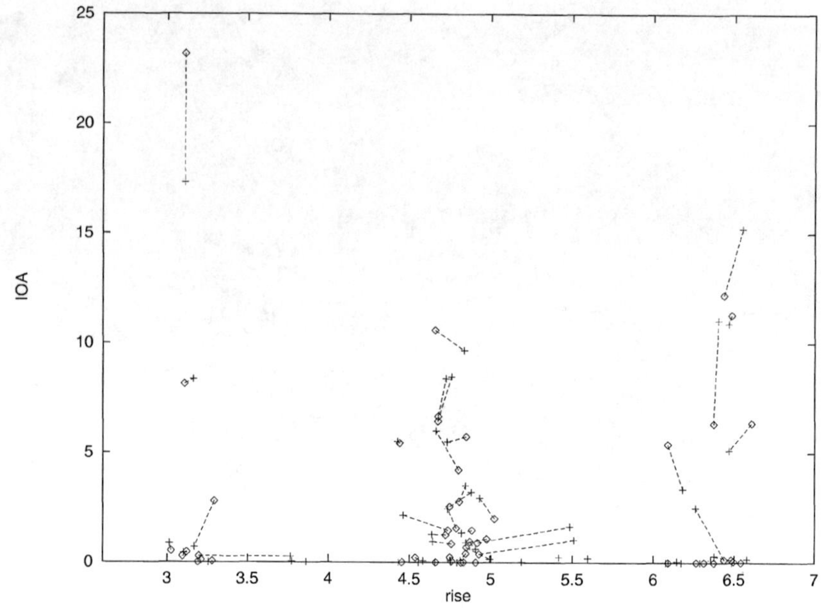

Figure 3.9. SSE packing interaction plot. The combined *rise* of the two secondary structures (x-axis) are plotted against their interaction as measured by the *IOA* overlap measure (y-axis). Data are shown for two small $\beta\alpha$ proteins `5nul` (diamonds) and `1fx1` (pluses). Corresponding interactions are linked by a broken curve. For example; $\alpha\alpha$ interactions lie close to rise=3 and three corresponding pairs can be seen (strong, medium and weak).

structure representation of moderate-sized proteins, there are not so many points to consider and a projection (or embedding) approach can be used.

The number of distances is especially low if only a single distance is considered between pairs. This can be used to produce a 3D representation of the network of SSE interactions but, given the limited number of interactions, it is usually possible to produce a good two-dimensional (2D) representation of the interactions. This can, of course, be checked by considering the size of the eigenvalues (as described in section 3.1). These 2D representations will be considered later but before that, some simple models will be considered that retain a token third dimension.

3.2.3 Layer architectures

In domain sized units, the secondary structures are typically between 10–20 Å in length and pack at roughly 10 Å apart. This makes 10 Å a convenient unit in which to describe their interactions in a simplified form. Further regularity is introduced in the form of the β-sheet which has a strictly set β-strand spacing of (just under) 5 Å. Together, these dimensions can be used to generate an idealized stick-figure to represent a protein. In this representation protein structures appear as layers of packed secondary structure (figure 1.4) typically, β on β (the β-sandwich class) or a β-layer between two α-layers (the alternating β/α class). The layered structure is clear in the preceding classes because of the regularity imposed by the hydrogen bonded β-sheets. However, this constraint is not present in the all-α class which adopt a less regular variety of forms.

As described in section 1.2, the units of globular proteins are secondary structures which pack together to form a hydrophobic core. Providing the protein main chain atoms are tied-up in one of the two secondary structure types, a core can be constructed using any mix of α or β layers (Chothia and Finkelstein 1990, Finkelstein and Ptitsyn 1987). Seldom more than four layers are ever seen in proteins and as these can be composed of only one of two secondary structures (i.e. no mixed layers), then the possibilities are few enough to enumerate.

- two layers: $\beta\beta$; $\alpha\beta$; $\alpha\alpha$.
- three layers: $\beta\beta\beta$; $\alpha\beta\beta$, $\beta\alpha\beta$; $\alpha\alpha\beta$, $\alpha\beta\alpha$; $\alpha\alpha\alpha$.
- four layers: $\beta\beta\beta\beta$; $\alpha\beta\beta\beta$, $\beta\alpha\beta\beta$; $\alpha\alpha\beta\beta$, $\beta\alpha\alpha\beta$, $\alpha\beta\alpha\beta$, $\alpha\beta\beta\alpha$; $\alpha\alpha\alpha\beta$, $\alpha\alpha\beta\alpha$; $\alpha\alpha\alpha\alpha$.

(These combinations allow for reversals since proteins do not distinguish top from bottom.)

This gives 19 possible combinations, but this is an overestimate since adjacent layers of α-helices are not always distinct. (The helices lack the strict registration imposed by the hydrogen bonding through the β-sheet.) Among these, not all possibilities are equally favoured in nature: amongst the three-layer options, the $\alpha\beta\alpha$ combination is very widespread while in the four-layer structures, the corresponding $\alpha\beta\beta\alpha$ structure is also encountered frequently. The occurrence of these different forms among the known protein structures will be considered in more detail in later chapters.

3.2.3.1 Finkelstein and Ptitsyn analysis

Finkelstein and Ptitsyn (1987) considered the possible arrangements of secondary structures in layers and 'digitized' proteins on a 2D grid (or frame) in which the α-helix occupied one grid square while for a β-sheet, each square was divided (vertically) into two β-strands. Using only the rule that β and α elements cannot occur in the same layer, they considered the surface area of contiguous arrangements of SSEs. This was measured as the perimeter of the assembly times a depth (length, L) plus twice the surface area on the grid. The value of L was a function of the surface area and the number of residues (N). This interdependency of the length on the secondary structure composition makes the model more complex than the simple assumption of a fixed length but allows more realistic modelling of actual proteins. The number of residues and surface area were combined into a measure of compactness which was greatest for cubic arrangements of elements (that is, those forming a square on the grid).

The model also gives the expected distribution in the degree of burial of the segments with some of the larger structures containing segments that are completely buried. Generally, proteins do not contain many segments that are completely buried—if they did then they might be too hydrophobic to fold or might find a better home in a lipid membrane. This gives two opposing trends: one towards compactness, the other trying to avoid excessive burial. Structures in the middle tend to have more elongated sections or, if they are large, break into distinct domains (figure 1.7). (See also chapter 5.)

3.2.4 Layer-based stick models

The models of Finkelstein and Ptitsyn (1987) are fixed on a secondary structure lattice and although they retain the flexibility of variable depth, they do not model the twist that is found to varying degrees in all layer-based structures. Although this aspect does not impinge on their analysis, for the purposes of comparing stick models to real proteins it is necessary to incorporate twist into the models.

3.2.4.1 $\alpha/\beta/\alpha$ layers

To represent the $\alpha/\beta/\alpha$ layers as a twisted model, a framework similar to that used previously for prediction (Cohen *et al* 1982) was used that consisted of a core β-sheet with a 20° twist between β-strands (spaced at 5 Å at their mid-points). The α-helices were placed above and below this sheet using a construction that preserved the local interactions with the sheet as previously used in the construction of ideal frameworks for transmembrane helices (Taylor *et al* 1994b), creating a realistic staggered packing between the helices. (See the Web site associated with the book for details and computer code for this construction.) Each helix lay, on average, 10 Å above the sheet and each secondary structure was 10 Å in length (figure 3.10(*a*)).

3.2.4.2 β/β layers

The model for the $\alpha/\beta/\alpha$ layer structures can also be used for stacked β proteins by, paradoxically, neglecting the β-strands (the middle layer) and reducing the scale by half. If the outer layers (previously α-helices) are taken as β-strands then the model is a good description of two twisted β-sheets packing against each other (Taylor 1993c). This is similar to that used previously in prediction by Cohen *et al* (1980) and more recently by Finkelstein and Reva (1991) (using a self-consistent field method) (figure 3.10(*b*)). The behaviour of these models under varying twist and their relationship to β-barrels will be considered in the next chapter.

3.2.4.3 $\alpha/\beta/\beta/\alpha$ layers

The four-layer $\alpha/\beta/\beta/\alpha$ model can be derived from the β/β layers by adding α-helices above and below the sheet using the same construction as was used to add helices to the single β-sheet.

All these layer models can be extended at their edges into a twisted ribbon (helical) structure, allowing any number of β-strands to be incorporated into any particular model.

3.2.4.4 β/α-barrel proteins

A β/α-barrel structure can be constructed along the lines of a squirrel-cage (an exercise wheel more commonly used for pet hamsters) in which the β-strands are represented by the rungs around the circumference (Lesk *et al* 1989, Scheerlinck *et al* 1992). To maintain a twist between the β-strands, however, the two sides of the wheel must have a relative displacement, which is most simply made by connecting each rung not to its opposing neighbour but to a position slightly further round (figure 3.10(*c*)). (A more detailed analysis of β-barrels will be considered in the next chapter.) This basic model can be 'decorated' with α-helices in a similar way to the $\alpha/\beta/\alpha$ layers, producing a framework for the alternating β/α-barrel proteins (figure 1.5).

Although this and the preceding 'flat' structures are highly idealized, they still retain a good fit to real protein structures (when represented in a similar stick form). Two of these are shown in figure 7.5. (The method used in this comparison will be described in chapter 6.)

3.2.4.5 Transmembrane models

A specialized protein architecture can be found in the bundles of packed helices that typically form integral membrane proteins. Neglecting their reversed hydrophobic polarity, these helices can also be modelled using the twisted lattice of sticks described above (Taylor *et al* 1994b, Bowie 2000) (figure 3.11).

3.2.4.6 All-α proteins

The transmembrane model (figure 3.11) can also be extended into a general layer structure that can be used as a model for some all-α proteins in which the helices all lie in a roughly (twisted) parallel manner. However, all-α proteins lack the hydrogen bonded constraint of any β-sheet to maintain absolute layers and as such have a much more varied and irregular set of possible structures. Some of this irregularity can be captured by greatly increasing the twist of the basic model (until some helices become almost orthogonal) but for small structures this is not a natural construct.

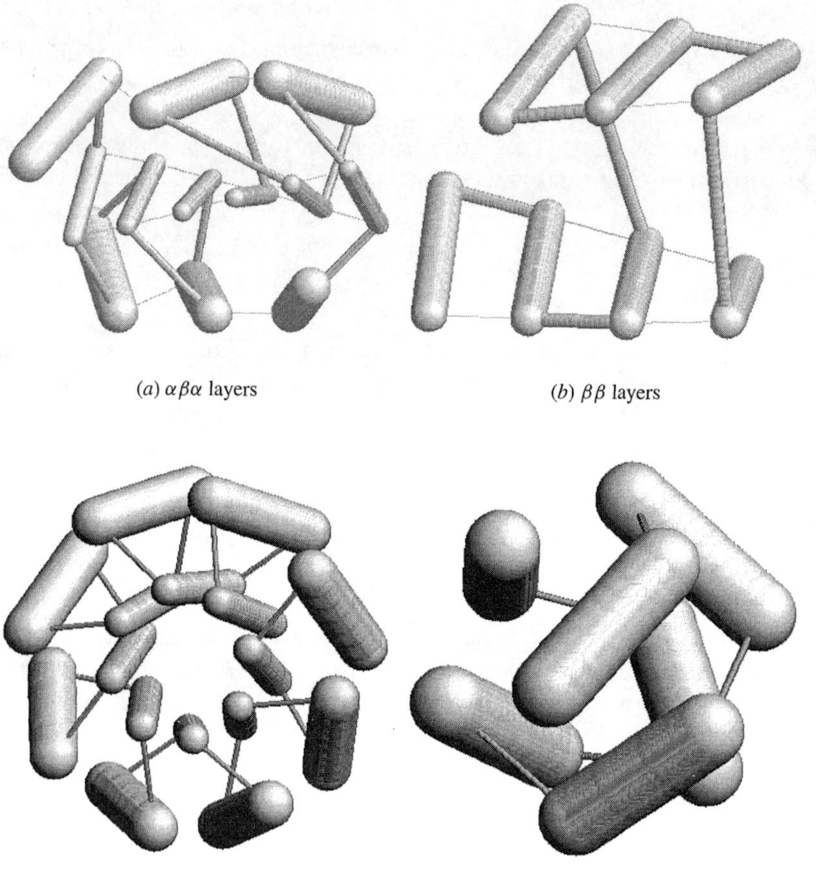

(*a*) $\alpha\beta\alpha$ layers

(*b*) $\beta\beta$ layers

(*c*) $\alpha\beta$ barrel

(*d*) all-α model

Figure 3.10. Stick-figure representations. Each of the major protein architectures are represented by their ideal 'stick' form (α-helices are drawn more thickly than β-strands). (*a*) $\alpha\beta\alpha$ layers. (Compare with figure 1.4(*b*) and figure 3.16.) (*b*) Two $\beta\beta$ layers or β-'sandwich'. Three strands pack over four—similar to the structure shown in figure 1.3(*a*). (*c*) Eight-fold $\alpha\beta$ (TIM) barrel. (Compare with figure 1.4(*a*) and figure 1.5.) (*d*) All-α model for six helices on the icosahedral frame of Murzin and Finkelstein (1988). The packing corresponds to the globin structure (figure 1.3(*b*)). In parts (*c*) and (*d*) the fold of the equivalent proteins is shown by a fine line. The figures were produced by the program RASMOL.

In the following section, a more flexible model based on semi-regular polyhedra will be described which captures well the packing seen in smaller proteins.

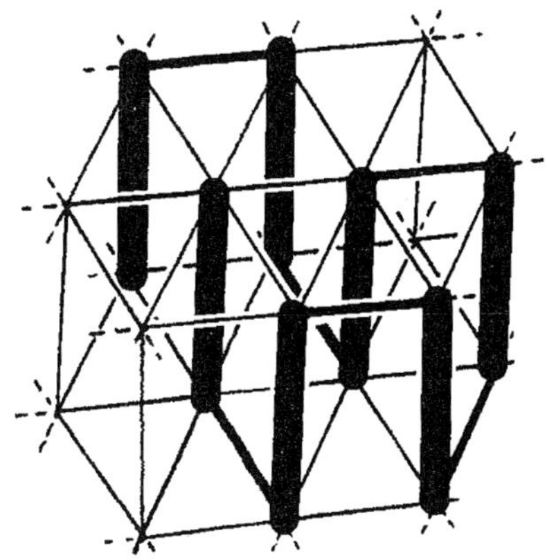

(*a*) lattice

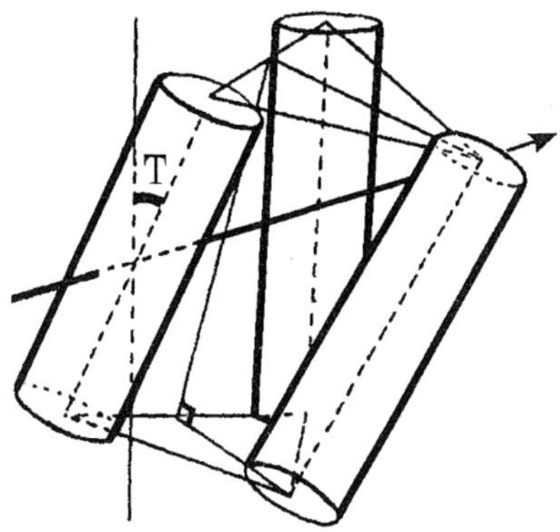

(*b*) twisted

Figure 3.11. Secondary structure lattice. (*a*) A simple hexagonal lattice was taken to represent the termini of helices on each face of the membrane. The helix end-points were typically 10–11 Å apart with 30 Å between faces. The fold indicated corresponds to the known structure of bacteriorhodopsin. (*b*) Geometric constructs used in the calculation of a twisted lattice. The central axis of the middle rank of the lattice is indicated by an arrow about which a twist angle T is applied between each successive helix. Helices in the next flanking layer of the lattice are then placed such that the separation between the ends of adjacent packing helices remains constant.

3.2.5 Polyhedra-based stick models

A useful model for the all-α class of protein was devised by Murzin and Finkelstein (1988) (M+F) who constructed idealized models for small globular proteins. If it is assumed that, to a first approximation, the core regions of α-helices are as long as they are thick, then two helices will have N- and C-terminal end-points that are equidistant both within a helix and between helices. This assumption allows very simple architectures to be constructed for bundles of packed helices in which all pairs of adjacent α-helices have equidistant end-points. Equidistant vertices, combined with an approximately spherical shape, generate a class of polyhedra that have equilateral triangles as faces and are sometimes graphically referred to as deltahedra (or more mathematically, as *simplicial polyhedra*). Different helical packings can then be constructed by placing helices along alternate vertices of these shapes.

An example of this is shown in figure 3.12 in which a path (fold) is traced over the M+F model for five helices. This path corresponds to the fold of a small (all-α) bacterial protein, the Cro repressor (PDB code `1r69`). When the real helices are fitted to the model positions there is a reasonable match but this

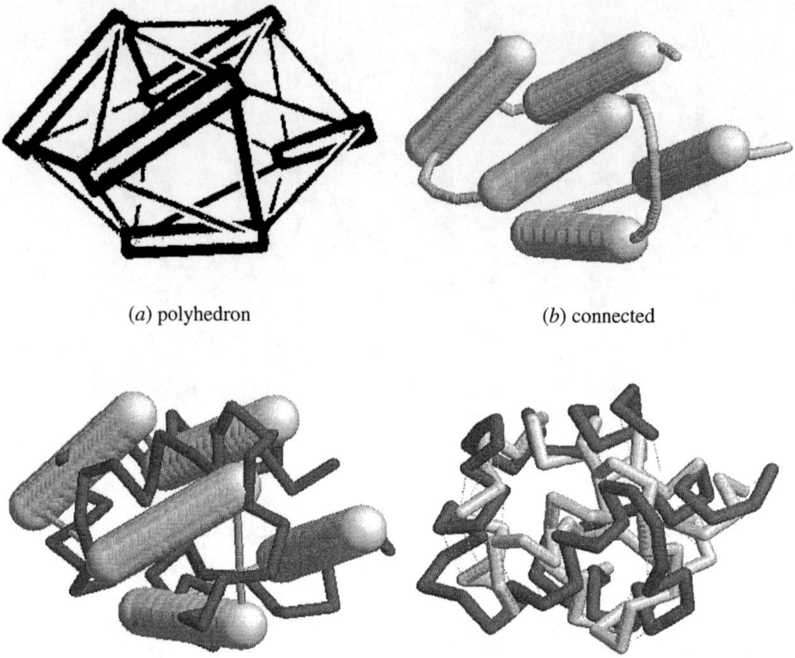

(*a*) polyhedron (*b*) connected

(*c*) fitted (*d*) α-carbon

Figure 3.12. Murzin and Finkelstein model. (*a*) Polyhedral model for five helices taken from Murzin and Finkelstein (1988) (with permission). The edges drawn double represent α-helices and a fold connecting them is traced with bolder lines. (*b*) The helical edges redrawn as 'sausages' and connected by a smooth trace. (*c*) A small bacterial repressor protein (PDB: `1r69`, α-carbon backbone in grey) can be fitted to this model (although the central helix is slightly misplaced). (*d*) When the sticks are expanded to α-carbon positions using the method of Taylor (1993a), the RMSd of the model to `1r69` (grey) is 4.2 Å (over 61 residues). Except for part (*a*) the figures were produced by the program RASMOL maintaining the same orientation as in (*a*) with the carboxy terminus rightmost.

is not a unique fit and other related folds give almost equal agreement. Murzin and Finkelstein (1988) describe an automated fitting procedure but this problem will be returned to in chapter 6 (along with a consideration of how unique any match can be). It is difficult to assess the match between sticks and α-carbon atoms and to make data of like type, the protein could be reduced to sticks but in this example, the stick model has been expanded to α-carbon positions using the method of Taylor (1993b). The two models can then be directly compared using minimal RMS superposition which gives a value of 4.2 Å over the full length of the protein (61 residues).

This example took a small protein in which the helices were almost the same length as the edges on the polyhedron. Helices can be much longer and generally they will extend well either side of the edge on which they lie. The series of models described by Murzin and Finkelstein (1988) (M+F) are all convex hulls: that is, the angle between any two triangular faces is always less on the inside. Although some pairs of faces are almost flat, the overall convex shape means that when helices are extended either side of their vertex, they do not bump into each other.

3.2.5.1 Deltahedra models

The most regular members of the M+F models are its smallest and largest members: the tetrahedron (two helices) and at the upper end, the icosahedron (six helices) (figure 3.13). These, and their relationship to the other polyhedra will be described in this section.

The tetrahedron is the simplest polyhedron that conforms to the Murzin and Finkelstein (1988) model and represents two helices crossing at right angles. As there are few proteins that consist of just two helices, the first model to be analysed by them was the three-helix model that is represented by the next platonic solid, the octahedron. Neglecting helix direction, there are two distinct ways of placing three helices along the edges of an octahedron, one of which is the mirror image of the other.

The model for four helices is not a platonic solid and can be pictured as an octahedron that has been split in the middle and opened (like a burger bun) to allow the insertion of more vertices. Linking the two halves with two vertices leaves a square opening and to add a third vertex (of equal length) this must be distorted into a rhombus. This also distorts the base of what were the perfect pyramids that formed each half of the octahedron. With its decreased symmetry, the number of ways of arranging four helices over this polyhedron are greatly increased to ten arrangements.

Maintaining a reference to the original octahedron, the next model (for five helices) can be pictured as deriving from an opening now being made on the other side of the bread bun, lifting the two original pyramids of the octahedron completely apart (like a ‘Big Mac’) with three more vertices being added. This operation restores some symmetry as the end pyramids are again completely regular. As a result, the number of ways of placing helices does not increase and, probably for some deep reason, remains at ten.

The final polyhedron considered by Murzin and Finkelstein (1988) is again Platonic: the icosahedron, which can accommodate six helices. Compared to the preceding polyhedra, this might seem to be something of a cuckoo but the same logic of splitting an adding vertices can be followed. This time the five-helix model must be split lengthwise (tip of pyramid to tip of pyramid) and opened to

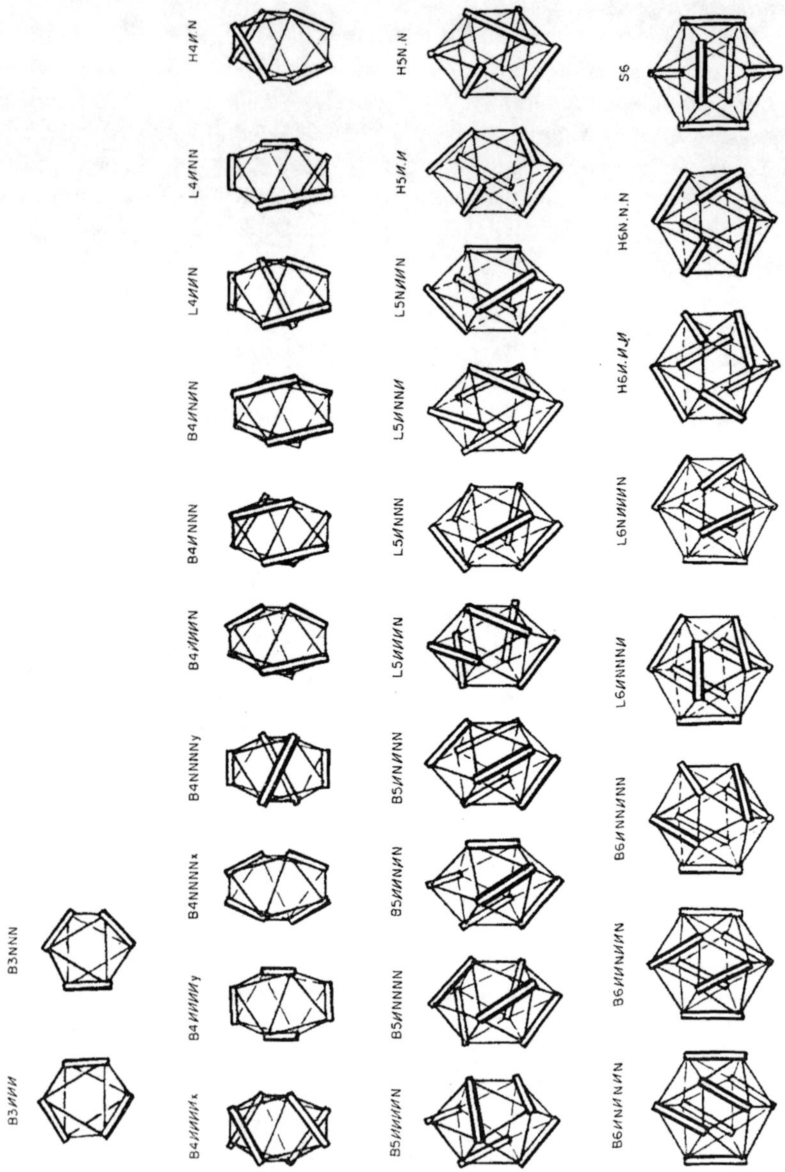

Figure 3.13. Murzin and Finkelstein models from two to six packed helices (thick curves). Reproduced from Murzin and Finkelstein (1988) with permission of Elsevier and the authors.

make the square base of the original octahedron into a pentagon. Filling the gap with, again, three new vertices generates the highly symmetric icosahedron, and because of the increase in symmetry (for some very deep reason) the number of ways of arranging helices remains at ten.

This progression of splitting and adding edges could be followed further but adding more vertices into the icosahedron gives rise to flat hexagonal faces. The most symmetric version of which has two hexagons linked by zig-zag vertices like a drum. Further shapes can be systematically generated by expanding any hub of a hexagon into a triangle (adding two new vertices), however, the resulting structures become increasingly hollow, which for a model of densely packed proteins is not desirable. (If continued with symmetric additions, virus-like shells would be obtained.)

Before leaving the description of α-helical secondary structure frameworks, it is interesting to compare the polyhedra models of Murzin and Finkelstein with alternative models which might provide alternative routes to modelling larger proteins.

3.2.5.2 Close-packed hexagonal lattice

An approach that is similar but different from the M+F polyhedra, is to take fragments out of a close-packed hexagonal (CPH) lattice. This is the lattice formed when spheres (traditionally cannon balls) are stacked and is the densest possible packing for spheres. As with the previous models, the smallest configuration (four cannon balls) is the tetrahedron. Adding two more spheres can produce a boat-like shape but now one face that previously was slightly convex, is flat. In this model, however, the original tetrahedral core need not be preserved and six spheres can be selected in the configuration of an octahedron—equivalent to the second polyhedron of Murzin and Finkelstein (figure 3.14(*a*))

Adding two more spheres to the octahedron produces a configuration not unlike the four-helix model of Murzin and Finkelstein. If this is viewed as two layers of spheres, with four in each layer forming a rhombus, then the difference is that the rhombi have a relative rotation in the MF figure where they have a relative translation in the lattice configuration (figure 3.14(*b*)). A similar comparison can be made between the five-helix models but now each layer is formed by a trapezium of five spheres. At this point the models begin to diverge as the lattice configurations must maintain contact between the spheres across the core whereas the M+F polyhedra become increasingly hollow.

Beyond six helices, neither model is obviously preferable. Extending the polyhedra in the M+F series leads to hollow shapes whereas the fragments from the regular CPH lattice either remain flat (two lattice layers) or, if they are more globular, increasingly incorporate fully buried positions. If their different solutions were to be analysed using the surface/volume approach of Finkelstein and Ptitsyn (1987), then some useful configurations might be identified.

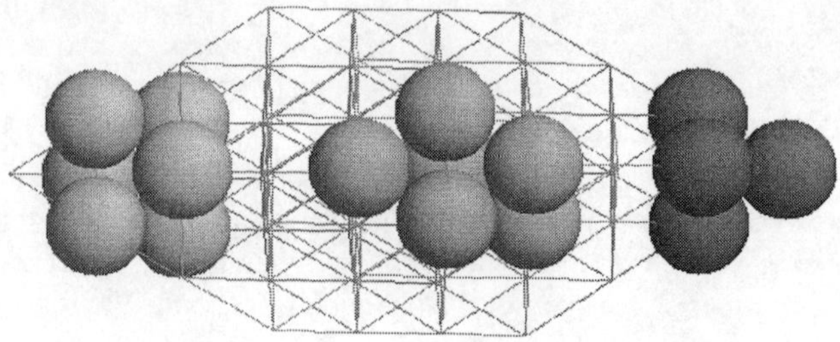

(*a*) 8, 6 and 4 point clusters

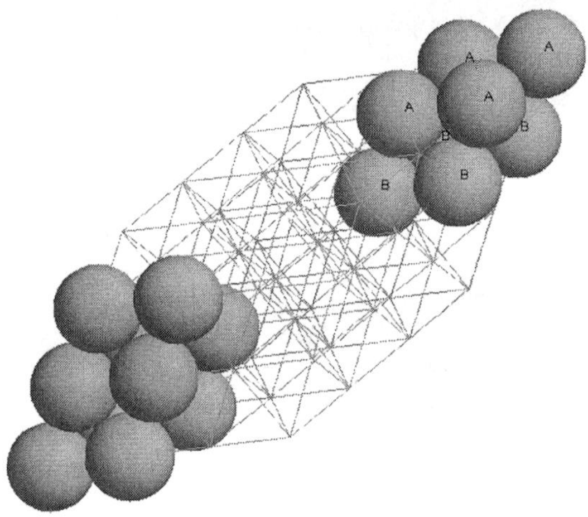

(*b*) 10 and alternate 8 clusters

Figure 3.14. Close-packed-hexagonal clusters. Clusters of points in the close-packed-hexagonal (CPH) lattice (feint lines) can be taken to represent different helix packing. (*a*) The four-point tetrahedron (right) models two orthogonal helices while the eight-point octahedron (left) models four. An intermediate six-point cluster is shown. (*b*) An alternate eight-point cluster is shown (right) composed of two layers (A and B) which can be extended by adding more positions to the layers (left). Such clusters may provide an alternative to the Murzin and Finkelstein models.

3.2.5.3 Bernal 'lattice'

The progression of solids obtained by adding successive spheres onto a core provides a similar but different approach to the CPH lattice. This is equivalent to representing each helix as a dumb-bell of two (touching) spheres so for each progression in size, two spheres must be added to the core[4]. Similar models arise in metal atom clusters (Dong and Corbet 1994).

The model for two helices is the tetrahedron formed from two packed dumb-bells and is, of course, identical to the equivalent deltahedron and CPH lattice models. Further helices can be added to this core—each centred above a face of the original tetrahedron. Because of the symmetry of the tetrahedron, there is only one way to add two spheres, creating a compact shape rather like the hull of a paper boat. Unlike the equivalent sized deltahedron (the octahedron) this boat configuration has a concave face, whereas all the pairs of faces on the deltahedra are convex.

When adding further spheres, three paths can be followed: either add more spheres over each face of the original tetrahedron (creating an un-protein like tetrahedral star), add one onto the core and one on the edge, or add both on the edge. The latter becomes quite elongated and is a fragment from an extended chain of tetrahedra. Except for the first, all these shapes are quite different from the Murzin and Finkelstein polyhedra and while their inclusion of concave faces may be undesirable, they have never been tested to see if they provide a better fit to proteins.

The accretion-of-spheres model encounters the same flat/buried dilemma as the CPH models but with less severity as it is not constrained to the exact layers of the CPH lattice. Its extension leads into the area of generalized crystallography, pioneered by Bernal to encompass quasi-regular solids and liquids (in particular, water) (MacKay 1986, Finney 1970). The addition of spheres around a tetrahedral core comes to a point where an axis of two spheres is surrounded by five spheres forming an almost perfect pentagonal ring. The gap in the ring is only 8.5% of the sphere diameter and it takes very little distortion (when spread over all edges) to make a reasonably regular configuration. Further spheres can be added to this new core and each addition can also be accompanied by a similar regularization to spread any errors throughout the edges. When this procedure is continued for a large number of points, the resulting configuration is equivalent to the large model of a liquid constructed by J D Bernal using straws and ping-pong balls (figure 3.15). Interestingly, the pseudo-fivefold symmetry that was present in the basic unit remains locally throughout the 'lattice' and is similar to quasi-crystals and Penrose patterns (Cahn and Gratias 1987, Hargittai and Hargittai 2000)[5]. Indeed, MacKay's 3D equivalent of a Penrose lattice may provide yet another model to investigate but will not be pursued here.

[4] These models are best appreciated using a combination of marbles and Blu-Tac (the slightly sticky adhesive putty) or following in the footsteps of J D Bernal, ping-pong balls and straws might be tried. More recently, the magnetic 'toy' called 'GeoMag' provides an easier way to construct these shapes.

[5] Chapter 7 of the latter reference (Hargittai and Hargittai 2000) gives a readable account of the prediction and discovery of five-fold symmetry, relating it to many of the topics mentioned here (liquids, Penrose patterns, generalized crystallography, viruses) and, in particular the people involved—most of whom (Bernal, Finney, Klug, MacKay, Penrose) were associated with Birkbeck College in London.

Fragments can then be drawn from the tetrahedral 'lattice', as for the CPH lattice, to correspond with the Murzin and Finkelstein (M+F) models. These have been discussed before for the smaller models where the options are limited, but for the larger models up to and beyond six helices, the possibilities are many and can properly only be investigated by computational enumeration of compact clusters. One interesting possibility, however, is to note that the icosahedral (six-helix) M+F model has a close match in a configuration of 12 spheres that surround a central vacant lattice point[6]. (Only one edge in this polyhedron deviates slightly from the average length.) The choice between all these alternatives, or combinations of them, can only be decided when tested against real protein data. This will be returned to in Part 3 where a similarity to compact random walks will be discussed.

3.2.5.4 Vertex mobility in polyhedra

As we have seen in comparing the classic crystallographic model of a hexagonal lattice with the pseudo-crystallographic clusters derived from sphere accretion, equivalent polyhedra can be obtained that differ only in the shift of one or more vertices. It can easily be imagined that twists and movements of the protein structure might mean that at one time the structure is a better fit to one then the other. This was indirectly realized by Murzin and Finkelstein (1988) who allowed some proteins to be matched to a polyhedron that had room for one more helix than was present in the protein. The resulting gap was often occupied by a large ligand, such as a haem. However, in principle, there is no reason why a protein should be matched to a polyhedron that has the 'correct' number of helices. A protein fitted to an oversized polyhedron will have a loose or open structure and it can be imagined that its suggestive collapse down through smaller polyhedra might imitate a folding process. Indeed, the protein could start as an almost straight line on a very large polyhedron and gradually condense through a number of steps to a unique compact polyhedron.

A more limited degree of movement can be realized for a single helix in an oversized polyhedron model. If there is one spare vertex (or edge), then any adjacent helix can move partly (or fully) into this space, vacating its own position that can now be taken by another helix. The process is similar to the jumbled pictures held in a square grid with one vacant space. Just as the picture fragments are moved around the grid to finally form the intact picture, so the helices might be moved around the polyhedron until their correct interactions are found. Unlike the picture square, they might then expel the vacant vertex (or vertices) and make a final collapse to a smaller polyhedron, so locking-in the conformation. This

[6] This approximation of an icosahedron has led this liquid lattice to be referred to as having icosahedral symmetry. However, later, it will be referred to as a tetrahedral 'lattice' to reflect its origins described in the recursive accretion of tetrahedra. This use should be distinguished from the structure of diamond or ice which is a true crystallographic lattice sometimes referred to as tetrahedral because of the tetrahedral valance of the atoms.

(*a*) tangible

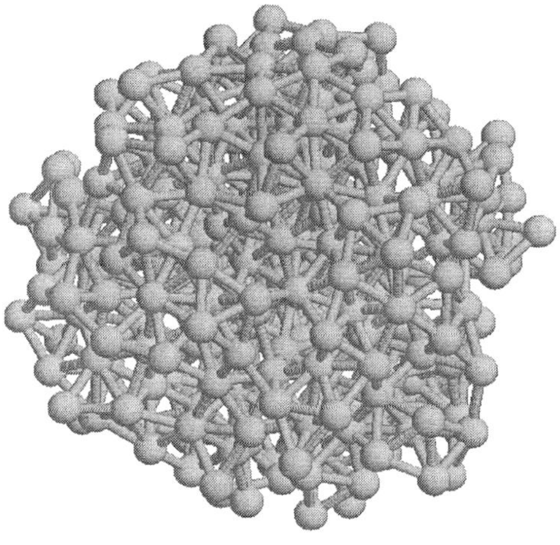

(*b*) virtual

Figure 3.15. The Bernal quasi-random 'lattice'. To investigate the properties of liquids, J D Bernal constructed a quasi-random 'lattice' based on tetrahedral packing. (*a*) The original construction made from ping-pong balls and straws. (*b*) A computer model based on the recursive packing of tetrahedra.

model is similar to the type of motion that is believed to occur in the molten globule state (Ptitsyn 1995) which is considered to be the penultimate phase of protein folding before the structure makes a small contraction and becomes fixed.

Distortions or movement in the polyhedra can also be modelled without recourse to using a vacant edge. If four points on the surface, A, B, C and D lie in two adjacent triangles forming a rhombus with no edge between A and D, then they can be transformed through a square to the alternative rhombus with no connection between B and C. If the edges $A - B$ and $C - D$ are occupied by α-helices, the transition results in a considerable change in helical packing, including a switch of chirality. The more open configuration at the square-packed transition state may be considered as equivalent to a molecular 'breathing' mode. A mathematical analysis of this type of vertex mobility has been carried out by Luo *et al* (1993).

3.2.6 Packing nomenclatures

3.2.6.1 Murzin–Finkelstein system

Given the variety of packing arrangements that are possible of the larger polyhedra (30 on the four, five and six helix models) Murzin and Finkelstein (1988) devised a simple nomenclature by which they can be identified easily. This was based on the main helix/helix contacts found in each arrangement. The 'strongest' contacts are those in which two helices lie side-by-side on adjacent triangles (linked by three vertices). These can either make a left-handed or a right-handed twist when viewed down the central connecting vertex and the two configurations can be represented graphically as: ∨∧ (left) or ∧∨ (right). The latter symbol was, of course, represented by the (backwards 'N') Cyrillic letter for 'I' in the original publication. (See the annotations on figure 3.13).

Using these interactions, pairs of helices can be chained together into a string of interactions. For example, given three helices A, B and C; if the AB interaction is type ∨∧ and the BC interaction is type ∧∨, then the full interaction is: ∨∧∧∨. Often these interactions form a closed circle so if there is also an AC interaction of type ∧∨, then the packing can be specified as ∨∧∧∨∧∨. Murzin and Finkelstein (1988) referred to these closed interaction cycles as barrels (B) and open chains as a folded leaf (L). Sometimes there is no continuous string of interactions and for this they introduced a hairpin (H) type. Adding in the number of helices completes their nomenclature, with the exception of a one special packing (S) on the icosahedron where there are no longitudinal packings. On the octahedron, there are only two packings of types: B3∨∧∨∧∨∧ and B3∧∨∧∨∧∨. For the H type packings, breaks in the string are required and for this a dot is used to mark the gap (for example in H5∨∧.∨∧).

3.2.6.2 An extended M+F nomenclature

While useful, the nomenclature of Murzin and Finkelstein (1988) is ambiguous on the starting position in a barrel-type interaction and for the packings with few longitudinal interactions, there is other information that could be included. Most importantly it was not derived automatically and for its application to the analysis of large numbers of proteins, it was desirable to have a computer algorithm to assign the packing class (Taylor 1991b). For this and other applications which consider the direction of the helices, the nomenclature of Murzin and Finkelstein was systematized and extended as now described.

For any pair of helices lying anywhere on the surface of a deltahedron, four interhelical distances can be specified in terms of the edges between their end-points (termini). For the closest packing helices, three ends will be separated by one edge length while two edges must be traversed to link the fourth pair of points. For example; in figure 3.12(a), the two helices to the front (helix 3 above and 4 below) lie in the \/\ configuration. This configuration can be represented by the ordered digits '1112', specifying the distances between the amino (n) and carboxy (c) termini of the two helices in the order: nn, nc, cn and cc. (i.e. $nn = 3_n, 4_n = 1$; $nc = 3_n, 4_c = 1$; $cn = 3_c, 4_n = 1$; $cc = 3_c, 4_c = 2$; with the latter vertices having no connecting edge). The packing of any other pair of helices can be specified by a similar four-digit number. Since these numbers are cumbersome, and we are only interested in a limited set, each numeric descriptor was assigned a letter (table 3.1). This system is more comprehensive than that of Murzin and Finkelstein (1988) but does not have the same graphic link to the actual packing as is found in their system. Because of this, a simpler set of descriptors is also given which corresponds more closely to the Murzin–Finkelstein system. Simply for typographic reasons, their characters have been rotated, transforming their 'N' (/\/) into Z and their \/\ into S (with a little smoothing). For little cost in complexity, the relative directions of the helices (antiparallel/parallel) were also encoded using upper/lower case.

The strings of packing descriptors used by Murzin and Finlelstein were reproduced by assigning a score to each packing based on the sum of the digits in the numeric descriptor (e.g. 1112 scores 5). This value was then used to sort all pairs of helices using single-linked cluster analysis into the lowest scoring ordered string[7]. Since some pairs will have the same score, precedence was given as defined by the order in table 3.1. The resulting string specifies the best-packed circular tour visiting all the helices. Keeping the same example as above (figure 3.12(a)), from table 3.1 the interaction of helices 3 and 4 (1112) can be found to have an 'a' type interaction (right-handed) or an 'S' type (antiparallel, right-handed) in the M+F-like nomenclature while the preceding pair is left-handed and takes an 'A' and 'Z' designation. Following this path around the best packed circuit gives the full and compact interaction strings as: `2A3a4A5a1c2`

[7] This cluster process is the same as that used in multiple sequence alignment to select the pairs of most similar sequences to be aligned (Taylor 1988).

Table 3.1. Helix packing types. Ordered numeric descriptors (see text) are ranked along with their corresponding letter symbol. Each symbol represents a unique packing configuration for a pair of helices on the polyhedral frameworks of Murzin and Finlelstein (1988). Packings A–D and a–d contain only one long end-point distance (length 2) while the packings below the line have two long end-point distances and make orthogonal interactions. The 'other' category has only one close approach which is an end–end interaction. Under 'M+F-like', a descriptor is given that corresponds to those described by Murzin and Finlelstein (1988) (see main text) in which Z corresponds to their ∧∨ and S corresponds to their ∨∧. Unlike the Murzin and Finlelstein descriptors, a distinction is retained between antiparallel (uppercase) and parallel (lowercase).

Numeric packing	Full left	Full right	M+F-like left	M+F-like right
1112	A	a	Z	S
2111	B	b	Z	S
1211	C	c	z	s
1121	D	d	z	s
2211	E	e	.	.
1221	F	f	.	.
1122	G	g	.	.
2121	H	h	.	.
1212	I	i	.	.
2112	J	j	.	.
other	X	x	.	.

and `ZSZSs`. This is a cyclic barrel-type packing as the first string begins and ends with the same helix (2) and the second string does not incorporate any '.'s. It does not correspond exactly to the M+F descriptor: B5∨∧∨∧∧∨∨∧∧∨ (figure 3.13, third down in the five-helix column) as their string is split at an arbitrary position whereas the string `ZSZSs` is broken at the nominally 'weaker' 's' type interaction.

3.2.6.3 A layer-based nomenclature

In contrast to the all-α polyhedral models, a packing nomenclature based on secondary structure layers is very straightforward. A basic nomenclature can be derived simply from the number of SSEs present in each of the secondary structure layers described earlier[8]. For example, in the three layer $\alpha\beta\alpha$ class,

[8] This nomenclature is not unlike the system used to classify steam locomotives based on the number of their leading bogie wheels, main drive wheels and training bogie wheels (excluding the tender). The 'Flying Scotsman' was a 4-6-0 class.

a `2-5-3` architecture is a five-stranded sheet with two helices on one side and three on the other side of the sheet (figure 3.16). Similarly in the four-layer $\alpha\beta\beta\alpha$ framework, layer occupation can be specified in the form: `0-3+4-1` which would be a β-sandwich of three on four strands with a helix packed against the latter.

Structures containing β-barrels are, at first sight, even simpler as all that needs to be recorded is the number of strands in the barrel. However, an important defining characteristic of barrels is the amount of stagger between the strands. This can be quantified in a shear number (McLachlan (1979a) and section 4.2 for a full description) and together the number of strands (N) and their shear (S) completely specify the overall barrel geometry. Following the nomenclature for flat sheets, if a barrel is surrounded by M helices, then the descriptor 'M-N.S' encodes the essentials of the structure. So, for example, the eight-fold $\beta\alpha$-barrel in triosephosphate isomerase (the 'classic' TIM barrel, figure 1.5) would be `8-8.8` as it has shear number 8.

An alternative to this nomenclature has been used (Taylor 2002a) (and will be reconsidered later) where it was of interest to capture partial barrels. These were recorded as `2-6.7` which is two helices packed against an incomplete barrel of six strands with a curvature best represented by the seven-strand barrel geometry.

All these nomenclatures do not record the specific arrangement of the helices on the sheets or the topology (fold) of the chain over the framework. These aspects will be reconsidered towards the end of this chapter but will be dealt with more fully in Part 3.

3.2.7 From 3D to 2D

3.2.7.1 Projection down to 2D

To construct a 2D representation of a protein structure a distance can be measured between all pairs of secondary structure elements and the resulting distance matrix projected into 2D using the methods described in chapter 2. However, the distance to choose to represent the separation of two line segments is not obvious. Various possibilities are the mid-line/mid-line distance or the closest approach of the lines, which is either the contact normal or an end/end or an end/line distance (Cohen *et al* 1982). Given the preceding analysis of packing, an alternative is to pick a distance that represents the separation in the middle of the line interaction, that is: the overlap region (figure 3.7). This can be specified as the mid-point/mid-point distance between the two overlapping line segments. A choice must still be made for non-overlapping segments and this can be taken as their closest end/end separation (anything else would involve some overlap as already defined). The SSE mid-points for a small $\beta\alpha$ protein (the chemotaxis Y protein) are shown in figure 3.16(a).

In this example (figure 3.16), the mid-points of the SSEs are almost co-planar and already provide a good 2D representation of the structure. However,

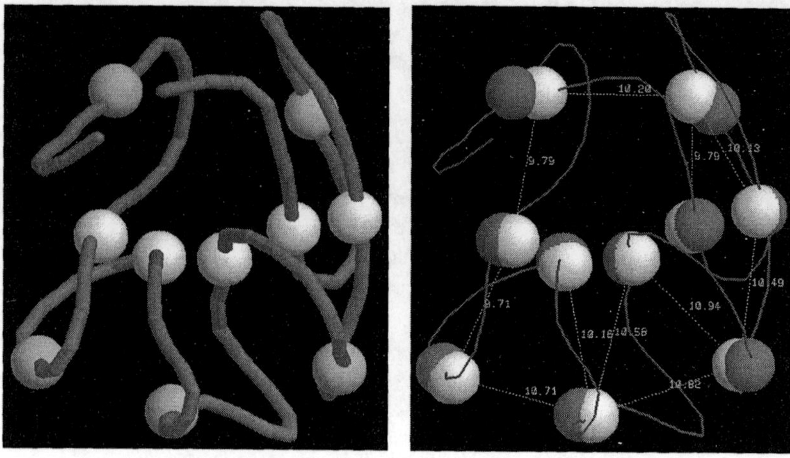

(*a*) mid-points (*b*) DG-refined

Figure 3.16. Simplified representation of `3chy`. (*a*) The smoothed backbone trace of the chemotaxis-Y protein is shown with the mid-points of the automatically defined line segments shown as spheres. The five central spheres lie on β-strands (white) with α-helices (grey) above and below. The three-layer 2-5-3 structure can be clearly seen. (See also figure 1.4(*b*).) The interactions between the strands of the sheet form a bend as a result of the imbalance of packing different numbers of helices either side of the sheet (compare figure 1.4). (*b*) The same orientation is maintained to show the revised mid-point positions (white) after their local neighbour distances have been set to 10 Å (5 Å for $\beta\beta$) and refined using distance geometry (DG).

in general, it can be useful to 'tidy up' this representation by making an exact 2D representation with idealized SSE separations and this is usually carried to the point where the SSEs are constrained to lie on a grid. (Although as seen in figure 1.4, this is not essential and does not capture barrel structures well.) For the current example, some automatic 'tidying' has been made using the method of distance geometry (DG) (section 2.2). For pairs of secondary structures where there are interactions defined by sufficient line segment overlap, the mid-point/mid-point distance was set to 5 Å for $\beta\beta$ interactions and otherwise 10 Å. With these values, the distance matrix for the mid-points was projected into two dimensions resulting in the configuration shown in figure 3.16(*b*).

When secondary structures are represented as symbols constrained to a grid, the representation is usually referred to as a 'topology cartoon' and has been used extensively to describe protein folds since some of the earliest analyses of structure (Sternberg and Thornton 1977b, Sternberg and Thornton 1977a, Nagano 1977). It has also formed the basis for semi-automatic (Flores *et al* 1994) and fully-automatic analyses of proteins at the 'topological' level (Sternberg *et al* 1985, Rawlings *et al* 1985, Rawlings *et al* 1986, Clark *et al* 1991). Although primarily employed for visualization, as the structure databases become larger, automatic methods are being increasingly employed in automatic topology matching (Koch *et al* 1992, Koch *et al* 1996, Gilbert *et al* in Press, Gilbert *et al* 1999). These methods tend to be restricted to matching just β-sheet topology as the more complex α-helix interactions are not adequately encoded.

3.2.8 From 2D to 1D

3.2.8.1 Topology strings

The encoding of protein architectures as layers of secondary structure allows the fold of the chain to be described as a series of moves between the layers. Concentrating on the three-layer $\alpha\beta\alpha$ architecture, each layer can be designated by the letters A, B and C (respectively) specifying a position coordinate in one dimension. Location in the layer was encoded as a numeric displacement relative to the first SSE to be found in the layer (specifying a second coordinate dimension). The third dimension encodes just the orientation of each SSE relative to the first SSE. This string encoding is similar to that devised by Flower (1998) but is directly rooted in a coordinate frame making it more computationally tractable. A chain path can then be encoded using three descriptors for each SSE. as shown in figure 3.17.

This encoding scheme might appear to depend on the orientation of the molecule, however, both the orientation and positions of the SSEs are determined relative to the first strand and, if in addition, the labelling of the layers is not predefined, then the scheme can be made orientation independent by assigning the label 'A' to the first α-layer to be occupied. Although the scheme is independent of orientation for the whole molecule: that is identical structures will have the same string, it remains sensitive to the starting point so two identical substructures within larger molecules need not have the same string. This difficulty will be considered further in section 7.4.

A typical topology string for a flavodoxin-like protein (`3chy` fitting form `2-5-3`) is: `+B0.-A0.+B-1.-C0.+B1.-C1.+B2.-C2.+B3.-A1`, where A, B and C are the three layers prefixed by their relative orientation to the first strand in the sheet and suffixed by their position relative to the first element in each layer. (Any empty positions in the α layers were ignored.)

3.2.9 Uniqueness of string descriptors

A typical protein structure consists of hundreds of residues, each comprising roughly ten atoms, each of which require three coordinate values (specified to at least one tenth of an ångstrom). This is a considerable volume of data (around 100 Kb) and it must be considered whether such complexity can reasonably be reduced to a string of around 30 alphanumeric characters. Much of the reduction

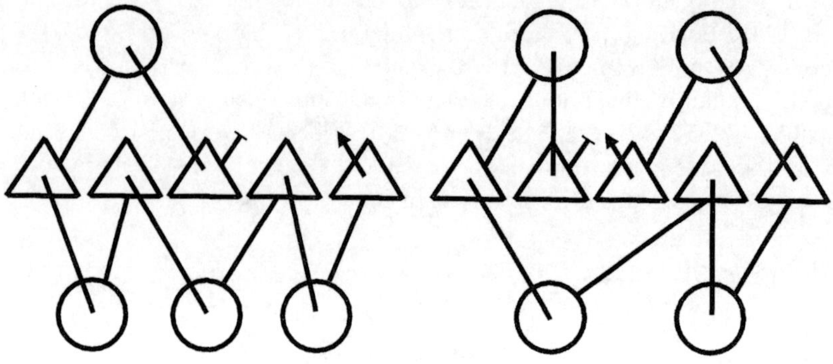

(*a*) +B0.-A0.+B-2.-C0.+B-1.-C1.+B1.-C2.+B2

(*b*) +B0.-A0.+B-1.-C0.+B2.-C1.+B3.-A1.+B1

Figure 3.17. Example topology strings for two small $\alpha\beta\alpha$ layer proteins. (*a*) Fitting form `1-5-3` (one helix above and three below a five-stranded sheet) with topology string '`+B0.-A0.+B-2.-C0.+B-1.-C1.+B1.-C2.+B2`' and (*b*) form `2-5-2` with topology string `+B0.-A0.+B-1.-C0.+B2.-C1.+B3.-A1.+B1`. In the diagrams, helices are depicted as circles and β-strands as triangles. In the topology strings, the three layers of secondary structure ($\alpha\beta\alpha$) are designated A, B and C respectively. Each SSE is given a label of three parts indicating orientation ('+', '−'), layer and position in the layer. The first strand in each layer is, by definition, at position 0 with others numbered relative to this. In the topology diagram negative numbers lie to the left, positive to the right. Similarly, in the strings, a positive orientation corresponds to a SSE approaching ('out of the page') in the diagrams. The fold in *a* corresponds to adenylate kinase but the fold in *b* has never been seen.

is gained by ignoring side chain atoms and encoding main chain conformation as one of two secondary structures (or neither). The final saving is made by assuming that secondary structures are regular.

3.2.9.1 External coordinate-based strings

For the layer-type nomenclatures the strings described give explicit coordinate positions for the secondary structures and all that is lost is the phase of the secondary structures and details of connecting loop conformations. As we will see later, the phase can even be reasonably estimated by orienting the hydrophobic residues towards the centre (Taylor 1991b). Using this information, small proteins, such as the chemotaxis Y protein (used as an example throughout this chapter) can be reconstructed from its topology string with a resulting error of just 5 Å over equivalent α-carbon atoms. A remaining source of error derives from the relative positioning of each SSE in a layer. For the more rigidly (hydrogen bonded) β-strands this is not significant but greater variation can be found in the packing of helices.

3.2.9.2 *Internal coordinate-based strings*

The position with the Murzin and Finkelstein (M+F) packing descriptors for the all-α proteins is less certain. These strings specify an internal coordinate system (like a distance matrix or torsion angles) rather than an XYZ set of coordinates in an external reference frame. In the original M+F system there can be strings that are cyclic ('barrel'-type) and non-cyclic ('leaf'-type), with the latter including incomplete ('hairpin'-type) strings. If the string is not cyclic then it cannot define a specific structure and this was overcome in the extended M+F system by making all strings cyclic—even if this incorporated a weak link, such as an end–end interaction (see section 3.2.6.2). If a packing string is a complete cycle of strong interactions ('S' or 'Z' type in table 3.1) then there will be limited flexibility in the ring whereas with weaker interaction types (those below the line in table 3.1) then flexibility will increase producing a less unique structure.

This supposition was given a limited test using the model described in figure 3.12 by adding sticks in their designated packing configuration but with a random angular displacement in any available degree of freedom that did not alter the packing class. Ring closure was attained by the simple method of repeating the packing many times until by chance the terminal helices lay close to their correct configuration. Twenty models were constructed with the fully packed circuit (`2A3a4A5a1c2` or `ZSZSs`) and a further twenty constructed with just a single constraint between the termini of the terminal helices. These results are shown in figure 3.18, and although difficult to judge in a static picture, the better constrained string has an RMSd of 0.40 Å to the M+F figure compared to 0.65 Å for the those with the weak terminal connection. Both these strings, and in particular the first, strongly preserve the topology of the fold indicating (at least for this limited example) that the internal coordinate-based cyclic strings can capture the fold in a reasonably unique way.

3.2.10 Predicting helix contacts

3.2.10.1 *Simple descriptions make prediction simple?*

The observation that protein folds can be described in a string of a few characters has led to the speculation that they should not be too difficult to predict. In the previous example, if the packing string `ZSZSs` could be deduced from the protein sequence, then the overall fold would be obtained and much of the details (secondary structures and side chain positions) would follow (or be greatly constrained). Attempts at predicting helix packing, however, have met with little success (Cohen *et al* 1979, Taylor 1991a) unless some additional constraints are available (Cohen and Sternberg 1980b, Taylor 1993b). As with much work in *ab initio* structure prediction, the possible pairs of interactions all look much the

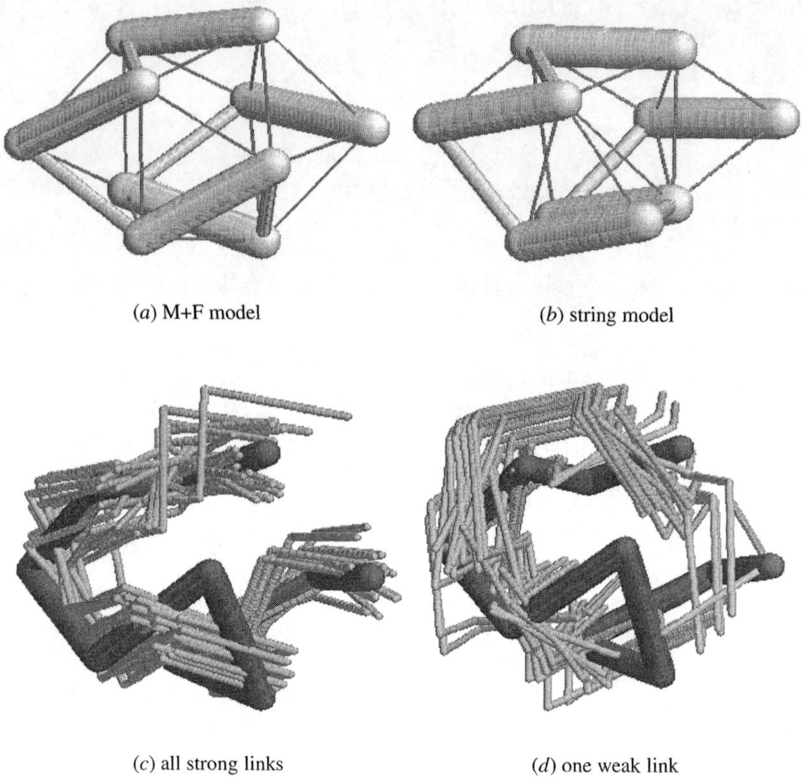

Figure 3.18. Uniqueness of string-based models. (*a*) Polyhedral model of Murzin and Finkelstein (1988) for a small all-α protein (PDB: `1r69`). The helical edges are drawn as 'sausages' connected by thinner tubes to show the path of the chain. The other edges (that define the packing) are drawn as fine lines (the carboxy terminus lies rightmost). (*b*) A model generated from the topology string '`ZSZSs`' by 'dialling-up' configurations as described in the text. (*c*) Twenty models (as in part (*b*)) superposed on the M+F model (part (*a*)) in grey. (*d*) As in part (*c*) but using the 'weaker' string '`ZSZS.`' which has just a single end–end interaction closing the ring.

same and there is little in the sequence to distinguish an 'S' from a 'Z' or parallel from antiparallel and equally little to even select the correct pairings in the first place! Despite this pessimistic outlook, it is still fun to try and some experiments will be returned to in chapter 9.

3.2.10.2 Observed packing classes

The topic of helix/helix packing was initially addressed by Crick (1953) from a theoretical/modelling perspective, not long after the structure for the α-helix had been proposed by Pauling (and confirmed by x-ray analysis). Crick's approach was based on a cylindrical projection of the residue positions from the helix axis forming a 2D net. When two such nets are superposed (with one being turned over to model the face-to-face interaction) there is a clear set of positions in which the top net can be placed so that its residues systematically avoid those on the bottom net. Crick called this model 'knobs into holes' and this simple model has provided the basis for most subsequent analysis.

On a helical net representation, each residue is surrounded by six other residues: two adjacent in sequence and two on adjacent turns at relative positions +3, +4 and −3, −4 in the sequence. This produces a net in which the residue packing approximates that of hexagonally packed spheres and Richmond and Richards (1978) have given a general analysis of this for various helices. From a simpler viewpoint, there are only six ways two such nets can pack if residue j from the top net is placed into the hole formed by residues $i, i+1, i+4$ in the bottom net, then the adjacent hole formed by $i, i+3, i+4$ can, in turn be filled by $j \pm 1$, $j \pm 3$ and $j \pm 4$. Each of these alternatives dictates a different packing angle between the helices but the exact value of this will depend on the radius taken for the α-helix, which in turn depends on the nature of the residues in the packing interface. Based on solvent accessible surface area calculations, Richmond and Richards (1978) made some estimates for this based on a construct called the 'cylinder of equal penetration' (CEP).

The curvature of the helical net around the cylinder means that the region of residue interaction is limited and that the 'holes' on the hexagonal net are not all equal. Since adjacent residues in the sequence splay apart, a larger 'hole' is effectively formed by the four residues: $i, i+3, i+4, i+7$. The difference in terms of the net superpositions is, however, simply a slight translation. The closer packing along the $i \ldots i+3$ and $i \ldots i+4$ directions means that the interaction can be viewed as the packing of ridges into grooves (Chothia *et al* 1981) (rather than knobs into holes) but the relative utility of these two descriptions is debated (Walther *et al* 1996) (see Efimov (1999) for a review). Despite an extensive survey of helix packing in globular proteins the packing classes are broad and there still appear to be no rules that are sufficiently firm enough to predict helix packing class from sequence.

3.2.10.3 Coiled-coil packing

Patterns in helix/helix packing become more regular as the interaction of the helices is increased and this reaches a maximum in the packing of long helices in fibrous proteins. There is extensive literature on this topic with 'rules' that are good enough to predict whether the coiled-coil of helices will be a double or a triple helix but, as stated in the introduction, this class of protein will not be considered in detail. For reviews see Cohen and Parry (1994) and Burkhard *et al* (2001). The periodic sequence of coiled-coils can be analysed using Fourier transform methods (McLachlan 1983, Finer-Moore and Stroud 1984).

Chapter 4

Sheets and barrels

4.1 β-sheet geometry

The analysis in chapter 3 treated β-structure at the level of its component strands (represented as sticks). This analysis adheres to one of the original definitions of secondary structure, stated by some of the 'founding-fathers' of protein structure analysis, as: *'the spatial arrangement of its main chain atoms without regard to the conformation of its side chains or to its relationship with other segments'* (Kendrew *et al* (1970), as quoted by Levitt and Greer (1977)). While the α-helix conforms to this definition, its application to a β-strand is ambiguous since the strand, is to a large extend defined mainly by the other strands in the sheet to which it is hydrogen bonded. It, clearly, cannot be applied to a β-sheet which is defined only in terms of the relationship between segments.

This implies that the β-sheet has a place in the structural hierarchy of proteins somewhere between a single secondary structure element (SSE) and a complete protein (or protein domain). Previously, the term *super-secondary* structure has been applied to recurring structures formed from a few to several SSEs, however, it is probably more consistent to apply this term also to all β-sheets, whether they recur or not. Although the β-sheet is properly a super-secondary structure, in this section we return to a lower level, to the residue level, for its analysis as it cannot be assumed that linear β-strands are the best units with which to represent and analyse the many shapes and forms of β-sheets.

4.1.1 Sheet chirality

The most obvious property of β-sheets is that they are all, to varying degrees, twisted and that the twist always has the same hand. Viewed along a β-strand, this twist is right-handed and, of course, viewed across the strands, it is left-handed. (It does not make any difference to the hand whether the strand is viewed N$\rightarrow$C or C$\rightarrow$N, or from which side the strands are viewed across.)

The origin of twist in sheets has been analysed by a variety of measures employing detailed atomic representations of structure. Avoiding the details, however, the reason for a preferred hand to the twist can be appreciated from considering an energy based Ramachandran plot (Hu *et al* 2003). This plot must be derived from energy calculations as an empirical frequency plot of observed values from the PDB simply restates the observation that sheets have a preferred twist. There is considerable variation in the results depending on the particular flavour of the potential function—even using complex quantum representations with water—but most indicate clearly that there are a greater number of low-energy ϕ, ψ angle combinations in the part of the plot associated with a right-hand twist along the strand.

In the normal Ramachandran plot of ϕ, ψ values (see figure 2.4), this area lies on the upper-right side of the $\phi = -\psi$ antidiagonal, with twist increasing from a mid-region point at $\{\phi = -120, \psi = 120\}$ towards a point at $\{\phi = -60, \psi = 180\}$. It can be estimated that roughly 3/4 of the allowed values lie on the side with right-hand twist along the strand. On the basis of the larger conformational space in this region, it can be argued on statistical grounds, that it will be more likely (and hence entropically favourable) for strand conformations to adopt these values. It must be remembered, however, that most of these calculations are carried out using a dipeptide, so they do not include the steric effect of neighbouring strands or hydrogen bonding. Other studies have considered twist in two-stranded β-hairpins and ribbons (Efimov 1991b, Ho and Curmi 2002) and large sheets (Salemme and Weatherford 1981a, Salemme and Weatherford 1981b, Salemme 1981). In these larger assemblies, the picture becomes complicated as sequential residues need not necessarily adopt the same ϕ, ψ values and in twisted ribbons, an alternation between two regions of the Ramachandran plot is observed (Efimov 1991b).

4.1.2 Geometric models for a twisted sheet

4.1.2.1 Simple helical models

Returning to the implications of these studies for simplified models, twisted sheets can be represented either, as in chapter 3, with straight strands running perpendicular to the helical axis or with twisted strands following the helix with the hydrogen bonds lying perpendicular to the axis. In the first model, if the strands are straight, then the gap between them increases with distance from the twist axis. As the strands should be hydrogen bonded at a constant distance, this is clearly not ideal. In the orthogonal model, with strands following the helical twist (Znamenskiy *et al* 2000), then as the number of strands increases, so does the radius of the helix that they must follow. Even for three strands, it does not take much progression along the strands before the hydrogen bonds start to slip out-of-register.

This problem was recognized in the theoretical studies of Salemme (1981) (see also the related citations above) who concluded that his ideal sheets were internally stressed in an attempt to reconcile these tensions. In the analysis of real

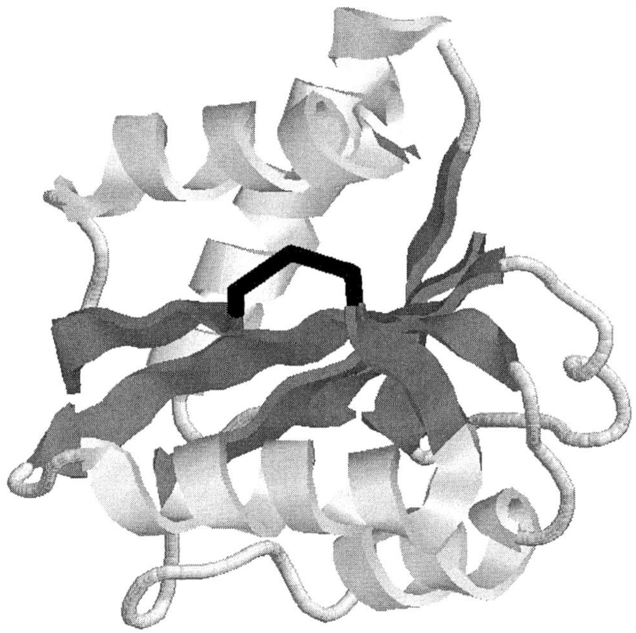

Figure 4.1. A beta-bulge. The edge strand of the sheet (grey) in a small $\beta\alpha$ protein flavodoxin (PDB: `4fxn`) contains a large β-bulge (black). The backbone of the protein is shown as a ribbon trace using the program RASMOL while the α-carbon positions are shown for the bulge.

structures, it was recognized that this stress was relieved by the introduction of breaks (called bulges) into the edge strands of β-sheets (Richardson *et al* 1978) (figure 4.1). Large edge bulges can even propagate into the sheet like cracks. The problems with both the simple models described can be solved with the introduction of β-bulges. An interruption in a straight strand can allow it to be shifted back into hydrogen bonding distance while a break in a helical strand allows it to 'catch-up' with the pace of hydrogen bonding set by the central strand.

To minimize the need for bulges, the two alternate models can be used in different situations. For a few long strands, a sheet model with strands running in the same direction as the helix axis will involve least distortion whereas for a large number of short strands in a sheet, the axis running across the strands will provide a better model as the deviation at the ends of the strands will not be large and remain constant whatever the number of strands. By contrast in the helical model, maintaining a fixed angle between strands will rapidly reach the point where the strand becomes a circle. As we will see later, this problem is less

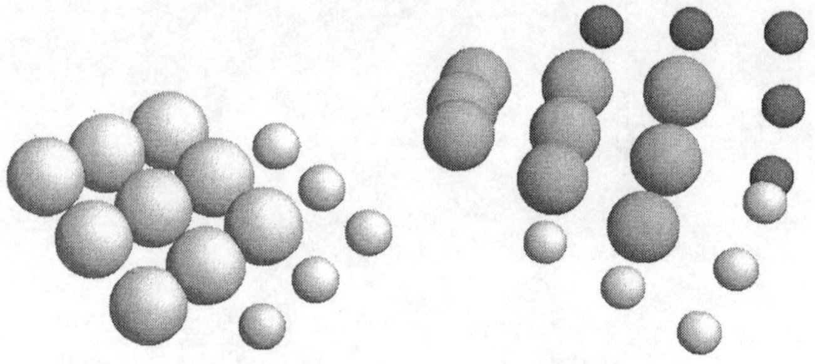

(*a*) (*b*)

Figure 4.2. Recursive growth model for a β-sheet. Residue positions in a β-sheet unit of nine residues are represented by large spheres. (*a*) A new unit (smaller spheres) is added onto a corner (generally, onto each available corner). The sheet has a slight twist of 0.1 radians between the three strands in each unit. (*b*) Using a unit with a higher twist (0.3 rad.) shows that successive additions (small light and dark spheres) do not result in identical edge locations. Within a given distance, these were averaged to give one position.

critical when there is stagger between the strands as this reduces the dimensions of the grid over which hydrogen bonding must be maintained.

4.1.2.2 *Recursive growth model for a β-sheet*

To test the limits of these two models, a large sheet can be made by the recursive addition of a basic sheet unit—similar to the approach used in a previous section to grow a tetrahedron-based lattice (see section 3.2.5.3 and figure 3.15). If the basic unit is constructed from a grid of three positions along three strands, then additional units can be added by overlapping the four corner residues (figure 4.2(*a*)). Because of the twist of the sheet, the distance between the residues on the ends of the strands is greater than those in the middle so when the corner of the (new) unit being added is placed in the centre of an (old) existing unit, the four residues cannot be matched-up exactly. While a least-squares RMS fit could be calculated, for just four positions, the simpler approach of aligning two local coordinate frames was used. These were calculated with X running along the centre–corner direction and Y running in the direction of the other (almost) orthogonal diagonal. Since X and Y are not exactly orthogonal in rectangular or twisted sheets, a Z-axis was calculated orthogonal to both X and Y using their vector product ($Z = X \otimes Y$). From this a set of three orthogonal vectors can be obtained by recalculating the Y-axis as $Y' = X \otimes Z$ and when all are scaled to unit length, the resulting rotation matrix can be used to place corners in a common coordinate frame. (This is quite a useful operation and full details of its implementation are described in the Web site associated with the book.)

If the strands have a twist between them, using the transforms described above, the mid-edge positions added from different corners will not overlap exactly (figure 4.2(*b*)). The positions of such duplicated residues were merged by averaging, up to a limit of 2 Å separation, beyond which they were accepted as independent residues provided they were not too close to sequential neighbours. The result of allowing the addition of two residues at what should be just one grid position mimics the insertion of a β-bulge. It can be seen in the resulting network in figure 4.3(*a*), that the sheet remains relatively undistorted only over a network area of, typically, 5×5 units. As with the addition of tetrahedra to a lattice, slightly greater regularity can be attained by spreading the distortions using a simple regularization algorithm.

When very large sheets are grown in this way, the β-bulges form break-points from which regular additions can continue in different directions. The resulting structure remains basically helical but with 'ragged' edges and even bifurcation in the super-helix (figure 4.3(*b*)).

4.1.2.3 Conic and quadratic surfaces

From the representations of β-sheets seen so far, it is clear that they have a saddle shape and it is not surprising that a variety of saddle shaped surfaces can be used to model them. The simplest saddle-shaped surfaces are generated from 3D extrapolations of conic sections, including the paraboloid (from a parabola), the hyperboloid (from the hyperbola) and, of lesser importance in this context, the ellipsoid (from the ellipse). These surfaces also exist in combinations, the most important of which (from our point of view) are the elliptic paraboloid and the hyperbolic paraboloid. Most studies have addressed the problem of closed β-sheets (barrels) which require the surfaces to be expressed in terms of x, y and z using a general class of quadratic functions. Like the ellipsoid encountered in the section on moments of inertia, these surfaces can be represented individually by what is called their normal form (equation (2.15)). While these will be considered in a subsequent section, the class of quadratic surfaces considered in this section are a function only of the x, y plane (giving a value of z for each x, y point). For example, the hyperbolic paraboloid is:

$$2cz = \frac{x^2}{a^2} - \frac{y^2}{b^2}. \tag{4.1}$$

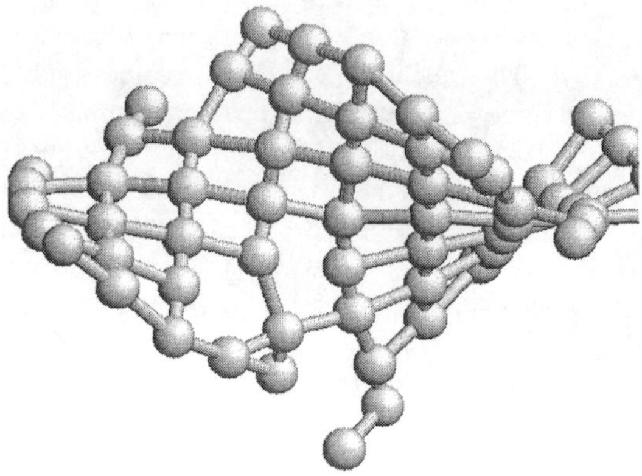

(*a*)

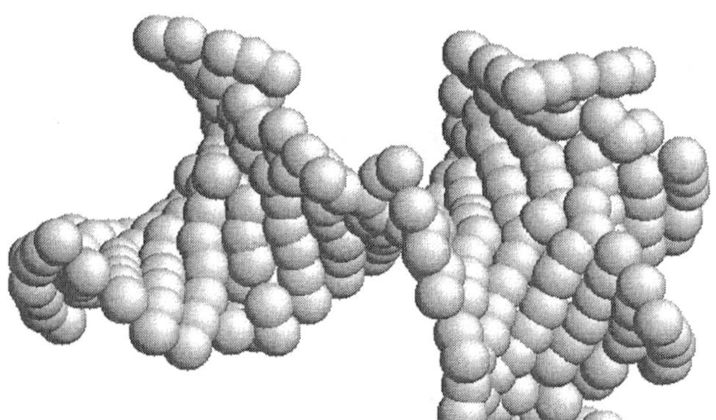

(*b*)

Figure 4.3. Recursive growth of a large β-sheet. Large sheets can be grown by the recursive addition of unit sheets (figure 4.2). (*a*) Showing close connections (bonds) between α-carbon positions (spheres). The geometry (negative curvature) of the surface means that bonding cannot be continuous. (*b*) Using larger spheres to show the sheet surface which is overall helical but with accumulated bulges and bifurcations giving the impression of a 'ragged-ribbon'.

Rather than use individual normal forms for each type of surface, a more general equation[1] for this set of surfaces is:

$$z = Ax^2 + 2Bxy + Cy^2 + 2Dx + 2Ey + F. \tag{4.2}$$

In this representation, each parameter ($A \ldots F$) can be associated with a feature of the curve For example, when $A > 0$ and all others zero, the surface is a simple U-shaped valley. With $A > 0$ and $C > 0$ the surface is now a depression—or more exactly, an elliptic paraboloid (figure 4.4(*a*)). More interestingly for β-sheets, if A and C have a different sign ($AC < 0$) then the surface has a saddle shape (hyperbolic paraboloid) (figure 4.4(*b*)). Similarly, if all parameters are zero except C, the surface is also saddle shaped but now with straight grid lines running over the surface (for constant x or y) (figure 4.4(*c*)). If $C < 0$, the twist along the x direction is left handed while for $C > 0$ it is right handed. A diagonal traverse of the surface returns to the curved lines of the parabolic grid obtained with $A > 0$ and $C > 0$. Combinations of A, B and C produce more complex surfaces (figure 4.4(*d*)) while D and E simply tilt the surface in x and y, respectively and F sets the base level.

The shift from surfaces with $AC < 0$ and $B = 0$ to $A = C = 0$ and $B > 0$ is effectively a rotation, which combined with the more obvious degrees of freedom associated with the D, E and F parameters, allows the surface to be matched to a set of points without shifting its reference frame. As with simpler linear regression, this can be formulated to give a least-squares fit of the surface to the point-set (D Moss, personal communication). Such a fit to the β-sheet of a small β/α-type protein is shown in figure 4.5.

The parameters in equation (4.2) were optimized to find the best fit of the surface to the α-carbon atoms of the β-sheet in the protein `1aps`. The surface is generated using the program `CURVE` (which can be found at: `www.biochem.ucl.ac.uk/ roman/surfnet/surfnet.html`).

To obtain comparable measures of fit between sheets in different orientations requires that the x, y plane should be a good approximation of the sheet to begin with. This can be obtained using the inertial moments described earlier to find the best plane to use as a base over which to calculate the surface (section 2.1.4.1). However, both the determination of this base and the fitting of the surface will be sensitive to what residues are selected as part of the sheet. This problem will be returned to later in the context of determining the extent of a β-sheet.

4.1.3 The surface of a twisted sheet

While the secondary structure definition of Kendrew and colleagues dismissed the consideration of side chains, this aspect cannot be neglected in the analysis of the

[1] In equation (4.2), the seemingly irrelevant factors of two on parameters B, D and E are normally included because a symmetric matrix of the parameters (with A, C and F on the diagonal) can be diagonalize to yield the parameters a, b and c of the normal form (equation (4.1)). This is equivalent to the extraction of the momental ellipsoid from the general equation of inertia in section 2.1.4.1.

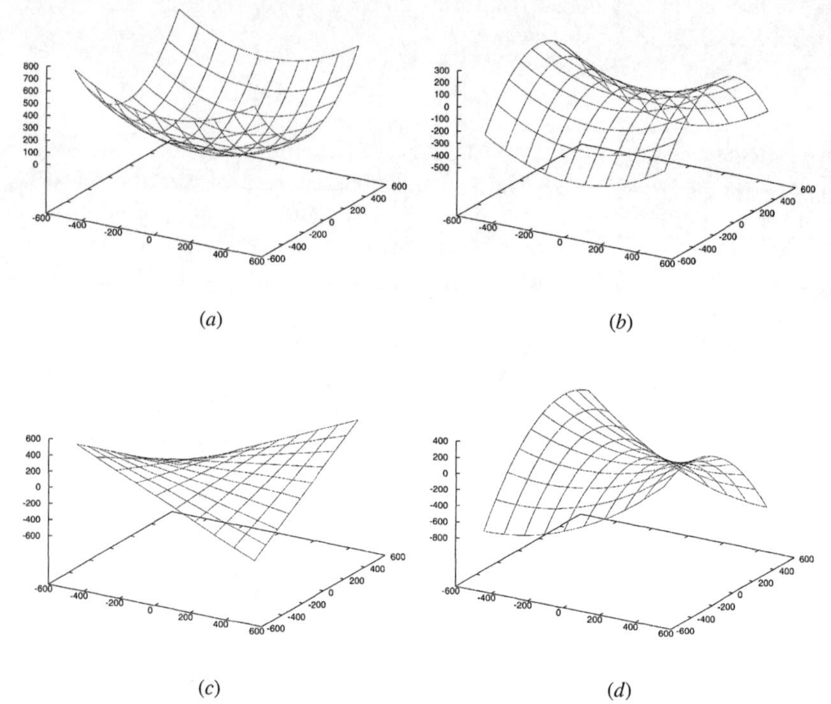

Figure 4.4. Quadratic surfaces as a model for a β-sheet. Changing parameters in the general quadratic gives rise to a variety of surfaces that can represent a sheet. (*a*) Cup shape (no twist), (*b*) saddle shape, (*c*) simple twist and (*d*) twist plus saddle. These parameters can be optimized to find the best fit to any given sheet (see figure 4.5).

packing of sheets with other secondary structures. Considering firstly the sheet by itself, as previously, details of side chain packing will be neglected in favour of simplified geometric models.

A distance of 10 Å was adopted in the previous chapter as a typical separation between pairs of packed secondary structures, irrespective of their type. This implies that the 'coating' of side chains over the backbone is roughly 5 Å thick. The model of a sheet that must be considered is therefore that of a slab of deformable material (side chains) fixed either side of a more rigid layer (main chain) that can be twisted but not compressed in either planar dimension. This can be represented nicely by two blocks of foam 'rubber' stuck either side of a piece of chicken wire. To give this mattress-like structure some dimensions, a typical sheet of five strands is roughly square and taking a separation of 5 Å between strands then we have a mattress roughly $25 \times 25 \times 10$ Å.

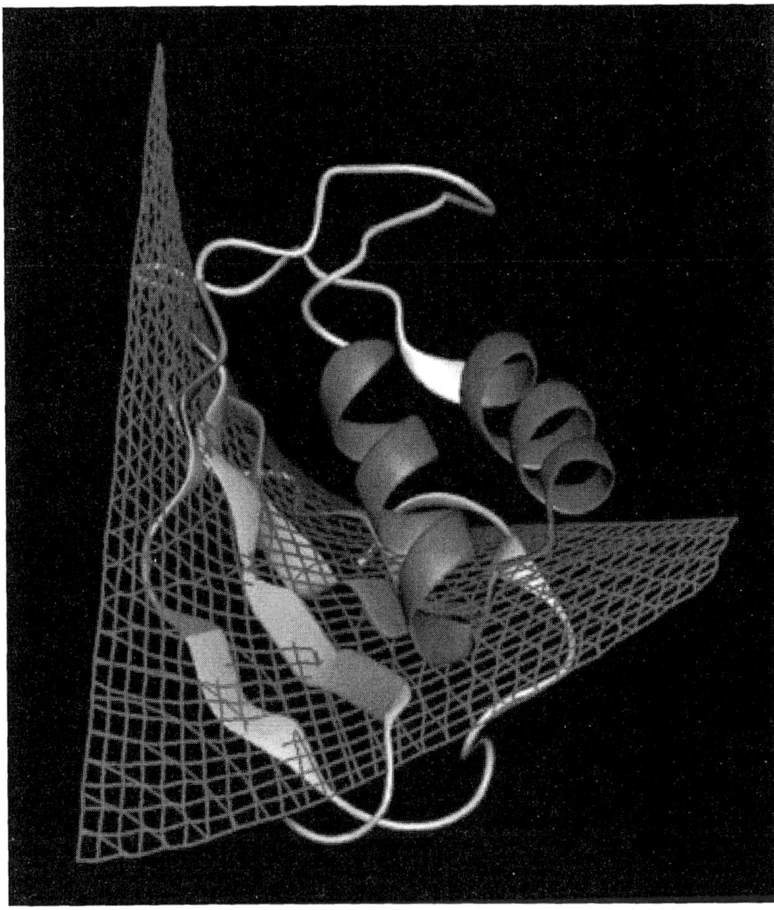

Figure 4.5. A hyperbolic paraboloid fitted to a β-sheet.

When this model of a sheet is twisted around a helical axis as described, then on each surface, two of the corners move towards each other while two move apart, creating a double saddle shape. This transformation can be analysed more quantitatively by a model of a β-sheet in which a series of sticks (β-strands), of equal length (L), twist about an axis ($\boldsymbol{a}$) at their mid-point (figure 4.6). The strands are equally spaced along the axis at a distance (D) with a twist τ. This is simply a statement of the model for a twisted β-sheet used in the previous chapter (figure 3.10(b)). To this model we now add a side chain position (R) at the end of each strand such that the line connecting R to the helix (at height h) is perpendicular both to the strand and to the gradient of the helix at the strand end-point (figure 4.7). If the pitch of the helix is α (its angle to the horizontal in

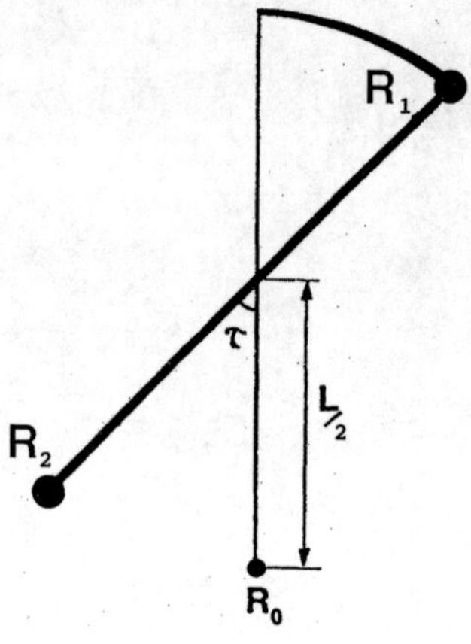

Figure 4.6. A stick model for a β-sheet. β-strands are represented as sticks of length L with end-points R_i twisting around a central axis by an angle τ with each strand. The twist is left-handed when viewed down the twist axis corresponding to a view across the strands as depicted.

figure 4.7), then the angle of the side chain vector to the horizontal is:

$$\theta = \frac{\pi}{2} - \alpha. \tag{4.3}$$

The angle θ can be found in terms of the twist per strand (τ) as its tangent will be the ratio of the height risen with each step (D) and the segment length (x) swept-out with each step. Since the circumference/segment = $\pi L/x = 2\pi/\tau$, then $x = \tau L/2$, or:

$$\alpha = \arctan\left(\frac{2D}{\tau L}\right) \tag{4.4}$$

giving:

$$\theta = \frac{\pi}{2} - \arctan\left(\frac{2D}{\tau L}\right). \tag{4.5}$$

This model can be used to calculate the variation with twist of the distances between four residues R_0, R_1, R_2 and R_3 which characterize a unit of the sheet surface (figure 4.7). The residue positions can be found by firstly rotating the residue vector relative to the strand axis then rotating this relative to the helix axis. For example, relative to a coordinate origin on the helix axis half way between the two strands, the residue position of R_1 (point $\boldsymbol{p}_1$) is: $\boldsymbol{p}_1 = \{p_{1X}, p_{1Y}, p_{1Z}\} = \{r, h, d\}$ (where $d = D/2$ and $r = L/2$). Rotating about X by θ gives, $\{r, h\cos\theta, d - h\sin\theta\}$ and rotating again around Z by τ gives,

$$p_{1X} = r\cos\tau - h\cos\theta\sin\tau \tag{4.6}$$

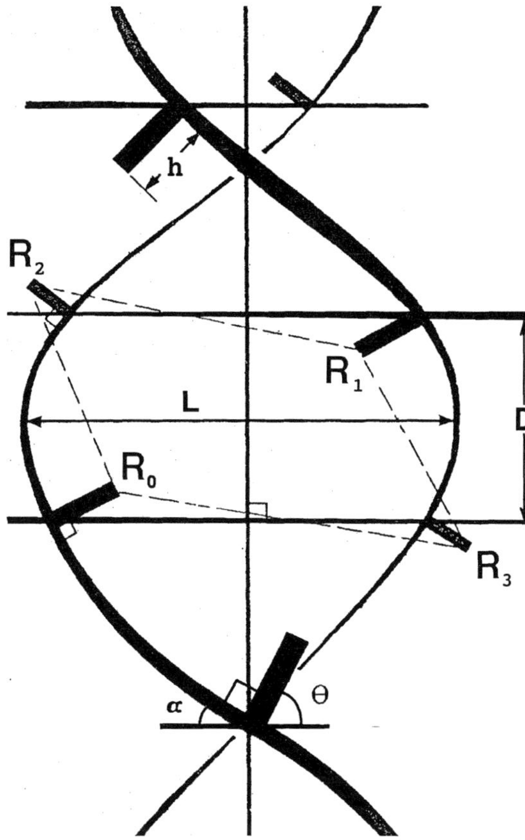

Figure 4.7. A stick model with residues above the sheet. The simple model in figure 4.6 is extended by raising the residues (R_i) above the sheet by a height h. The angle of the helix to the horizontal (pitch) is α and the residue vectors remain orthogonal to the helix and to the strands, making an angle θ to the horizontal. The twist per strand is τ, as in figure 4.6.

$$p_{1Y} = r \sin \tau + h \cos \theta \cos \tau \tag{4.7}$$

$$p_{1Z} = d - h \sin \theta. \tag{4.8}$$

Each of the other residue positions can be then be swung into position using similar transformations (figure 4.8). However, for a number of points, these operations are more easily implemented 'blindly' using a rotation matrix (see the appendix). The interresidue distances can then be found (by Pythagoras) from the point vectors.

By doubling the helix radius and adding residue positions along the axis the model can be extended for little extra effort to include three strands each with three residues above and below. When twisted (figure 4.9), the surface of the

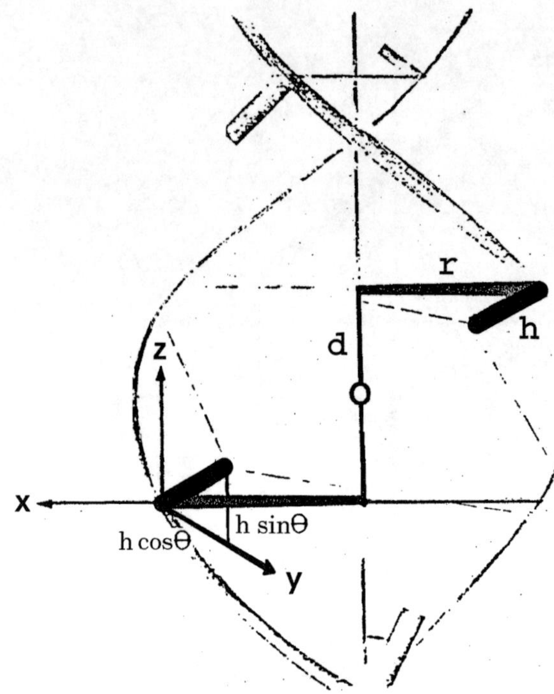

Figure 4.8. The coordinate frame for the sheet model. Components of figure 4.7 are emphasized to show the coordinate system used in the text. Placing the origin on the axis between two strands (marked by a circle) a residue can be reached in three steps of: d along the axis, r (the helix radius) along the strand and h the height of a residue above the sheet. When perpendicular to the helix trace, h can be decomposed into sine and cosine components of θ, all of which is further rotated by τ around the central axis.

sheet on both side becomes saddle shaped with two quite different curvatures on the up-curling and the down-curling diagonals between the corners. These radii can be calculated from the positions of the corner residues and the central residue. If the distance between the corners is $2c$ and the distance from each corner to the central residue is b, then the line between the corners passes at a height a over the central residue with $a^2 = b^2 - c^2$. By a simple construction, the radius of the circle passing through all three residues forms a right-angle triangle with sides r, c and $r - a$, so by Pythagoras,

$$r^2 = c^2 + (r - a)^2. \tag{4.9}$$

Expanding and simplifying,

$$r = \frac{c^2 + a^2}{2a} \tag{4.10}$$

and substituting for a,

$$r = \frac{b^2}{2\sqrt{(b^2 - c^2)}}. \tag{4.11}$$

The distance between the α-carbons of two residues along a strand on the same face of a sheet (L in our model) is reasonably constant at around 6.5 Å, while the distance between the strands (d) is again fairly constant at 4.7 Å. In our current model, however, the strands splay apart (as discussed above) so a smaller value of $d = 4$ was taken with $L = 6.5$. This leaves the height of the residue centre (h) to be determined and if the whole sheet is 10 Å thick, then the residue should have a radius of 5 Å, giving $h = 2.5$ Å. (These are the values used in figure 4.9 but the spheres are drawn smaller.) Of these three dimensions, the height of the residue 'side chain' centroid (h) is the least well determined as it depends on how much residue interpenetration occurs between the two packed surfaces in sheet/sheet packing. To allow for this a series of values around $h = 3$ is considered in most of the calculations below.

Using these dimensions (plus a series of values for h), the two radii across the diagonals of the sheet were calculated. These are referred to below as the up-up radius (R_{uu}) between the inwards curling corners and the down-down radius (R_{dd}) between the outward curling corners. The value for these two radii are plotted in figure 4.10 for differing degrees of interstrand twist. Both radii start at a very large value (infinite for a flat sheet) and decrease to a point at one radian where R_{uu} is only 2.5 Å, at which point the residues on the up-up corners would have met. Typically $R_{dd} = 2R_{uu}$. An interesting point is where the radius of the inner points (encompassing the up-up corners) equals 5 Å (at $\tau = 0.6 = 34°$) as this is the interstrand twist for a sheet to neatly curl around a single α-helix.

Another striking feature of the twisted surface is that it moves from a flat rectangular unit to a twisted parallelogram with an associated decrease in area. This area (A_{par}) can be calculated from the vector product of the two edges of the parallelogram. However, because the surface is not flat, a better estimation (A_{tri}) can be made from the sum of the areas of its two component triangles (figure 4.11). These alternatives are:

$$2A_{\text{par}} = |\boldsymbol{s} \otimes \boldsymbol{t}| \tag{4.12}$$

and

$$2A_{\text{tri}} = |\boldsymbol{s} \otimes \boldsymbol{r}| + |\boldsymbol{t} \otimes \boldsymbol{r}| \tag{4.13}$$

where $\boldsymbol{r} = \boldsymbol{p}_1 - \boldsymbol{p}_0$, $\boldsymbol{s} = \boldsymbol{p}_1 - \boldsymbol{p}_0$ and $\boldsymbol{t} = \boldsymbol{p}_2 - \boldsymbol{p}_0$. These areas are plotted in figure 4.12(a) against interstrand twist.

Unexpectedly, the area of the parallelogram (A_{par}, full curves) rises initially which is due to the stretching-out of the down-down corners as the residues tilt outwards. By contrast, the area calculated from the two-triangles method (A_{tri}, broken curves) either rises slightly or falls monotonically depending on the length of the residue vector (h). At high twist values ($\tau > 0.5$) both areas fall with the

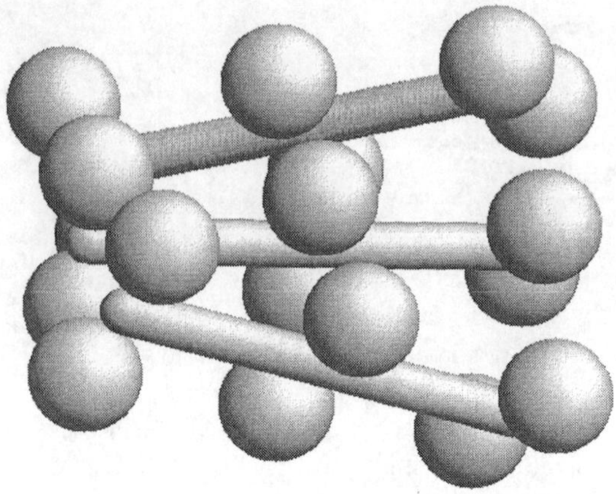

Figure 4.9. A simple model of a twisted β-sheet. Using the geometric construction shown in figures 4.7 and 4.8, a twisted sheet was made of three strands with three residues above and below the sheet on each strand. For simplicity, the stagger of the residues above and below the sheet was not included as all measurements on this model were made on the same surface (see figure 4.13 for a more realistic model.) The in-curling corners (top-right and bottom-left) are referred to as $up - up$ while the out-curling ones (top-left and bottom-right) are called the $down - down$ corners.

greatest decrease found with the parallelogram-based measure. Both areas start at 26 Å^2 (6.5×4) and at high twist have dropped to around 18 Å^2, a decrease of 30% of the surface area of the sheet. If this observation applies to real β-sheets (that is: atomic models), then it may be sufficient to explain why sheets twist since loss of surface area (in particular hydrophobic area) is the main energetic drive in proteins. It cannot of course, explain the preferred hand to the twist seen in real sheets as the model considered here is achiral.

With a simple model it is not easy to reconcile the differences in the change in sheet area measured by the different methods. Although the two-triangles method may give a more accurate estimate of the curved surface, when the two triangles close together with increasing twist (like butterfly wings) the gap may be so narrow that in a more realistic atomic model, it would become inaccessible to solvent and thus should not properly be considered a part of the surface.

4.1.3.1 Residue packing in twisted β-sheet

From the analysis in the previous section, it might be expected that β-sheets would be driven to increasingly higher twists to minimize their surface area. This, of course, does not happen because the residue side chains have bulk and will eventually clash. If we assume, as in the model above, that residues have a 2.5 Å radius then two residues at aligned positions on adjacent strands will bump

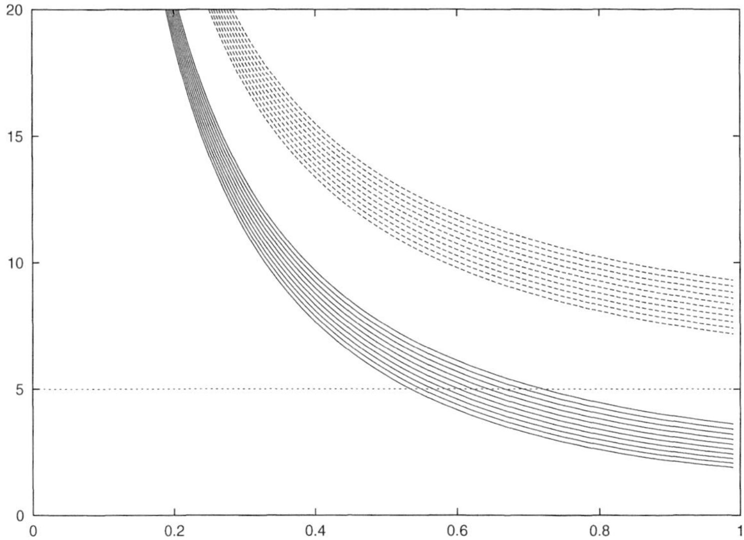

Figure 4.10. Surface curvature changes with twist. As the model for a thick sheet is twisted from flat to one radian (X-axis), the radius curvature between the corners changes from infinite to less than 5 Å for the up-up curling corners (R_{uu}, full curves) and under 10 Å for the down-down corners (R_{dd}, broken curves). For each radius, a series of values are shown for different sheet thicknesses from 2.5–3.5 Å. (See figure 4.9 for a description of up-up and down-down corners.)

(interpenetrate) by 1 Å while two adjacent residues on the same strand (and the same face of the sheet) will have a 1.5 Å gap between them. On twisting the sheet, the bump will be avoided and the gap will close until at high twist, the residues along the up-up diagonal will start to bump. This implies that there is an ideal position with minimal steric repulsion.

Taking the assumed radius of 2.5 Å, the root-mean-square (RMS) bumping error can be summed over all residue pairs and plotted for varying twist (τ) and residue height (h). Similarly, any gaps between residues with centres over 5.0 Å apart can be similarly summed as a RMS error. Both measures, but especially the latter, were calculated only over residues that are adjacent on the unit parallelogram, counting the short (up-up) diagonal but not the long (down-down) diagonal. The expected minima can be seen in figure 4.12(b) which for the bumps is long and low but rises sharply for twists over 0.6 (radians). The gaps close to a minimum around the same region.

Residues packed with no bumps and no gaps would, ideally, be hexagonally packed with equal distances between all adjacent residues. This can also be monitored on the model as the RMS error over the difference between all pairs of adjacent residues. This measure has the advantage that it does not use an assumed

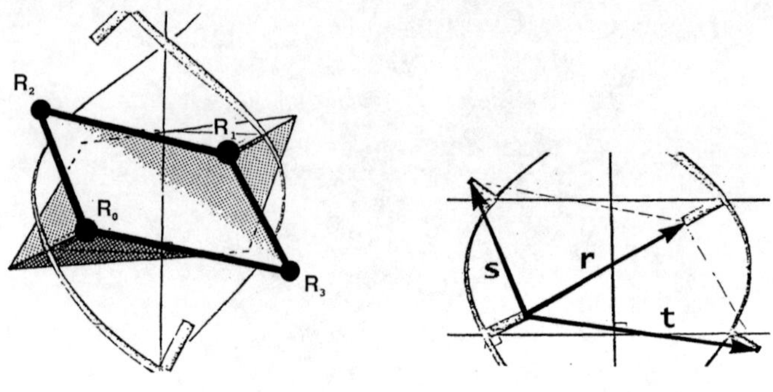

(*a*) (*b*)

Figure 4.11. Surface measures on the twisted sheet. (*a*) The interresidue distances on the unit sheet from R_0 to R_1, R_2 and R_3 were monitored. (*b*) Their corresponding interresidue vectors $\boldsymbol{r}$, $\boldsymbol{s}$ and $\boldsymbol{t}$ were used to calculate the surface area of the unit from their vector product.

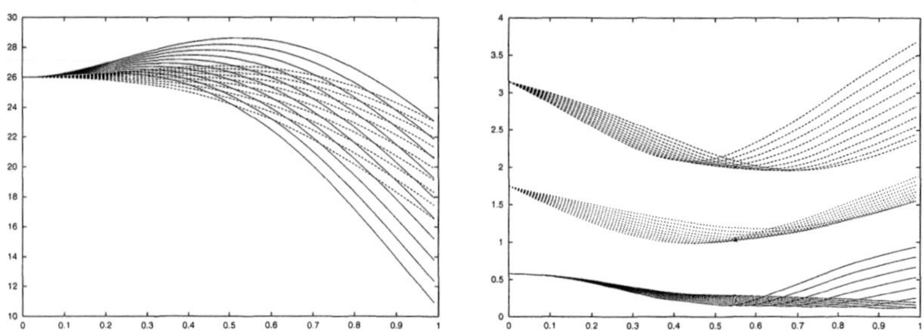

(*a*) surface curvature (*b*) residue packing

Figure 4.12. Twist induced surface changes. As the model for a thick sheet is twisted from flat to one radian (X-axis), (*a*) the surface area decreases as measured by a parallelogram approximation (full curves) and a double-triangle approximation (broken curves). In concert, (*b*) the spacing between residues on the surfaces also changes with gaps (broken curves) and bumps (dotted curves) becoming smallest around 0.5–0.6 radians. At this point, the residue spacings also have minimal deviation from hexagonal packing (full curves). For each quantity in parts (*a*) and (*b*), a series of values are shown for different sheet thickness from $h = 2.5$–3.5 Å.

residue radius and, indeed, can provide a value for the residue radius when ideal packing is attained. This distance-difference error, as it must, follows both the bump and gap error with a range of minima from 0.4–0.7 radians, depending on the value of h.

Although the minima on all these functions are shallow, a combined ideal value that penalizes excessive bumps and gaps can be formulated as: $\delta d(g^2 + b^2)$, where δd is the RMS distance-difference error and g and b the RMS error on gaps and bumps, respectively. This function attains a minimum when $\tau = 0.5$ radians and $h = 3.75$ Å, at which point the mean residue separation in 5.8 Å. Iteratively substituting the mean residue separation for the residue radius (bump/gap distance) in the model converges on a radius of almost 6 Å with $h = 2.25$ and $\tau = 0.67$ (38°). The values of h and τ are correlated and between these two values of h (2.25–3.75) there is a series of $\{h, \tau\}$ combinations with almost equally low error. For $h = 3$ the minimum occurs with $\tau = 0.58$ (33°) and a residue radius of 5.8 Å (figure 4.12(*b*)).

4.1.3.2 An atomic model for a twisted sheet

The analysis of the previous sections suggests that ideal residue packing can be attained in a twisted sheet and that this might be driven by a decrease in surface area. To properly assess the latter in particular, requires a more detailed atomic model.

In the preceding analyses, the relationship between the top and bottom surfaces of the sheet were irrelevant (measurements were only made on one surface). To construct an atomic model, the shift between the surfaces, corresponding to the pleat of the sheet, must be included. Although this is a simple translation of $\pm L/4$, it changes the distance of the residues from the twist axis giving three distinct helices of residues at $\pm L/4$, $\pm 3L/4$ and $\pm 5L/4$—all of which have different pitches and hence different values of θ (equation (4.5)). To make a rough α-carbon model, guide positions were constructed half-way between the new residue positions and their old points of 'attachment' to the strand. This α-carbon backbone 'sketch' was then regularized to give equal spacings between the α-carbon-atoms (of around 3.8 Å) (figure 4.13(*a*)). Using only the residue spacings, this new model was allowed to find its optimal side chain packing as described for the more symmetric model in the previous section. Because of the now slightly greater spacing between the residues (being shifted away from the axis), their mean spacing was 5.9 with $h = 3$ and $\tau = 0.52$ (30°).

Using the method described in section 2.1.2.2, the program `ca2main` can be used to convert the sheet α-carbon backbone into a main chain backbone. Throughout this the residue positions have remained unchanged and they now lie in a realistic position relative to the constructed β-carbon atoms. Finally, to make the complete model into a realistic protein, the residue positions can be incorporated into the structure as the sulphur atoms of cysteine side chains (figure 4.13(*b*)). This format allows the model to be presented to a program

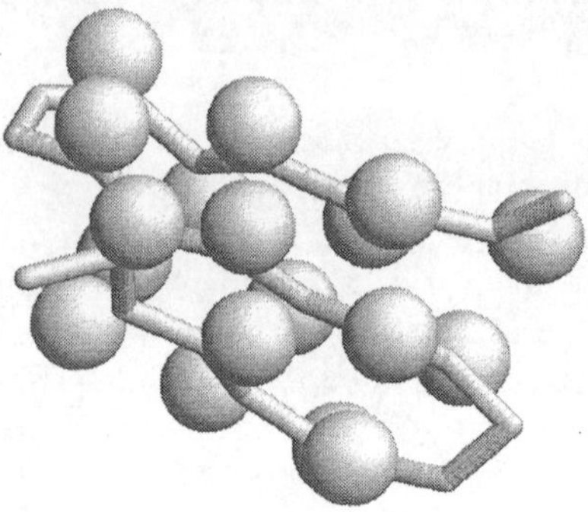

(*a*) α-carbon with centroid

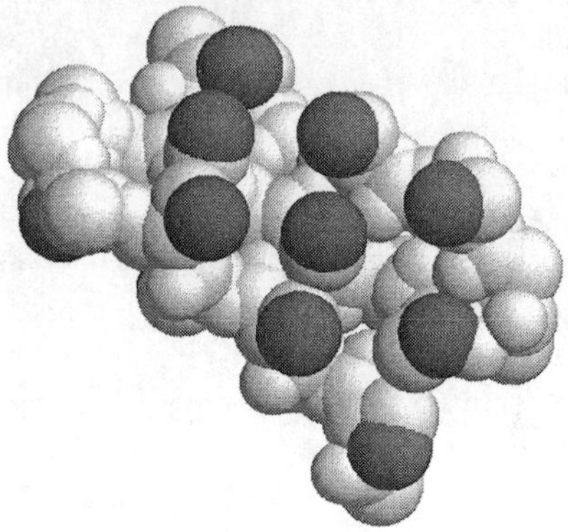

(*b*) all-atom (poly-Cys)

Figure 4.13. Realistic models for a twisted β-sheet. (*a*) Stagger was introduced between the surfaces of the simple twisted sheet model (figure 4.9) giving α-carbon atom positions. This model has been expanded in (*b*) to give all main chain and β-carbon atom positions (using the program `ca2main`). The residue centroids (above the α-carbon positions in (*a*) were retained as sulphur (grey) side chain atoms (making cysteine residues).

to measure the solvent accessible surface area of the atoms (Lee and Richards 1971). Of particular interest is the area of the central residue position as the sheet twists. Although this is only a single atom, its environment will be the same as any residue in a sheet that is not on the edge and as such residues will tend to be more hydrophobic, it will give a measure of potential hydrophobic surface burial.

The program `access` of Lee and Richards (1971) (as modified in turn by Tim Richmond, Mike Sternberg and Simon Hubbard), allows the probe radius of the solvent (water) molecule to be varied as well as the radius of any other specified atom. Advantage was taken of this to measure the accessible surface area for a series of probe radii and two different radii for the sulphur atom that represents the side chain. The smaller radius was 2.0 Å (16.7 Å^3) which is typical of a single methyl group while the larger was taken as 2.5 Å (65.4 Å^3) which is more representative of a typical hydrophobic residue. For both small and large side chains (figure 4.14(*a*) and (*b*), respectively) the exposed surface area of the central reside dropped markedly over the range of twist from $\tau = 0.2$ to $\tau = 0.8$. Around the value of ideal hexagonal side chain packing ($\tau = 0.6$) the area has decreased around three-fold from the flat sheet configuration. Beyond $\tau = 0.6$ for the larger radius, the central residue is almost completely buried.

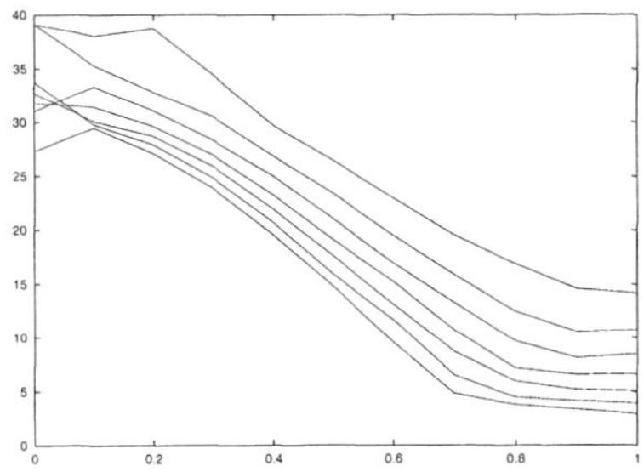

(*a*) $r = 2.0$

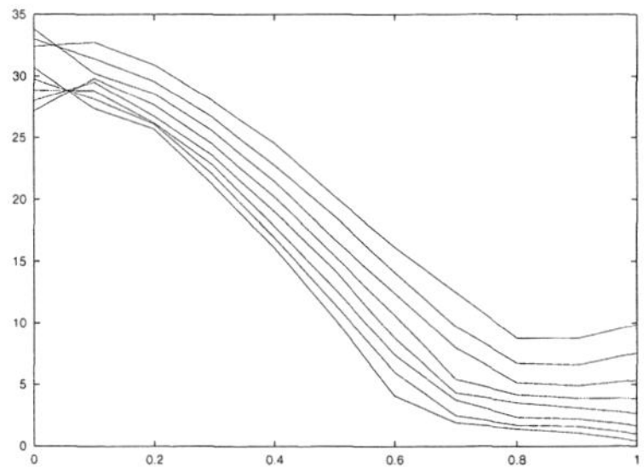

(*b*) $r = 2.5$

Figure 4.14. Surface area changes on an atomic model of a twisted sheet. The twist of the atomic model for a twisted sheet in figure 4.13(*b*) was varied and the solvent accessible surface area of the side chain sulphur atom (grey) of the central residue measured. A large area decrease was seen for two different radii (r) of this atom (*a*) and (*b*) for a variety of probe (water) radii. (See text for details.)

4.1.4 Sheet bend and curl

The representation of a β-sheet using a general hyperbolic paraboloid (section 4.1.2.3, figure 4.4) introduced the possibility that sheets can bend as well as twist. In the fitting of this surface described previously, the bend was not specifically associated with any particular direction on the sheet but using the stick model elaborated over the previous sections, it is simple to introduce a bend both along the β-strands and across the strands (in the H-bonding direction). With just three points considered in each dimension, this is not sufficient to define a parabola, however, there is no particular reason why a parabola should be adopted and it is preferable to use a simple circle defined by three points, giving a radius of curvature in each direction. To distinguish these two dimensions, curvature along the strand will be called ***bend*** while curvature across the strands will be referred to as ***curl***. (This term should not be confused with the curl of a vector.) The term 'bend' is naturally associated with a one-dimensional object (like a β-strand), but as bend across the strands can only be associated with a sheet, the term 'curl' as commonly applied to a non-planar sheet of paper seems appropriate. Together, 'bend' and 'curl' will be referred to as 'coil'.

The resulting thick sheet with varying degrees of twist, bend and curl can now be recursively built into extended sheets by the recursive growth algorithm described in section 4.1.2.2 and assessed using the packing of surface side chains as described in section 4.1.3.1. In addition, residues from each new unit are only added if they do not clash with any previously existing residues. (A clash was declared if any central or surface layer 'atom' came within 3 Å of any other

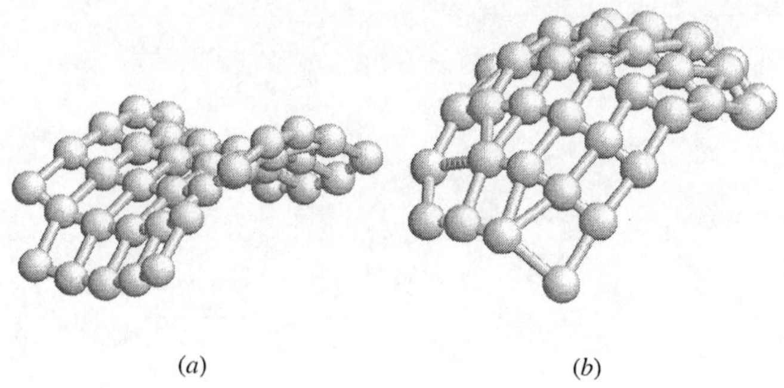

Figure 4.15. Growth of β-sheets with positive and negative curvature. The recursive growth model for a β-sheet (section 4.1.2.2) was applied to a unit sheet with (*a*) positive bend and negative curl angles and (*b*) positive bend and positive curl. (Both models had a twist angle of zero).

atom.) Starting with a flat sheet, bend is easily introduced without any side chain bumping. With curl, however, since the side chains are bumping each other at the start, things get worse on one side of the sheet while they are relaxed on the outer surface. Combining bend and curl produces effects that are equivalent to the combinations of parameters A and C in equation (4.2) (figure 4.4(*a*) and (*b*)). When both bend and curl are positive, a section of a sphere is obtained (figure 4.15(*a*)) (similar to the cup-like elliptical paraboloid) and when they have opposite sign, a saddle shape is obtained (figure 4.15(*a*)), similar to the hyperbolic paraboloid.

More natural shapes (for β-sheets) are obtained when twist is introduced and a series of sheets are shown in figure 4.16 for twist $= 0.3$, bend $= \pm 0.2$ and curl $= \pm 0.1$. (Less curl was used because of its greater propensity to induce bumping.) Most of these sheets fall into the category of 'ragged' twisted ribbons, however, when both bend and curl are positive, their in-turning spherical trend (positive curvature) counteracts the out-turning hyperbolic trend (negative curvature) of the twist to form a closed barrel-like structure. This can be seen in figure 4.16(*b*) and if growth is allowed to continue, the two edges close together. In this particular example, with arbitrarily chosen parameters, the edges do not meet in exact hydrogen bonding register. In the following section we will address the problem of the choice of parameters that generate exact barrels.

A more mathematical treatment of the transition from a twisted ribbon to a barrel has been made by Louie and Somorjai (Louie and Somorjai 1982, Louie and Somorjai 1983) using the construct of a helicoid which can be continuously 'deformed' into a closed (circular) form called a catenoid which is similar to a hyperboloid of one sheet.

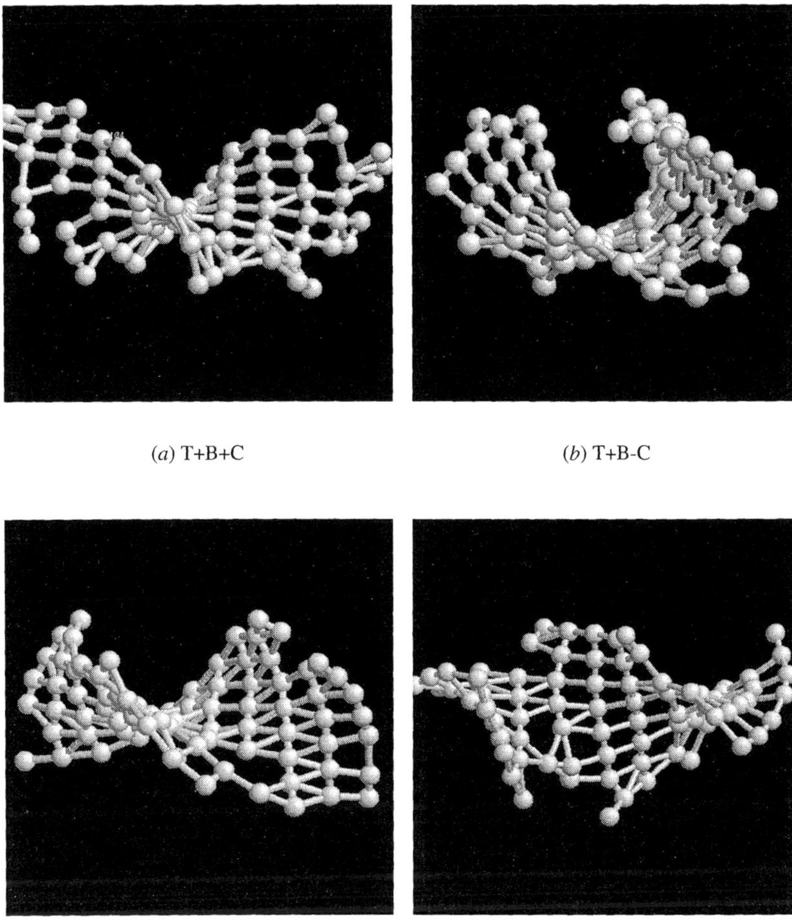

(*a*) T+B+C (*b*) T+B-C

(*c*) T-B+C (*d*) T-B-C

Figure 4.16. Growth of β-sheets with twist, bend and curl. The recursive growth model for a β-sheet (section 4.1.2.2) was applied to a twisted unit sheet with different bend and curl angles. The twist angle, $T = 0.3$; bend, $B = \pm 0.2$ and curl, $C = \pm 0.1$. Their combinations are designated as $T \pm B \pm C$ in each of the four frames (*a*)–(*d*). Most produce 'ragged ribbons' but in (*b*) a less ragged sheet is formed.

4.1.5 β-sandwiches

4.1.5.1 β-sheet stacking model

A model was introduced in the previous chapter for the stacking of two β-sheets (figure 3.10(*b*)). In our more detailed analysis in this chapter it is clear that the model for a thick twisted β-sheet described previously has implications for the way in which two β-sheets can pack together. The different radii on the up-up and down-down diagonals means that the sheets cannot fit neatly on top of one another, however, with the introduction of coiling (bend and curl), it is possible to match the radii. This means that the two sheets twist together about a single axis which implies that the local axes of twist through each sheet are no longer parallel (figure 4.17).

To quantify this relationship, the construct used for a single twisted sheet in section 4.1.3 can be adapted almost unchanged. Instead of a pair of strands twisting around a central axis (figure 4.7), each strand in this previous construction becomes a sheet. The line segments that were previously in the direction of the side chains now point to the adjacent strands in the sheet. This means that the sheet now runs over the surface of the barrel and not through the axis. What was previously another helical path of strand end-points on the opposite side of the axis, now forms the basis for the construction of an equivalent sheet on the other side (figure 4.18). However, with a 90° shift in view, what was left-handed becomes right (and *vice versa*) so the coordinates must be reflected to preserve the natural twist of the sheet. It can be seen in figure 4.18(*b*) that not only is the left-hand twist across the sheet preserved but there is also a new left-handed twist introduced between the sheets. This is a property observed in real sheets (Chothia *et al* 1981, Chothia and Janin 1981).

The dimensions of the construct in figure 4.18 must also change from its previous use, with the radius of the helix (r) being half the distance (D) between the two sheets taken as 12 Å, the separation of strands within a sheet (h) is 4 Å and the separation of residues along the sheet (on the same face) is 6.5 Å. Given a twist angle, the coordinates of two unit (3×3) sheets were calculated from this construction and the residue positions used to seed the recursive growth of larger sheets. In order to construct a thick sheet, the dummy residue centroids were calculated using the twist and bend angles extracted from the unit (stacked) sheet. (Since the line of hydrogen bonds is straight, the curl angle is zero.) These were simply 'dialled-up' using the model for a single twisted sheet, then this sheet was superposed with minimum RMS deviation over the α-carbon positions of each sheet in the stack and the residue centroids transferred.

4.1.5.2 Packing in stacked β-sheets

With a relatively low twist angle of 0.2 rad, the model of sheets twisting around a cylinder generates a well packed sandwich with flat packing surfaces both parallel and perpendicular to the helical (cylinder) axis (figure 4.19). This can even extend to very large sheets without serious distortion (figure 4.20).

As the twist angle becomes larger, however, the linear hydrogen bonding direction (with zero curl) becomes packed against an increasingly curved surface as the local twist axes of the sheets now cross at a higher angle (figure 4.21(*a*)).

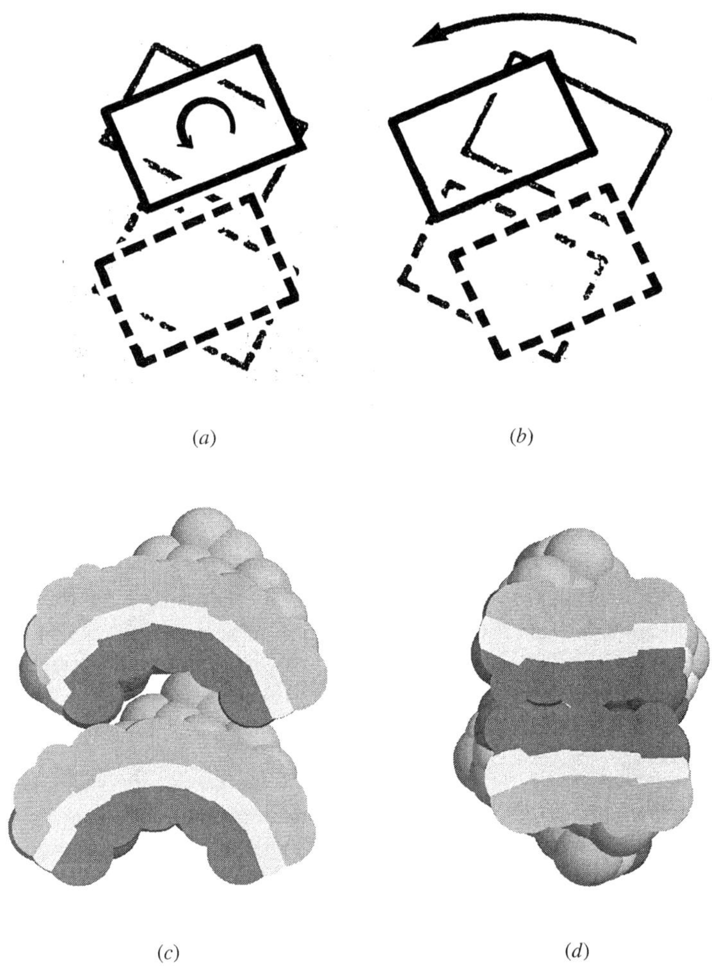

Figure 4.17. Two models of sheet stacking. (*a*) Two sheets twisted about their own local axis are stacked and (*b*) two sheets twisted about a common axis. The corresponding models for these are shown in (*c*) in which the first model (*a*) is sectioned along the diagonal while the second model (*b*) is sectioned across the strands. A clear cavity can be seen in (*c*) whereas in (*d*) good packing is maintained in the core.

This leads to less favourable packing with bumping at the edge of the sheets and a cavity in the core (figure 4.21(*b*)).

To avoid these defects, the curl angle can be changed to match the curve of the surface. When this is done, the packing in the core of the sheet becomes

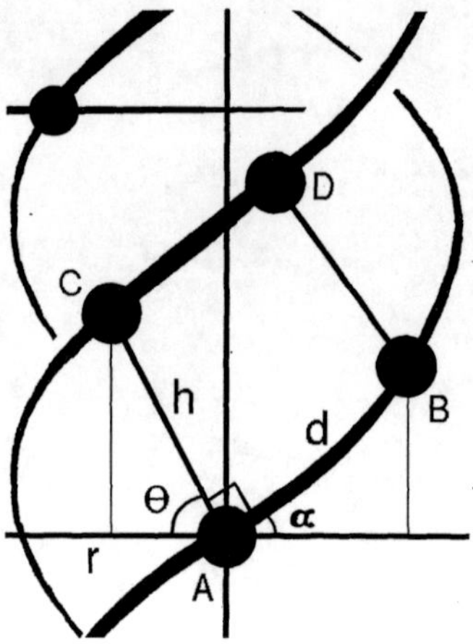

(a)

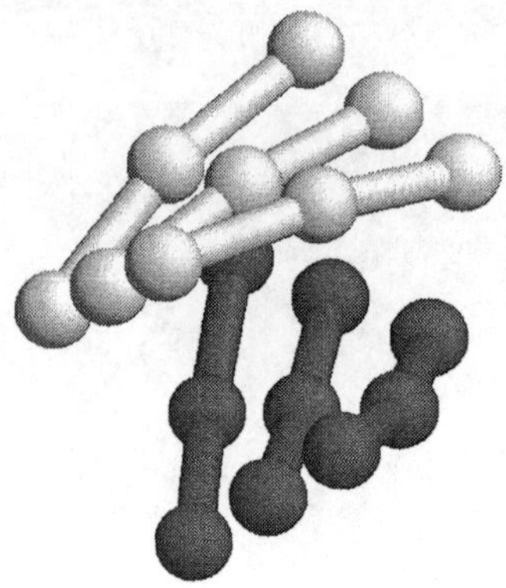

(b)

Figure 4.18. A geometric construction for stacked β-sheets. (*a*) Residues ***A*** and ***B*** lie (at a distance d) on a β-strand twisting around a central axis. They are hydrogen bonded (at a distance h) to residues ***C*** and ***D*** (respectively) on an adjacent strand. The strands make an angle α to the horizontal. (Compare figure 4.7.) (*b*) Two 3×3 sheet fragments based on the construct in part (*a*). A left-hand twist emerges between the sheets.

approximately flat again (figure 4.21(*d*)) while the edges of the sheet now curl away from each other (figure 4.21(*c*)). The introduction of curl into a relatively untwisted sheet, however, is not likely to be favourable because of the already close contact between the residues along the hydrogen bonding direction and it is more likely that adaptions will be made to fill the relatively small cavity in the core of the uncurled sandwich.

4.1.5.3 Alternative stacking models

With the exception of the last refinement, the model investigated above maintained linear hydrogen bonding (zero curl) but allowed strands to bend. With another 90° shift in view, the reverse situation could have been created—with linear strands and curling sheets. These options correspond to the two considered in section 4.1.2.1 for the simple twisted sheet. The option with linear strands corresponds closely to the model used for linear strands segments (sticks) in section 3.2 (figure 3.10(*b*)).

As with the simple twisted sheet, both these models are idealizations at the ends of a continuous range and there is no reason to presuppose that zero curl or zero bend should be preferred to any intermediate values. Indeed, neither is

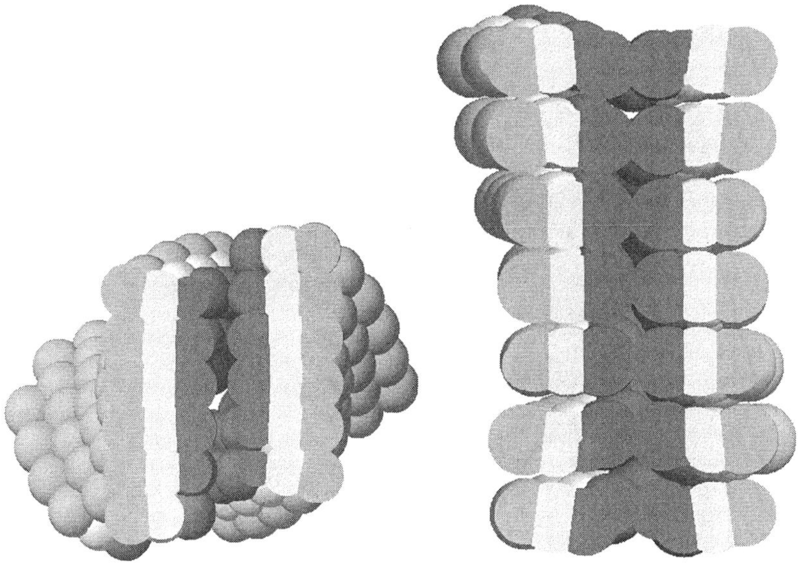

(*a*) (*b*)

Figure 4.19. Sections through stacked β-sheets . Using the geometric model in figure 4.18, a β-sheet was grown using the recursive tiling algorithm (section 4.1.2.2). The resulting stack was found to be well packed in the core when sectioned (*a*) across the β-strands and (*b*) parallel to the strands.

there any reason why the two sheets should have the same twist-bend-curl (TBC) angles or that they should pack at an angle dictated by the geometry of a helix over a cylinder. To find complementary surfaces in this phase-space of seven angles (TBC $\times 2 + 1$) would be a formidable task but as many of the TBC combinations can be discarded on the basis of internal residue packing within the sheet it may reduce to a tractable volume.

In the above analysis (figure 4.21), the curl angle was adjusted to reduce the cavity inside the core of two stacked sheets and avoid the edges bumping. If the curl angle were shifted in the other direction, the size of the cavity increases, becoming cylindrical and the edges of the sheets bump to such an extent that they might easily be joined to each other forming a closed barrel. The parameters that allow this to occur in a regular way will be described in the following section.

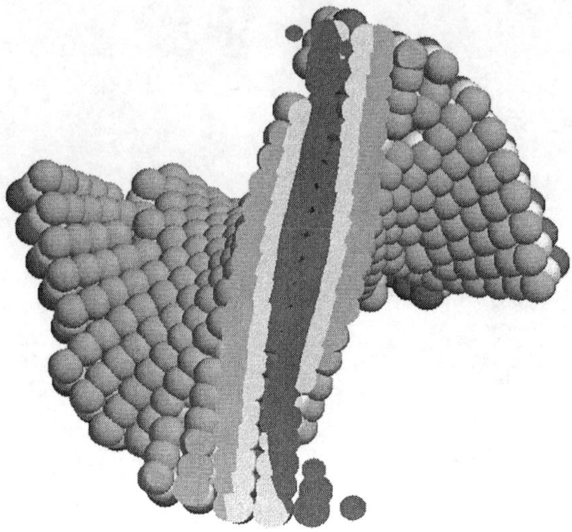

Figure 4.20. Extensive sheet stacking. A very large pair of sheets can be stacked using the parameters calculated from figure 4.18.

4.2 β-barrels

4.2.1 Hyperbolic surfaces

Early approaches to the analysis of β-barrels in proteins concentrated on the regular barrel seen in triosephosphate isomerase (TIM) (figure 1.5) which has eight parallel strands in the barrel. The TIM barrel has a markedly splayed shape at its outer rims which, combined with an assumption of straight β-strands, led to the quadratic surface called the *hyperboloid of one sheet* being adopted as an obvious model (Novotny *et al* 1988, Lasters *et al* 1988, Lasters 1990). In its normal form, this surface has the equation:

$$1 = \frac{x^2}{a^2} + \frac{y^2}{b^2} - \frac{z^2}{c^2} \tag{4.14}$$

(which is identical to the ellipsoid bar one change of sign). This model was tested on a variety of protein β-barrels by Novotny *et al* (1988) and found to be better than a number of other models, including catenoids and the elliptical cylinder. However, they went on to give the axis of the hyperboloid an additional twist (producing a new surface that they called a *strophoid*) and used this for fitting by a least-squares procedure.

Taking the β-strands as line segments, Lasters *et al* (1988) also used an iterative procedure to find the fit of the best hyperbolic surface to these. Their

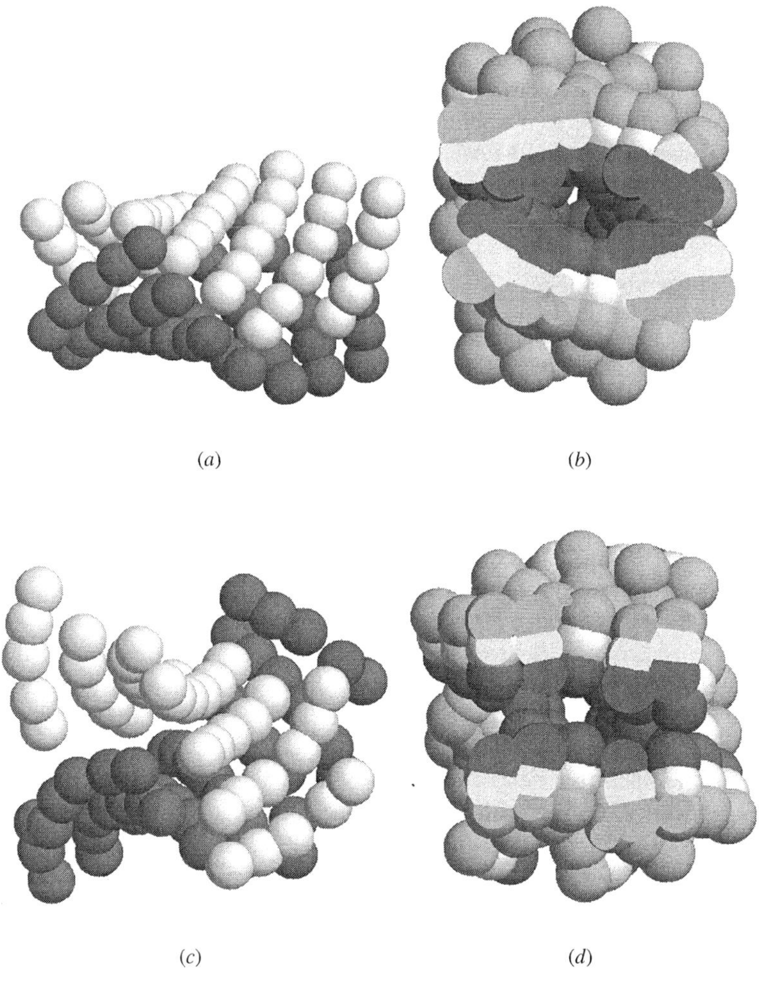

Figure 4.21. Packing between sheets with high twist. (*a*) A pair of sheets (light and dark) derived from the model in figure 4.18 (with curl angle zero) when seen in section (*b*) have a cavity in their core. This can be reduced by introducing curl (*c*) and (*d*) but the model now deviates from figure 4.18.

fit was parametrized as the interstrand twist and separation (−26°, 4.5 Å) and the angle the strands made with the barrel axis (−35°). More interestingly, the cross section of the hyperboloids fitted to a selection of barrels showed some marked deviations from circular, ranging from semiaxis lengths (*a* and *b* in equation (4.14)) of 7.1, 7.1 (circular) to 8.3, 5.6 in the TIM barrel. The external curvature of the barrel (c in equation (4.14)) was typically 10 Å.

Still concentrating on the eight-fold parallel barrel, in later works, Lasters and co-workers re-parametrized their model to incorporate a more detailed atomic model, allowing them to examine the requirements to form an exactly hydrogen bonded barrel (Lasters *et al* 1990). When these structures (derived from a variety of proteins) were then energy-minimized, it was found that the β-strands deviated only slightly from their starting positions as straight strands.

4.2.2 Shear and stagger

It was clear from the earliest description of the eight-fold barrel in TIM (Banner *et al* 1975) that the strands around the barrel have a considerable relative stagger in their hydrogen bonding register. This is nicely captured in the hyperbolic model described previously and for the TIM barrel where there is a stagger between the strands of one unit/strand which means that the first and last strands are 'shifted' eight positions relative to a barrel in which the strands remain parallel to the barrel axis. While this is a neat relationship, there is no reason for the stagger to equal the number of strands and a cylindrical barrel can be constructed with any degree of stagger (within the physical constraints of stereo-chemistry).

It was first noticed by McLachlan (1979a) that the specification of the number of strands and their relative stagger completely determines the overall structure of a β-barrel. Again concentrating on the $\beta\alpha$-barrel, this idea was initially developed by Lesk *et al* (1989) and more generally by Chou *et al* (1990) and Murzin *et al* (1994a) (Murzin *et al* 1994b) to incorporate all possible barrels. Most of these analyses concentrated on the geometry of the strands relative to the barrel axis, however, with a view to generating barrels using the recursive algorithm (section 4.1.2.2) employed widely above, it is of greater interest to extract the local twist, bend and curl angles for the basic (3×3) unit sheet.

The basic model for a β-barrel as employed by McLachlan (1979a) has two sets of orthogonal helices on a cylinder. Starting at any residue, the two helices can be traced in opposite directions (one along a strand, the other across H-bonded strands) until they meet again on the same strand (figure 4.22(a)). On their respective journeys, one helix has stepped across N strands while the other has hopped over S residues. Along these two paths, both have dropped the same height: which along the strands is $Sh\cos\alpha$ and across the strands is $Nd\sin\alpha$, where h is the distance between strands and d is the distance between residues on a strand. This gives a shear number (designated S by McLachlan) as:

$$S = \frac{Nd\sin\alpha}{h\cos\alpha} = Nd\tan\frac{\alpha}{h}. \tag{4.15}$$

More importantly, given S and N, this relationship allows the helix angle α to be found as:

$$\alpha = \arctan\frac{Nd}{Sh}. \tag{4.16}$$

In the orthogonal direction, together both helices make one circuit of the cylinder, so:

$$2\pi R = Nd \cos\alpha + Sh \sin\alpha \tag{4.17}$$

giving the radius R in terms of the two given values h and d, and the chosen number of strands and their stagger. While useful, this value for the radius assumes that the distance between two residues is measured on the surface of the cylinder whereas in most applications, the distance is normally the point–point separation of the residues which will be the chord of this arc. For N strands, the angle stepped around the cylinder for each strand is $2\pi/N$ (the angle ψ in figure 4.22(*b*)), with a simple construction bisecting this segment, it can be seen that the chord length (x) is $2R\sin(2\pi/N)$. The line of H-bonds connecting the residues (with separation h) makes an angle α to a horizontal on the cylinder, so the segment length between strands is $h/\cos\alpha$. As we want to make the chord length equal to the segment length, then:

$$2R \sin\frac{2\pi}{N} = \frac{h}{\cos\alpha} \tag{4.18}$$

giving

$$R = h/\left(2\sin(2\pi/N)\cos\alpha\right) \tag{4.19}$$

which is the expression given for the radius by McLachlan.

The barrel model of McLachlan is based on a cylindrical construct which differs considerably from the hyperbolic models described in the previous section that were found to provide a good description of barrel architecture. An important difference is that the McLachlan barrel model is unbounded and the strands can continue to twist around the cylinder for ever and still maintain their hydrogen bonding distance. By contrast, the hyperbolic models must be bounded as the separation between the strands continually increases away from the equator. These models can only be constructed because the stagger between the strands restricts the extent of the hydrogen bonded network that needs to be simultaneously connected. These two approaches were reconciled by Murzin *et al* (1994a) in a rather complicated analysis using differential geometry that established a relationship between the number of strands and their shear number with the splaying (hyperbolic) nature of the barrel. A slightly simpler model based on recursive addition of a basic unit will be considered below.

4.2.3 Cylindrical β-barrels

4.2.3.1 Constructing cylindrical barrels

The model of McLachlan is similar to that used in figure 4.18 for the model of stacked twisted sheets with the only difference that the line of H-bonds linking the strands now bends around the cylinder surface (figure 4.22(*a*)). Only two distances are given: the hydrogen bonded separation h and the spacing of residues

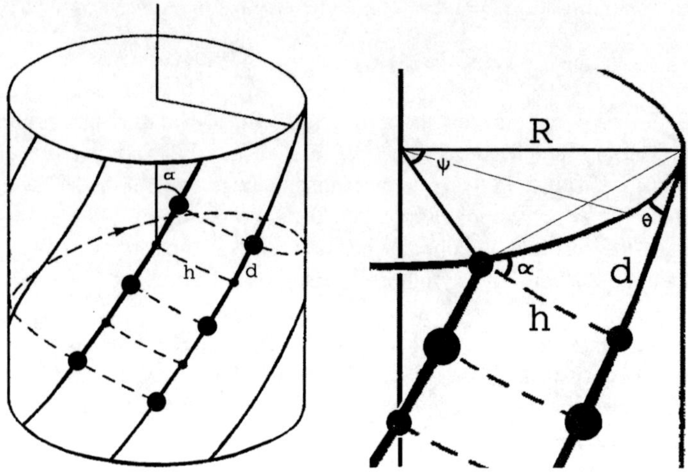

(*a*) (*b*)

Figure 4.22. McLachlan model for a cylindrical β-barrel. (*a*) β-strands (full curves) run in helices around a cylinder, connected at right angles by hydrogen bonds (dashed curves). One of these is shown as an extended trace linking a strand to itself four residues further along. (*b*) An enlarged portion of part (*a*) showing the three angles (α, θ, ψ) referred to in the text.

along the strand d. To be consistent with the constructions used in previous sections, the spacing along the strands will be taken as that between two residues (6.5 Å) since it is only possible to align alternate residues on the same side of the sheet. The hydrogen bonding separation was taken as 4.0 Å above which is shorter than the observed (4.7 Å) to compensate for the divergence of straight strands and although the strands do not diverge on a cylindrical barrel, the same value will be retained. From these values, the angle α can be calculated from McLachlan's relationships for shear (equation (4.16)) Since the exact length of the edges of the unit sheet is not important, the simpler equation for the radius (R) can be used (equation (4.17)) rather than the version corrected to give exact values (equation (4.19)).

Taking any residue as central in a 3×3 grid, a coordinate frame can be defined with the helix axis along Z and the perpendicular line connecting the central residue to the helix along X. Following a helix of hydrogen bonds, the angle turned around the central axis between each strand (ϕ) is $\pm h \sin(\alpha)/R$. Similarly, along a strand, the angle turned (ψ) is $\pm d \cos(\alpha)/R$. Relative to the central residue (***a***), the coordinates of four positions need to be found. These are the next hydrogen bonded residue (***b***), the next position along the strand (***c***) and the diagonally opposing points (***d*** and ***e***), the coordinates of which are:

$$\begin{aligned}
a_x &= R, & a_y &= 0, & a_z &= 0;\\
b_x &= R\cos\phi, & b_y &= R\sin\phi, & b_z &= h\cos\alpha;\\
c_x &= R\cos\psi, & c_y &= R\sin\psi, & c_z &= d\sin\alpha;\\
d_x &= R\cos(\phi+\psi), & d_y &= R\sin(\phi+\psi), & d_z &= h\cos\alpha - d\sin\alpha;\\
e_x &= R\cos(\phi-\psi), & e_y &= R\sin(\phi-\psi), & e_z &= h\cos\alpha + d\sin\alpha.
\end{aligned} \tag{4.20}$$

From these positions, the coordinates of the four remaining points can be found by symmetry, giving the full nine points in the grid (***P***) as:

$$\boldsymbol{P} = \begin{bmatrix} \boldsymbol{d}, & \boldsymbol{b}, & \boldsymbol{e} \\ \{c_x, -c_y, -c_z\}, & \boldsymbol{a}, & \boldsymbol{c} \\ \{e_x, -e_y, -e_z\}, & \{b_x, -b_y, -b_z\}, & \{d_x, -d_y, -d_z\} \end{bmatrix}. \tag{4.21}$$

The grid of points (***P***) can now be used in place of the simpler twisted unit described previously (section 4.1.2.2) and used to generate sheets. Unlike the simple twisted sheet, all the corners of this basic unit are identical and so superpose exactly ensuring that the geometry of the barrel is faithfully regenerated. An example of a barrel with eight strands and shear number eight is shown in figure 4.23 and a larger one in figure 4.24.

4.2.3.2 Residue packing in cylindrical barrels

The criteria used to assess the residue packing in simple twisted sheets described above can also be applied to barrel structures. To do this it is necessary to add to the dummy side chains. This can be done most simply by using the previous model and just 'dialling-up' the twist, curl and bend angles. The result differs slightly from adding side chains as normal vectors to the barrel surface but once these are overlapped and equivalent positions averaged, the effect is much the same. As before, to avoid edge effects, the packing was calculated only over a core unit cell. In some barrels (or coiled sheets in general) the inward pointing residues from different parts of the barrel can clash across the centre of the barrel. This effect will not be considered below and only the interactions of the adjacent residues will be taken into account. To assess this packing, the scoring function described above (section 4.1.3.1) will be used which penalizes overspaced interactions as well as steric clashes with the most favoured configuration of side chains approximating hexagonal packing.

In contrast to the simple twisted sheet considered above, once any coiling (bend or curl) component is introduced, there is now an asymmetry between the two surfaces of the sheet. With their different radii from the axis, the packing on the outer sheet can be more spaced compared with the inward facing residues in the centre of the barrel. Packing scores were calculated for both faces for a variety of β-barrels with different numbers of strands and stagger and the value of the scores plotted as two surfaces (figure 4.25(*a*)). It can be seen that when one

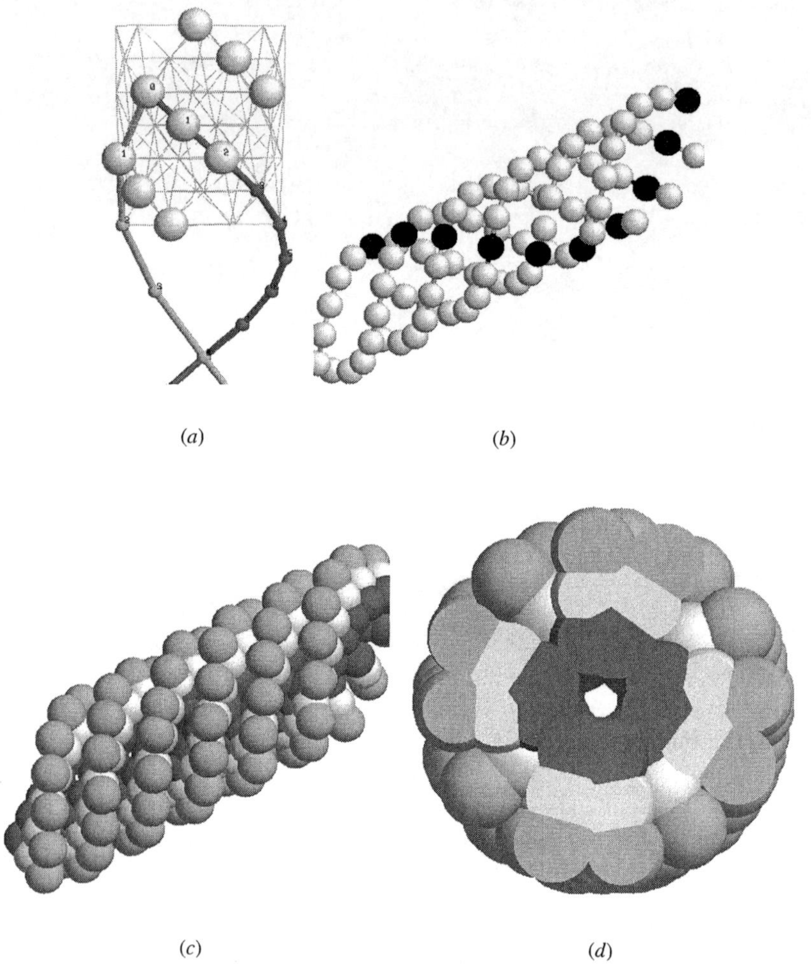

Figure 4.23. An eight stranded cylindrical barrel. The geometry of McLachlan (figure 4.22) was used to construct an eight stranded barrel with shear number (stagger) eight. (*a*) The 3 × 3 unit sheet on the cylinder surface with strand and H-bond helices. As only alternate residues (on the outer face) are shown, these helices rejoin after eight steps in the H-bond trace (black) but only four steps along the strand (grey). (*b*) The recursive growth model extended the unit sheet into a cylinder, one strand of which is shaded black. (*c*) The inner and outer residue centroids (grey shades) added to the model and (*d*) an orthogonal section through the cylinder.

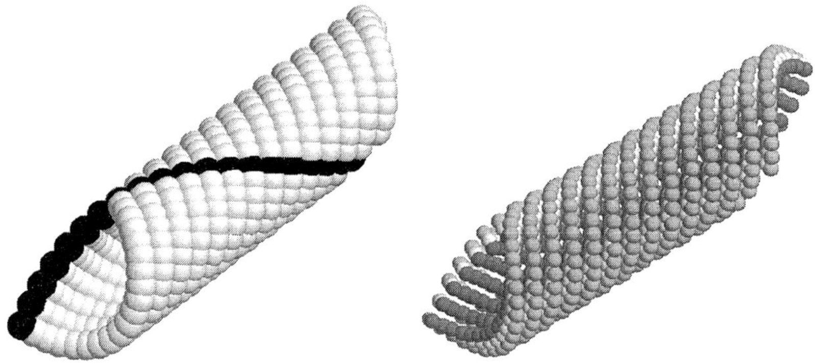

(*a*) (*b*)

Figure 4.24. A large cylindrical barrel. A twenty strand sheet with shear number 14 was constructed in the same way as the smaller barrel in figure 4.23. Parts (*a*) and (*b*) are depicted in the same manner as parts (*b*) and (*d*) in that figure (but with a slightly larger sphere radius in part (*a*)).

surface is well packed the other is not and a combined packing was calculated as the product of the packing scores for the two surfaces (figure 4.25(*b*)). This revealed a valley shaped surface rising slowly along the diagonal of $N = S$ (strands=shear). This can be clearly seen when minima for each number of strands are connected as a line (figure 4.25(*b*), full curve). A similar result was obtained in the analysis of optimal packing by Murzin *et al* (1994a).

4.2.4 Hyperbolic β-barrels

4.2.4.1 Constructing hyperbolic barrels

As was seen above, most observed β-barrels are not cylindrical but have a distinct splaying at their ends and are better approximated by a hyperbolic shape. In our twist-bend-curl (TBC) model this is equivalent to setting the bend angle to zero giving straight strands. The twist and curl angles can then be calculated to find values that allow the formation of barrels. These can be found in an equivalent manner to cylindrical barrels since, starting from the same residue, a strand will meet a helix of hydrogen bonds again after both have dropped the same height (down the barrel axis). With N strands spaced at a distance h, staggered by S intervals of distance d, then the angle of the strands to the vertical (defined by the barrel axis) is the same as in equation (4.17). In the cylindrical construction, both paths are helical but in the hyperbolic construction the strands become linear while the helix of H-bonds becomes a more complex function over the hyperbolic

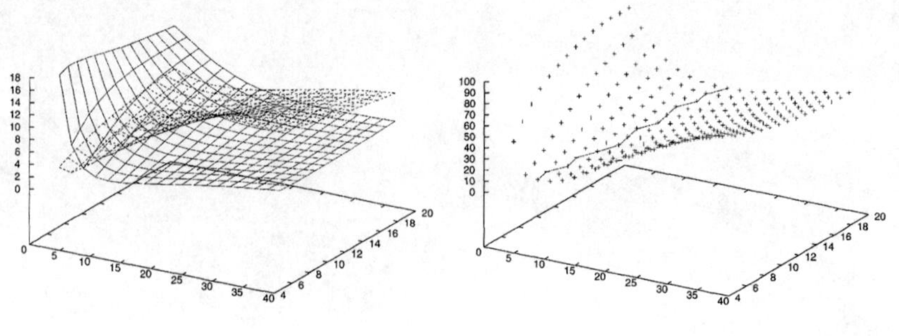

(*a*) (*b*)

Figure 4.25. Residue packing in cylindrical barrels. The residue packing scores used on simple twisted sheets (section 4.1.3.1) were applied to barrels of differing number of strands (N) and shear (S) (X, Y plane). (*a*) Packing scores are plotted separately for the inner and outer sheet surfaces and (*b*) combined into a single (product) score. The minimum value for each number of strands on this surface follows the line of $N = S$.

surface (figure 4.26(*a*)). The resulting positions of the unit 3×3 sheet were then used to recursively grow sheets. For the eight-stranded barrel with stagger eight, a relatively regular barrel is formed but continues to extend beyond the barrel region without closure (figure 4.26(*b*)).

The packing in hyperbolic barrels can be analysed in the same way as cylindrical barrels using the packing score functions developed for simple twisted sheets in section 4.1.3.1. This produces a similar valley to the cylindrical construction but with steeper sides more like a 'U'-shaped valley. The flatter valley floor results from the more even distribution of packing between the inner and outer layers permitted by the splayed-out shape of the hyperbolic barrel. The shear number producing the optimal packing for each number of strands rises at more than double the rate for cylindrical barrels.

4.2.4.2 *Extended hyperbolic barrels*

The recursive addition of units to form large cylindrical barrels simply results in longer cylinders but with hyperbolic barrels, something must change to accommodate the increasing radius of the barrel moving away from the middle (equator). As was seen above with the 'ragged ribbons' (figure 4.3(*b*)), it is possible for the sheet to bifurcate and, as was seen in figure 4.26(*b*), unclosed extensions can be made. The closest approximation to a closed extended barrel is with low shear number, neglecting $S = 0$, the two stranded barrel constructed with $S = 2$ is close to cylindrical. With $S = 4$, the cylinder begins to bifurcate at the ends. With increasing shear, this trend continues: producing increasingly fungal-like shapes (figure 4.27).

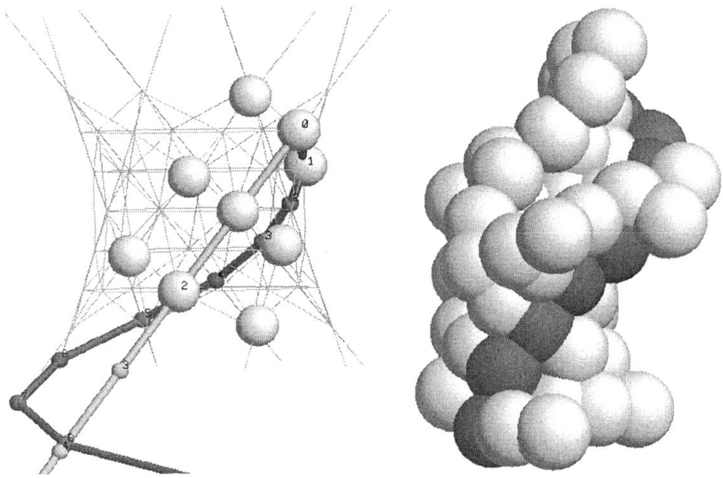

(*a*) (*b*)

Figure 4.26. An eight stranded hyperbolic barrel. The modified geometry of McLachlan (figure 4.22) was used to construct an eight-stranded hyperbolic barrel with shear number (stagger) eight. (*a*) The 3×3 unit sheet on the cylinder surface with strand and H-bond helices. As only alternate residues are shown, these helices rejoin after eight steps in the H-bond trace (black) but only four steps along the strand (grey). (*b*) The recursive growth model extended the unit sheet into a barrel, one strand of which is shaded grey. The top of the barrel contains β-bulges and is not closed.

As the shear number increases, the radius of the surface (parallel to the axis) approaches the radius of the barrel (perpendicular to the axis). When these radii are equal, a single sheet can form two orthogonal barrels and, in theory, this can extend indefinitely as a periodic minimal surface (MacKay 1986). Such exact regularity is not observed using the recursive growth algorithm, however, the space is densely packed with roughly equal portions of inner and outer surface visible. The shapes may be semi-regular and assuming this is so, they will be referred to below as aperiodic minimal surfaces. Two examples are shown in figure 4.28(*a*) and (*b*).

As the shear number is further increased, an abrupt transition is made to a very regular form in which barrels formed by both surfaces 'stack' in two regular super-helices that grow out of a single inversion region. Higher shear numbers lead to an increased difference in the radius and pitch of the two super-helices until a point is reached where the larger has zero pitch and terminates as a circle creating a structure rather like a bishop's crozier.

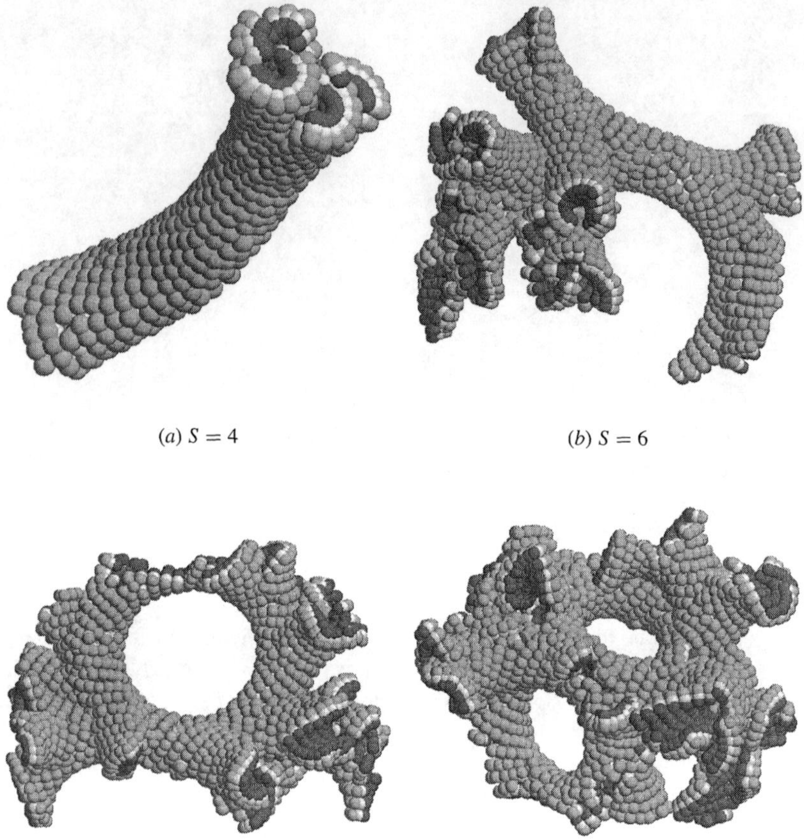

(*a*) $S = 4$ (*b*) $S = 6$

(*c*) $S = 8$ (*d*) $S = 10$

Figure 4.27. Extended hyperbolic barrels constructed with a unit sheet based on a 14-stranded barrel and different shear numbers (S) from $S = 4$ to $S = 10$ (S must be even) in frames (*a*)–(*d*), respectively. The outer surface of the sheet is shaded light grey and the inner surface dark grey.

It is not obvious that this variety of forms has any correspondence with nature but extended β-structures do form the basis of many biological materials (particularly in the insect and spider worlds) and it is not impossible that some of the strange structures encountered above may have unique properties that have been exploited in those mini-worlds. A structure that comes quite close to that seen in figure 4.28(*c*) can be found in the adenovirus tail-spike or fibre-shaft (PDB code: `1qiu`). However, this protein consists of stacked ring of β structure, rather than the spirals seen in figure 4.28(*c*) and (*d*).

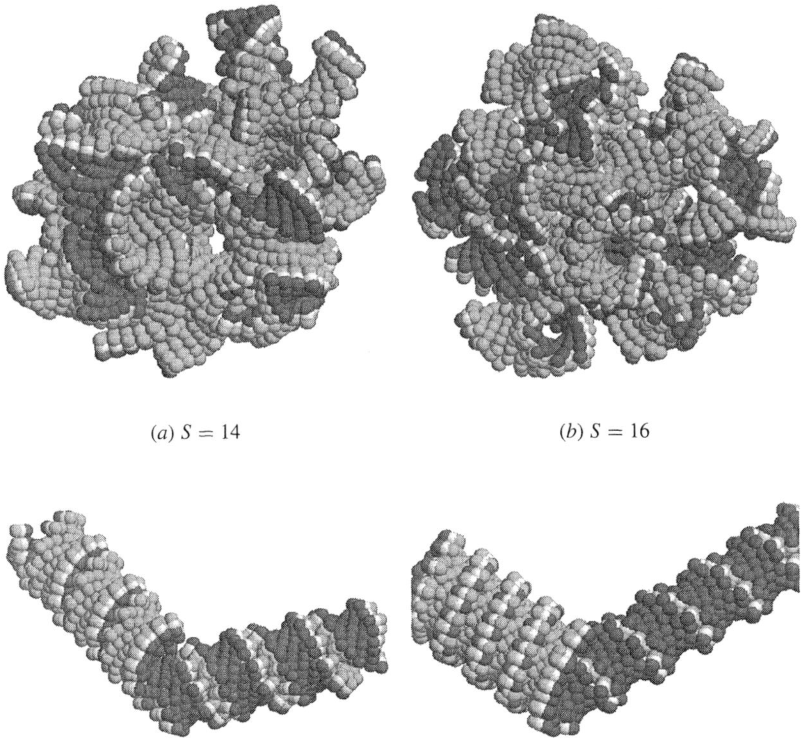

(*a*) $S = 14$ (*b*) $S = 16$

(*c*) $S = 18$ (*d*) $S = 24$

Figure 4.28. Aperiodic minimal surfaces and spirals in extended barrels. Continuing the series begun in figure 4.27, the spaces between the extended sheets contract to the point around $S = 14$ or $S = 16$ (*a*) and (*b*) where the connections between barrels are indistinguishable from the barrels themselves. (Referred to in the text as being like a periodic minimal surface.) When $S = 18$ there is an abrupt change to a cylindrical structure with an inversion point where the inner surface becomes the outer (*c*) and (*d*). The sheets are shaded as in figure 4.27.

4.2.5 Optimized barrels

From an analysis of observed β-barrels in proteins, Murzin *et al* (1994b) concluded that the preferred twist in barrels is around 20° with both curl and bend (coiling angles) being around zero. Using a model similar to the basic unit (3×3) sheet described above, they optimized the twist bend and curl angles towards these values for ideal barrels with different numbers of strands and stagger. The results of these optimizations are plotted in figure 4.29 along with the corresponding

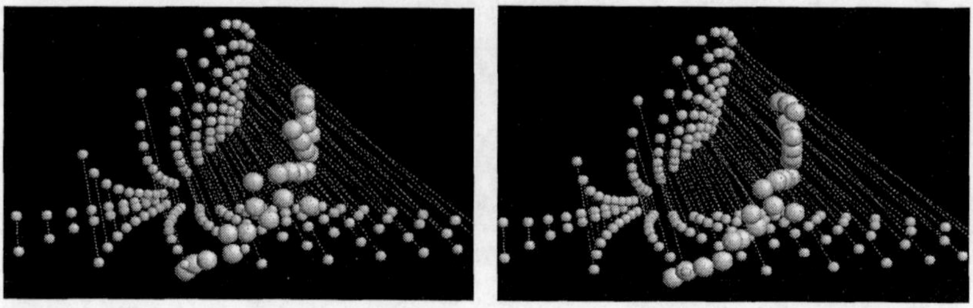

Figure 4.29. Barrels in TBC-space. The unit sheets on which each barrel is based is defined by three angles twist, bend and curl (TBC) and all the barrels considered in this section have been plotted in the 3D phase-space of these angles (shown as a stereo-pair). The hyperbolic barrels all have zero bend and so form a base-plane with each point (ball) representing a barrel with a different number of strands (N) and shear number (S). Each of these is connected by a dashed line to its equivalent cylindrical barrel (same N and S) which form a leaning wall to the left. Between these, lie the optimized barrels of Murzin and colleagues (plotted as larger spheres).

values for the simpler cylindrical β-barrels and hyperbolic β-barrels. The latter all have zero bend angles and so lie on a plane while the cylindrical barrels rise as a twisted surface above this. The optimized barrels lie mostly between these two surfaces (some have a small negative bend) forming a transition from hyperbolic towards cylindrical as the shear number increases.

PART 2

CLASSIFICATION

For the want of a bond, a strand was missed,
For the want of a strand, a sheet was missed,
For the want of a sheet, a fold was missed,
All for the want of a hydrogen bond.

Adapted from the nursery rhyme *the Horseshoe Nail*
(with apologies to Anon.)

Chapter 5

Networks and domains

The methods in the previous chapter have decomposed proteins into idealized components based on secondary structure. In this chapter we consider the more messy problem of identifying these structures in real proteins. The structures considered range in size from collections of a few secondary structures through large β-sheets and β-barrels to full protein domains. Two distinct but related problems are considered. The first involves ways to define the extent of a β-sheet, while the second extends this to the problem of defining protein domains. Both definitions are entangled but, being the more restricted problem, the β-sheet is considered first.

5.1 Hydrogen bond networks

In the previous chapter, each β-strand was taken as being defined independently and these were then assembled into the larger structures of sheets and barrels. However, neither level of structure takes precedence and the identification of a sheet or barrel equally defines what we consider to be a β-strand. There is a cooperativity in the definition and classification of structure that means that we need to consider the emergence of these structures from their defining networks of hydrogen bonds. At this level the β-sheet is a highly cooperative structure and once two strands begin to 'zip' together into a rudimentary sheet, their hydrogen-bonding partners come together forming a 'ladder' of hydrogen bonds (HBs) between them (Aszódi and Taylor 1994b). When the main chain polar groups point inwards to form the ladder, the planar nature of the peptide bond constrains their adjacent polar groups (across the peptide bond) to point in the opposite direction so orienting themselves in position to form another ladder with an additional strand. This cooperativity also extends in either direction along the strands: as the ends of the two chains are brought closer, it becomes increasingly likely that further HBs will form between them.

Although pictured above as a dynamic process, the cooperative properties of β-sheets also need to be considered in the definition of secondary structure in static coordinate sets. For example, by considering only local information, such as sequential torsion angles, it is possible to miss the definition of a β-strand even though it might lie between two other strands that are hydrogen bonded to it. An early method for secondary structure definition took advantage of the cooperative nature of the β-sheet structure by identifying 'ladders' of hydrogen bonds that run between two strands (Kabsch and Sander 1983). This method was based on an estimate of hydrogen bonding calculated from the relative positions and orientation of the amino (N–H) and carboxy (C=O) groups. Although one of the most successful methods for secondary structure definition, this method is of limited use for the models considered in this work that derive mostly from rough α-carbon backbone traces. While these could be converted to full main chain models (using `ca2main`), the resulting inaccuracies in the hydrogen and oxygen positions would be too great to provide a useful guide to hydrogen bonding for the Kabsch and Sander (1983) method.

5.1.1 α-carbon-based β-sheet definition

While the Kabsch and Sander method of secondary structure definition (DSSP program) has proved very useful over the years when applied to protein structures solved by crystallography, it becomes more limited when the coordinate data derives from rough model or even structures solved by NMR. The limitation derived both from the relatively exact definition of a hydrogen bond (based on main chain atom positions) to the relatively local view (ladder interactions) taken in the recognition of larger structures. In this section, both these limitations are extended to consider a definition of β-structure based on just α-carbon distances within a 3×3 unit β-sheet. As with the definition of line segments (section 3.1) the methods in this section operate on the smoothed protein structure (with three cycles of smoothing). Previously, smoothing was used to give a more even-handed treatment of α-helix and β-strands but here it is used to eliminate the pleat of the β-sheet so that bend of the strand can be measured more easily and compared with the studies in the previous section.

For each residue i in the smoothed protein, its nearest and second nearest neighbours (j and k) were found, such that: $d_{ij} < h$, $d_{ik} < h$, $|i - j| > 3$ (where d is an interatomic distance). To prevent links being established along a strand, it was also required that $|i - k| > 3$ and $|j - k| > 5$. This allows a minimal three-residue turn between adjacent strands and a minimal five residue turn between strands adjacent but one. When all these conditions are met, the three residues potentially align in a β-sheet as j–i–k. The same conditions, with the exception of the last, were then reapplied to the two sequentially adjacent triplets $i \pm 1$, $j \pm 1$ and $k \pm 1$, for which the signs were adjusted to minimize the interatomic distances. The resulting set of nine residues thus lie in the expected arrangement of a unit (3×3) β-sheet and this was recorded in a matrix of pairwise links ($\boldsymbol{B}$, initially zero) by adding 1 to each of the pairs across the sheet (B_{ij},

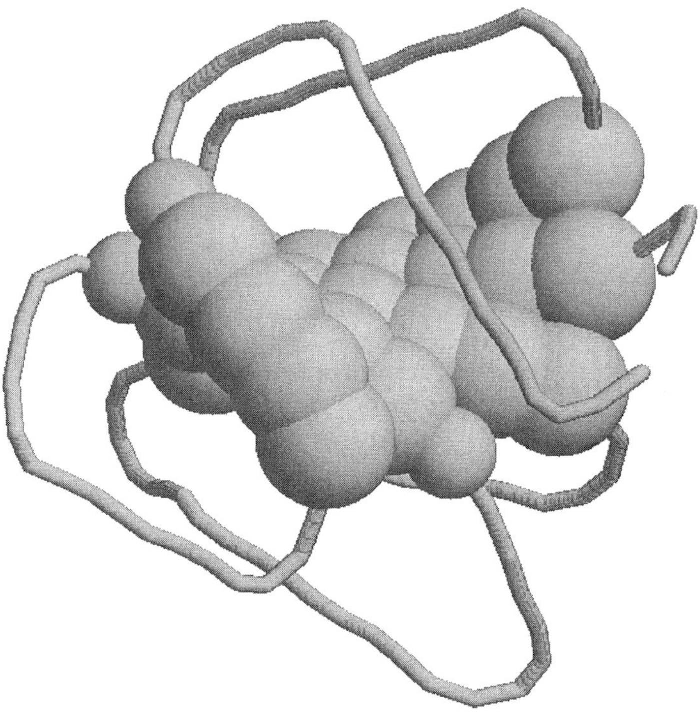

Figure 5.1. Automatically defined β-sheet. The algorithm described in section 5.1.1 was applied to a small $\beta\alpha$ protein (PDB: `3chy`) and the matrix of pairwise links ($\boldsymbol{B}$) calculated. The number of links to each residue are indicated by the size of the sphere on the (smoothed) α-carbon position. The β-sheet region (and only that region) is clearly identified.

B_{ik}, $B_{i\pm1,j\pm1}$, $B_{i\pm1,k\pm1}$) and along the strands ($B_{i,i\pm1}$, $B_{j,j\pm1}$, $B_{k,k\pm1}$), choosing signs as before.

The cut-off distance h was chosen as 7.5 Å, being a point midway between the separation of hydrogen bonded β-strands and the separation of stacked β-sheets—typically 5 Å and 10 Å, respectively. (The value of h will be further investigated later.) After processing each residue, the strongest pairwise links in the matrix $\boldsymbol{B}$ have a maximum value of six for residues that lie in the centre of a large sheet, dropping to one for corner pairs (Figure 5.1).

5.1.2 β-sheet dimensions

The matrix $\boldsymbol{B}$ defined above can be analysed as a weighted graph to derive some properties of the sheet. To facilitate this, the matrix of pairwise connections was recast as a linked-list in which each residue in the protein could maintain connections up to nine neighbours. (Normally only two would be required or sometimes three for bifurcating hydrogen bonds so nine is probably excessive). In this formulation, the sequential neighbours were distinguished as additional links.

Most basic of these properties are the sheet dimensions. The strand lengths are easily found as runs of linked residues scanning sequentially through the protein sequence. In the orthogonal direction, the width of the sheet can be found from the number of residues hydrogen bonded in a row. Beginning on a bonded residue, its links can be traced in one direction until an edge is found (a residue with only one link). From this the row can be traced in the opposite direction until another edge is found and the number of residues on this path give a width. As this is only one particular path, the process can be repeated for all residues and the maximum width or average width recorded. A combination of these two quantities can give a rough idea of how fully connected (rectangular) the sheet is.

Special allowance must also be made for bifurcating bonds as each path from the bifurcation may have a different length. This can be treated generally using a backtracking algorithm in which every link from each residue is traced to its end (including all other links along the way). If each residue had nine links this would lead to a combinatorial explosion of paths but fortunately bifurcation is limited. (Nevertheless, in the following calculations, large proteins were avoided.)

In figure 5.2, the mean strand length is plotted against the mean sheet width for a sample of over 1100 proteins with β-sheets (and under 300 residues). This is based on the definitions of β-strands defined using the line segment method described in section 3.1 and so has a bias towards odd numbers of strands. A clear preference can be seen at five residues with most data falling within ± 1. The spread in width is slightly greater but a mean width is around four strands with most of the data lying within ± 2 of this. The lack of data close to the five residue position occurs because the lengths of adjacent strands are correlated. (For example; a single strand cannot extend beyond the sheet.)

5.1.3 TBC angle distribution

Having a record of the neighbouring residues in a network allows unit (3×3) sheets to be identified and their twist-bend-curl (TBC) angles measured. For a 3×3 sheet, the latter two angles of the triplet are uniquely defined on the central residue in the unit, however, the twist angle is not unique and can be defined twice between the two edge strands and the central strand using the outer residues or can be measured four times based on the four residues in the quadrants of the unit sheet. This latter value has the advantage that it is not correlated with bend

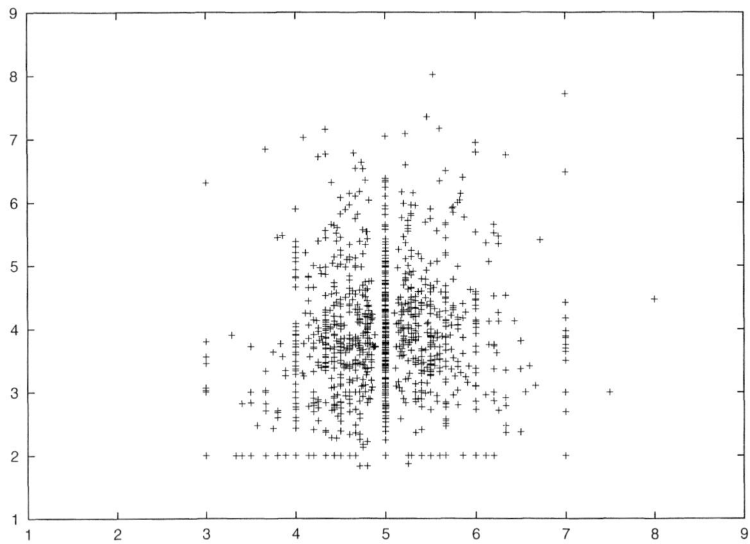

Figure 5.2. The size of β-sheets were determined from the pseudo-H-bond networks for a sample of proteins and plotted as mean strand length (X-axis) against mean sheet width (Y-axis). (The lack of points either side of integral strand lengths results from the cooperativity of the adjacent positions in the sheet definition.)

and using this value, an average was taken over the four quadrants, providing the corner residues were present and within 5.5 Å of their bonding partners. When these conditions failed to be met, the average was taken over the remainder.

These values were gathered over a random sample of 500 proteins that have β-structure extending to at least a 3×3 unit and plotted in a three-dimensional space (figure 5.3). The mean values of the angles are $T = 0.377$ (22°), $B = 0.188$ (11°) and $C = 0.266$ (15°). This confirms the expected twist angle for strands but gives a higher than expected coiling angles (B, C) compared to the zero values assumed by Murzin *et al* (1994a). However, these are absolute values and the curl (C) can be signed depending on whether it is with or away from the strand bend. In an untwisted sheet, the former makes a cup shaped sheet (positive curvature) while the latter makes a saddle shape (negative curvature). Taking this sign into account the mean curl value is close to zero but the distribution is bimodal, being almost a perfect reflection about $C = 0$.

The mean TBC values above can be used in the model for a twisted sheet described in the previous chapter to construct an average sheet. However, because this model used the residue spacing on the same face (6.5 Å), the bend angle must be doubled. With these values, two sheets were constructed with $\pm C$ to represent the bimodal spread in the curl angle. Positive curvature combined with twist is the requirement for cylindrical sheets while negative curvature with twist produces

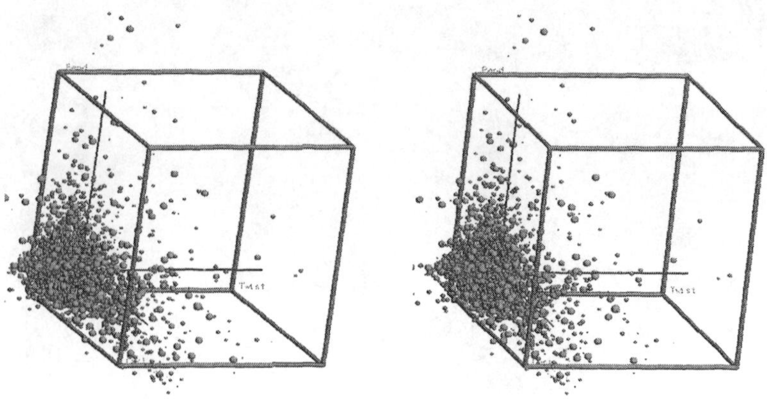

Figure 5.3. TBC angles in β-sheet. The twist-bend-curl (TBC) angles were determined for each unit sheet identified in a sample of proteins and plotted as a point in TBC-space (shown as a stereo-pair using the display program RASMOL). Each side of the box is one radian with the origin at the bottom-back-left corner. A line is drawn through the mean value in each dimension.

hyperbolic sheets. The sheets resulting from the average values exhibit a clear trend towards these forms (figure 5.4).

As the sheets in figure 5.4 are almost β-barrels, a specific look was taken at the TBC angles in three known barrels with different numbers of strands and stagger. One is narbonin (PDB codes: `1nar`), a TIM barrel with a typical hyperboloid shape (strands = stagger = 8), another is a superoxide dismutase (`1qupB`) which has a more cylindrical shaped barrel (strands = 8, stagger = 4) while the third is a viral protease (`1havA`), also with a cylindrical type barrel (strands = stagger = 6) (figure 5.5).

It is clear from the plot that there is considerable variation in the individual TBC distributions, to the extent that it would be difficult to tell from just these angles what type of barrel was present. While, some trend towards negative curvature ($C < 0$) can be seen for the TIM-type barrel, it is clear that a more direct approach is needed to identify barrels in HB networks.

5.1.4 β-barrel identification

Besides following just hydrogen-bonded links through the network, a step along the strand can be included. Again starting at any residue, these can be followed combinatorially and of interest are those paths that come back to where they started. By counting the steps across the network and along the strands, the nature of β-barrels can be determined by their number of strands and their stagger (which can be combined into a single shear number). In this approach as it stands, there

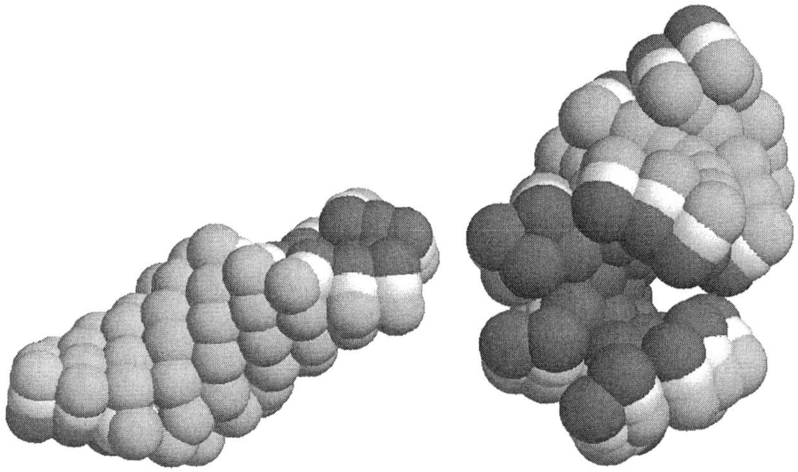

(*a*) positive curl (*b*) negative curl

Figure 5.4. β-sheets generated from average TBC values. Using the mean TBC values (figure 5.3), sheets were generated using the recursive tiling algorithm (section 4.1.2.2) with (*a*) the positive mean curl value and (*b*) the negative mean curl. The former tends towards a cylindrical barrel while the latter is close to a hyperbolic barrel.

is nothing to prevent the algorithm from following parts of the chain that are not bonded in a sheet. Rather than forbid this behaviour entirely, a limit was set that up to four residues could be followed between strands. This means that barrels that are not fully closed will still be recognized and these will be referred to below as '*part*' barrels as distinct from '*full*' barrels. Examples are shown in figure 5.6.

The sample of 1100 proteins with β-sheets (under 300 residues) used above was analysed and the distributions plotted to show the frequency of barrels observed for each strand number and stagger. A third dimension was also plotted which was the number of additional residues included to make a part-barrel. (Figure 5.7). Although this was not a large sample of proteins, a striking feature is the large number of part-barrels identified relative to full-barrels. Another interesting feature is the almost complete absence of seven-stranded barrels. (Some have been seen in larger proteins that were not in the sample considered here.)

Before considering a larger sample of proteins, some limitations of a simple exhaustive search over the HB network would need to be addressed. For large β-proteins, the combinations of pathways can become very large and in larger proteins there is often multiple sheets. The former problem can be overcome by introducing some heuristic methods to limit the extent of recursion, however, the

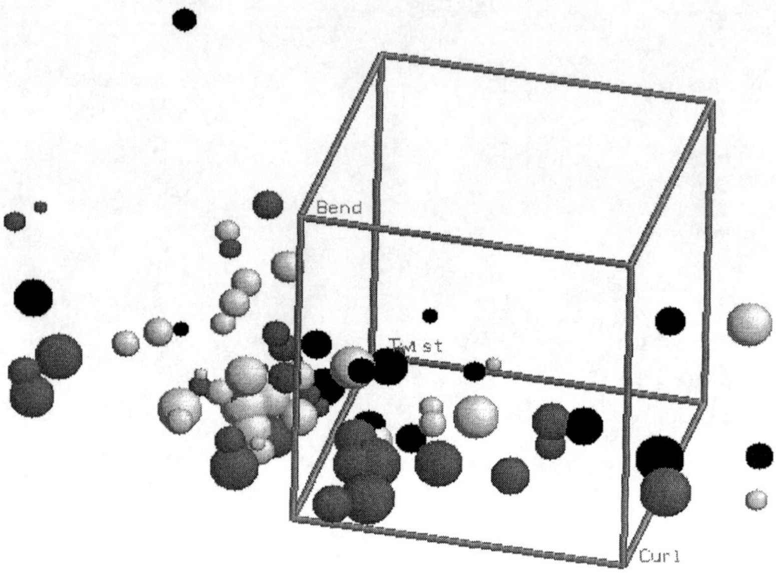

Figure 5.5. TBC values for three proteins with different sheets are plotted in TBC space. As in figure 5.3 the box is a unit (radian) cube but the origin is now near-bottom-left. The data are from a sheet $N = 8$, $S = 8$ (white), $N = 8$, $S = 4$ (grey) and $N = 6$, $S = 6$ (black). Each data point is drawn as a sphere with a radius proportional to the strength of the links in its unit sheet as measured in the matrix $\boldsymbol{B}$ (section 5.1.1 and figure 5.1).

latter problem is not so simple. While added book-keeping can keep account of the numbers of sheets and barrels, there will always be situations where a decision must be made to separate two weakly linked sheets. To help with this decision, the strength of the links in the network can be used to make the divide along a line of weakest links.

This approach will be developed in the following section where it is extended into a general method for defining domains in proteins. Once a protein has been divided into domains then the network recursion becomes much more limited and can be used to determine the extent and nature of the β-sheets and β-barrels.

5.1.5 β-sheet classification

Before looking at the more general problem of domain definition, it is valuable to assess the implications of the above studies for the classification of β-sheets and β-barrels. There has been a long tradition of classifying β-sheets as either (topologically) flat or as a barrel and the latter class has been further subdivided into common barrels (TIM barrel like) or orthogonal barrels. This latter distinction we have associated above with barrels that have a hyperbolic

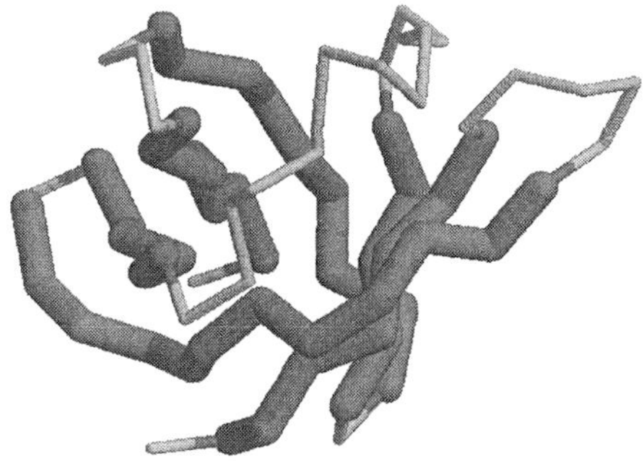

(*a*) Full

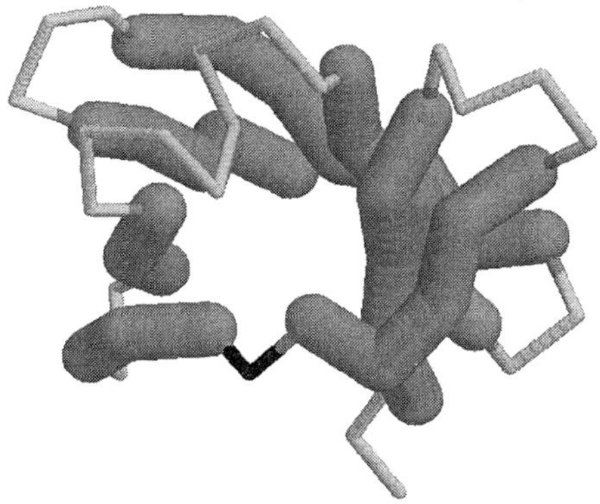

(*b*) Part

Figure 5.6. Full and part barrels. Two domains are shown from an all-β protein (PDB: `1a1q`) with β-strands drawn thickly and coloured grey. One domain (*a*) forms a fully hydrogen-bonded barrel while the other (*b*) has a break in one strand (black link) and does not form a barrel. Despite this, the two domains have the same fold (RMSd = 5.6 over 53 residues) and are probably evolutionarily related. The domains are shown in the same orientation.

surface and those that have a cylindrical surface, which is in turn associated with positive and negative curl angles, respectively.

These definitions represent points on a continuum and cylindrical barrels can be distorted smoothly into hyperbolic barrels and both types of barrels can be converted through part-barrels into a single flat sheet or into two stacked sheets. For example, the contrast drawn between a part-barrel and a full-barrel in figure 5.6 used two barrels drawn from the same structure. These two structures

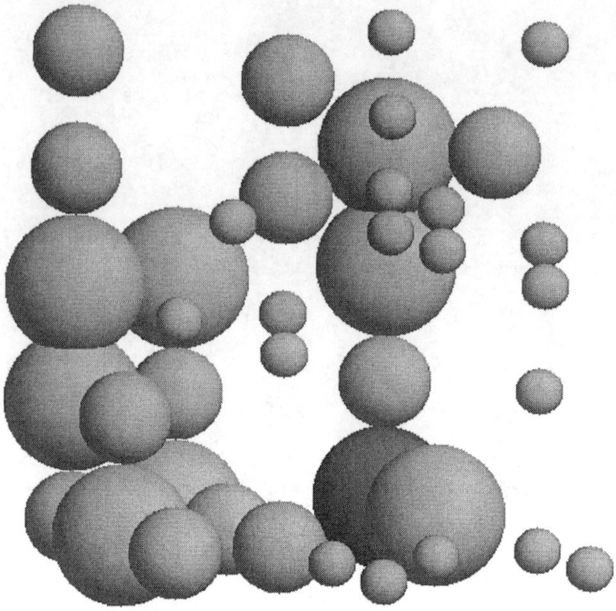

Figure 5.7. Occurrence of full- and part-barrels. For a selection of proteins, their barrel type is plotted as number of strands and shear number (X, Y plane) with the number of residues needed to form a part-barrel along the Z-axis. For each class, a sphere is plotted with a radius proportional to the observed frequency. The bottom layer are full barrels and the common $N = 8$, $S = 8$ (TIM) barrel is shaded darker. (There are almost no $N = 7$ barrels.) There is a significant number of part-barrels.

have the same overall fold (the views in figure 5.6 have the same orientation) and have almost certainly derived from an ancient gene duplication[1]. This shows that there is no evolutionary barrier between full- and part-barrels.

The prevalent distinctions in classification come not from geometric constraints but from an overview of known protein structures. However, this can be circular since the preconceived classes of sheet structure are those used to classify novel structures. This problem will be returned to again later in chapter 7 but before we sink into a morass of relativism, in outline, one approach is to use hydrogen bond networks to tentatively define sheets structures that can be used to tentatively define domains that can be used to tentatively define the fold of the domain. This approach can then be iterated in the hope that 'tentative' becomes 'definite'. If there is no convergence, then like the quantum wave function, all

[1] The protein is a serine protease and although this particular example (from a virus) has quite deviant structures in each half, it is more typical for the two barrels to be well formed with strands = stagger = 6. Indeed it was the first example of this protease (chymotrypsin) that led McLachlan to derive the strand/stagger nomenclature in order to illustrate that the two barrels had an equivalent structure.

options can be kept open and only realized when we need to make a firm statement about some aspect of the protein or its relationship to others.

5.2 Protein domain definitions[2]

It has been clear since the determination of the earliest protein structures (Phillips 1966) that there is a level of structural organization that is greater than the folding of the chain into simple secondary structure components. The exact definition of these structural domains, however, has remained problematic since there is a continual progression from proteins that slightly divide into two lobes to those that form clearly distinct folded regions separated by a flexible linking segment of chain (figure 1.7(*a*)). A component of (sequentially) local organization is partly a element in the idea of a domain but is not sufficient as some domains are formed from discontinuous segments of chain that are distant in the chain (figure 1.7(*b*)). Secondary structure, in particular the β-sheet also influences the definition of a domain since β-sheets are rarely split into separate domains. However, although one sheet would not normally be in two domains, two or more sheets may be in one domain, so again, this structural feature does not provide a sufficient definition. A concept sometimes taken as a rough working definition of a domain is that, if excised, the domain should remain folded as a stable structure. While difficult to test (either experimentally or computationally), an implication of this concept is that the excised domain should contain a hydrophobic core and should therefore be larger than, roughly, 40 residues. (For reviews see; Janin and Wodak (1983), Janin and Chothia (1985)).

5.2.1 Physical methods

All the above principles have, in various combinations, been taken to form operational definitions of domains. Local compactness was taken as the principle aspect in the early method of Rose (1979) (using the density of ellipsoids to define cores) and more recently has been extended by Holm and Sander (1994b) in a way that captures the relationship between compact units. Swindells (1995a) concentrated more on the requirement of having a hydrophobic core in each domain (Swindells 1995b), extending cores outwards from their deepest components and, where necessary, pruning and fusing these into larger units. Some older methods such as Rose (1979) and the more recent method of Siddiqui and Barton (1995) have focused on minimizing the number of chain-breaks needed to separate domains while also measuring the degree of association between the separating units while Rashin (1985) and Islam *et al* (1995) have employed solvent area calculations. Sowdhamini *et al* (1996) have also captured many of these ideas but at the level of secondary structure elements. Whatever

[2] This section is reproduced in part from Taylor (1999c) with the kind permission of Oxford University Press.

their primary guiding principle, most of these methods apply corrections to their initial definitions on the basis of the remaining (secondary) principles. Typically, the primary method generates alternative definitions that can then be selected using the secondary principles, which for example, may involve counting the resulting breaks in the chain and secondary structures. Often, these secondary filters become a complex weighted combination (as for example in the method of Siddiqui and Barton (1995)).

The methods described above generally take the approach in which a predefined idea of a domain is imposed on the structural data. In the language of systems analysis, this would be called a 'top-down' approach and the inherent danger in its application is being unable to recognize when the data does not fit the conceptual model. An alternative approach is to reverse the direction and let the idea emerge from the data, in what is sometimes called a 'data-driven' or 'bottom-up' approach. In this section, a 'bottom-up' method for structure domain definition is described that is based on a very simple idea that does not embody any specific knowledge of protein structure. The method also has few parameters, so allowing their effect to be systematically investigated and, perhaps most importantly, it can be easily extended to the simultaneous definition of domains using homologous structures and also to a probabilistic model.

5.2.2 An Ising model

The basic method is similar to an Ising model in which the structural elements of the model change state according to a function of the state of their neighbours. Although Ising models are typically applied to two states on a two-dimensional lattice, the approach has also been applied to the one-dimensional protein chain in the Zimm and Bragg (1959) model of helix-coil transitions. (See Thouless (1992) and Bruce and Wallace (1992) for reviews of the approach applied to magnetic and more general phenomena, respectively). In the current method, the neighbours are defined by spatial proximity in the three-dimensional protein structure, and the states are multivalued labels. In this implementation, the approach has affinity to the analysis of protein structure using connection topology (Aszódi and Taylor 1993) discussed in chapter 8.

5.2.2.1 Basic method

Each residue in the protein chain is assigned a numeric label. If a residue is surrounded by neighbours with, on average, a higher label, then its own label increases, otherwise it decreases. This test and reassignment is applied repeatedly to each residue in the chain.

Representing the sequence of labels as $S = \{s_1, s_2, \ldots s_N\}$, for a protein of length N, then the iteration can be stated as:

$$s_i^{t+1} = s_i^t + \mathcal{U}\left(\sum_{j=1}^{N} \mathcal{J}(s_i^t, s_j^t) \right) \qquad \forall i = 1 \ldots N. \tag{5.1}$$

At each iteration t, the new state of residue i (at $t+1$) is determined by the influence of all other residues (j) modified by the function $\mathcal{J}$ which is referred to as the coupling function. Where the function is simple multiplication, then the state revision can be represented by a matrix multiplication as in the Zimm and Bragg (1959) method. The function $\mathcal{U}$ takes the sum over the neighbours and transforms it to either +1 or –1 for positive and negative sums, respectively.

5.2.2.2 *Coupling function*

The coupling function ($\mathcal{J}$ in equation (5.1)) calculates the inverse distance (p_{ij}) between the α-carbons of residues i and j and returns a negative value if the state label of i (s_i) is less than that of j (s_j). An upper limit (radius, r) on the proximity of the neighbours was imposed on those taken into the calculation. Explicitly, the function evaluates the expression to which it is equivalenced below:

$$\mathcal{J}(s_i, s_j) \equiv \begin{cases} p_{ij} & \text{if } s_j > s_i \quad \text{and} \quad d_{ij} > r \\ -p_{ij} & \text{if } s_j < s_i \quad \text{and} \quad d_{ij} > r \\ 0 & \text{otherwise} \end{cases} \tag{5.2}$$

where, d_{ij} is the interatomic distance between the α-carbons of residues i and j, p_{ij} is the inverted distance r/d_{ij} and r the neighbourhood radius. The inverted distances constitute a matrix ($\boldsymbol{P}$) which will be of further use later.

Some trials were made with different functions, in particular, with no inverse weighting (giving a simple majority 'vote' in equation (5.1)) and with inverse squared weighting. The results for both were remarkably similar to the basic method but the latter appeared to undervalue the contribution of neighbours while the former increased the sensitivity of the result to the choice of cut-off radius (r). This behaviour is typical of Ising models in which the details of the lattice and the form of the coupling function make little difference to the global properties (Bruce and Wallace 1992). The choice of the cut-off radius will be considered below but its use as a scaling factor (in equation (5.2)) does not affect the result since only the final sign of the sum is considered in equation (5.1).

5.2.2.3 *Label assignment*

The most obvious choice for label assignment is the sequential residue number itself. This naturally embodies the desired property that sequentially adjacent residues will be predisposed to belong to the same domain. Other schemes will be considered below but unless otherwise stated, simple residue numbering should be assumed.

5.2.3 Model evolution and domain extraction

The recursive iteration of equation (5.1) results in compact regions evolving towards the same residue number. However, if there are two compact regions linked by a long exposed segment of chain (to take an extreme example), then each domain will evolve towards a local value and these labels will extend and meet half-way along the linker. At this point, neither side will have sufficient 'leverage' to 'force' the other to adopt its label and the system will cease to evolve (typically oscillating, or 'flickering' at the point of label discontinuity). For the extraction of domain definitions it is necessary that this stage in the evolution is detected and the iteration terminated, allowing the assignment of residues with a common label as a domain. To do this some minor technical difficulties need to be addressed.

5.2.3.1 Stopping the iteration

Because of the potential for domain boundaries to 'flicker', the iteration cannot simply be terminated when there is no further change in labelling. This problem was overcome by keeping an average over two cycles and monitoring the squared deviation of this between successive cycles (summed over the length of the sequence). Any simple oscillation will thus be averaged out: for example; if a residue position alternates between 5 and 7 on successive cycles, then the difference in successive averages ($5 + 7, 7 + 5, \ldots$) will be zero. The iteration was stopped when the mean squared deviation of the average between successive cycles was less than 10^{-6} or if this degree of convergence was not obtained, then the calculation was stopped after a number of iterations equal to half the number of residues in the protein. This gives sufficient opportunity for both the amino and carboxy termini to evolve to a common label if they lie in the same domain.

5.2.3.2 Refining unique labels

In the basic method, the labels evolve in discrete unit steps. This admits the small possibility that two independent domains might converge to the same value by chance. This possibility can be minimized by using a smaller step (increment/decrement) size but if a small step size is selected at the outset, then the evolution of the system will be very slow. A compromise was made by following the initial evolution of the system by a further set of $N/2$ iterations (where N is the number of residues in the protein) in which the step size decreased linearly from one to zero. After this, the value of the labels within a domain generally agreed to better than 10^{-4}, greatly reducing the chance of two domains having the same label within the error of convergence.

5.2.4 Conforming to expectation

The method as described to this point, when applied to a variety of representative proteins, performed remarkably well, especially considering that it embodies no encoding of any feature specific to proteins. However, as discussed in the Introduction, there are some assumptions in the received definition of a domain that need to be taken into account to produce a definition that conforms to expectation. Principal among these is the expectation that:

(i) the chain should not pass too frequently between domains;
(ii) small domains should be ignored or avoided;
(iii) secondary structure, in particular β-sheets, should not be broken.

5.2.4.1 Reclaiming short loops

Examination of some of the initial test results revealed that most of the frequent chain crossings between domains resulted simply from short loops ‘dipping’ in and out of the adjacent domain. These could easily be ‘corrected’ by resetting their label to that of the flanking domain: however, situations can be imagined where it is not obvious which loop should be reset; as illustrated by the following example in which two domains (with labels 5 and 7) mingle: $\{\ldots 5555577755577777\ldots\}$. Simple smoothing (taking an average over a window) cannot be used as this would alter the residue labels, however, a solution was found by using a form of smoothing based on the median, rather than the mean, in which the position in the centre of a window is replaced by the median of the values in the window (Bangham 1988). This method, when iterated to completeness, levels all peaks (or troughs) less than half the window size but these are flattened (or filled) only with existing values so no new domain labels are created by the process. A window size of 21 was taken, eliminating all excursions of 10 or less residues.

5.2.4.2 Reassigning small domains

As in other studies (Jones *et al* 1998, Siddiqui and Barton 1995), domains of 40 or less residues were not accepted. These might simply be ignored (marked as unassigned regions) but it was considered better to see if they might be joined onto another existing domain. This was done using a variant of the core calculation in equation (5.1) in which each residue in the small domain was directly assigned the (weighted) mean value of its neighbours, as follows:

$$s_i^{t+1} = \sum_{j=1}^{N} (s_j^t \cdot p_{ij}) \Big/ \sum_{j=1}^{N} p_{ij} \qquad \forall s_i^t < 0 \tag{5.3}$$

(where p_{ij} is an element from the matrix of reciprocal distances $\boldsymbol{P}$). Reassignment was made only for residues that shared a common label with less than 40 others and this was 'flagged' by setting the label of all such residues to -1. (Hence the condition $\forall s_i^t < 0$.) Although not explicitly stated, as before, the sum was taken only over residue pairs closer than the cut-off radius r and, in addition, residues in the process of being reassigned ($s_j^t < 0$) were also excluded.

After the reallocation of small domains, the balance between the larger domains might have altered. This potential disequilibrium was allowed for by taking the new set of labels (S^{t+1}, calculated by equation (5.3)) as the starting point for another complete domain assignment calculation and the whole exercise was repeated until no small domains remained or to a limit of five times. This limit was sometimes reached as some small domains are truly isolated and remain 'unclaimed' by any of their larger neighbours. In this situation, it was considered unnecessary to introduce any further steps to 'force' their reallocation.

5.2.4.3 *Keeping β-sheets intact*

The basic method deviated most seriously from expectation in a propensity to divide large proteins that were dominated by a single β-sheet. This tendency was most apparent in the eight-fold alternating β/α-barrel proteins, which often have weakly packed strands and helices as a result of the stagger in hydrogen bonding around the barrel (section 4.2). Solutions to this problem have been found previously through the use of the recorded secondary structure information (extracted from the protein structure databank) or based on calculated hydrogen bonding. In the current method, a self-contained solution was sought that depended (as does the basic method) on the use of α-carbon coordinates alone.

A bias was given to maintain the integrity of β-sheets by setting the initial label of their component residues to a common value. This value was derived from the analysis of the H-bonded network described in section 5.1. However, as discussed there, it is still not always clear from these networks whether to split a sheet or keep it intact. This subproblem was solved in the same way as the larger domain problem by using the basic Ising method itself but substituting the matrix $\boldsymbol{P}$ of inverted interatomic distances (in equation (5.1)) for the matrix $\boldsymbol{B}$ H-bond scores (section 5.1.1). However, it was found that this approach also was still prone to split weakly linked sheets into domains so the variation employed to reassign small domains was used instead in which each residue takes the weighted mean label of its neighbours, again, substituting the matrix $\boldsymbol{B}$ for $\boldsymbol{P}$ (in equation (5.3)).

$$s_i^{t+1} = \sum_{j=1}^{N} (s_j^t \cdot b_{ij}) \Big/ \sum_{j=1}^{N} b_{ij} \qquad \forall\, i = 1 \ldots N. \tag{5.4}$$

Unlike the reassignment of small domains, equation (5.4) was iterated to convergence using the stopping criteria employed in the basic method (section 5.2.3.1). No neighbour cut-off was applied as this is already inherently encoded in matrix $\boldsymbol{B}$ and equation (5.4) was evaluated only for linked residues

($\sum_{j=1}^{N} b_{ij} > 0$). This approach has the desired property that the entire network of linked residues is still not forced to adopt the same label and weakly (possibly spuriously) linked sheets can remain distinct. It should be remembered that this procedure only provides a set of starting labels to which the basic method is applied (as described previously) and this still has complete freedom to reassign any of the initial labelling.

It was also considered whether an equivalent bias should be applied to α-helices, however, long helices often pack against more than one domain and it seemed more natural that these should be allowed to partition freely as dictated by the basic method.

5.2.5 Setting the granularity level

The basic method has only one parameter which is the neighbourhood cut-off radius (r). The value of r affects the average size of the resulting domains (and can be associated with the correlation length in the application of Ising models to critical-point phenomena). When r is small the resulting domains tend to be smaller but the relationship is not direct and even when r is infinite, clear domains will still remain separated. Almost all the methods discussed in the introduction have parameters that affect the granularity of the result but none have any mechanism for objectively setting this property, other than to optimize the parameters to give a result that approximates the definitions recorded in the literature. These, of course, will vary from author to author and, despite some attempts at homogenization, remain a heterogenous standard.

One approach to this fundamental problem is to get two different (ideally independent) views and when these agree, then it can be assumed that some 'truth' has been found that is independent of any particular method. An approach along these lines was made by Jones *et al* (1998) using three methods of domain identification. Unfortunately it was found that, except for the most obvious examples, these were never in full agreement (to better than 85% of equivalently assigned residues). An alternative approach is to use a single method but applied to homologous proteins. However, this is limited by the availability of homologues with sufficient structural difference to provide independent solutions. To circumvent this problem a 'fake homologue' was created for each protein and the current method applied to both. This allowed the value of r to be varied and the level of granularity was accepted at the value where the two solutions agreed best.

5.2.5.1 Creating a 'fake homologue'

A simple way to create a structure with the same fold but differing in detail is to smooth the path of the chain. This technique has often been used to help visualize the fold of the chain, originally by Feldman (1976), and more recently by Aszódi *et al* (1995a). Smoothing destroys almost all the specific details of protein geometry, however, for the current method this is not a disadvantage as it does not rely on any of these characteristic geometric features. For each consecutive triplet of α-carbon coordinates, the central atom was replaced by the average coordinates of the triple (see section 2.1.1.1 and figure 2.2). This procedure was repeated five times giving a structure that was substantially different from the native coordinates but still recognizable. Although not directly comparable, the root-mean-square deviation between the smoothed and native chain was typically around 5 Å, which is equivalent to that found between analogous structures (having the same fold but no significant sequence similarity).

5.2.5.2 Comparing domain agreement

Comparing the domains assigned with the smooth and native chains, it was apparent that the smooth chains required a slightly larger cut-off radius to give roughly comparable results. This compensates for the reduced interatomic contact in the smooth chain, especially in regions of α-helix packing where the helices have been reduced to almost straight lines. Values of $r + 3$, $r + 5$ and $r + 7$ were tested and a bonus of three was found to be sufficient.

Following (Jones *et al* 1998) a matrix of common residue counts in all domain pairings was compiled. The best overall count was then extracted from this matrix, however, where Jones *et al* (1998) appear to assume that these values lie on the diagonal, a more general (but still 'greedy') algorithm was used in the current work which has previously been described in the alignment of multiple sequences (Taylor 1987). Since the smooth and native structures have the same number of residues, the number in agreement was reported as the percentage of the length of the protein.

5.2.5.3 Finding the best granularity level

The simplest algorithm to find the best agreement between the two structures is to vary the cut-off radius and monitor the percentage domain agreement. However, this is computationally expensive and a more restricted search strategy was adopted. From trial runs it was found that most solutions lay in the mid to lower part of the range of $r = 10 \ldots 20$. A start point was taken at $r = 14$ and a search expanded with alternating lower and higher values, in unit steps, and terminating when r fell below 10. If at any point during the search, the two domain assignments had 90% or more coincidence, then the search was halted and the current solution accepted. Otherwise the best agreement point was recorded and if at the end of the search, this was 85% or better then its solution was accepted. If during the search, both structures were reported as single domains, and no other solution agreed to better than 85%, the structure was taken to be a single domain. Structures for which no solution was found (either as single or multiple domains) were marked for individual consideration.

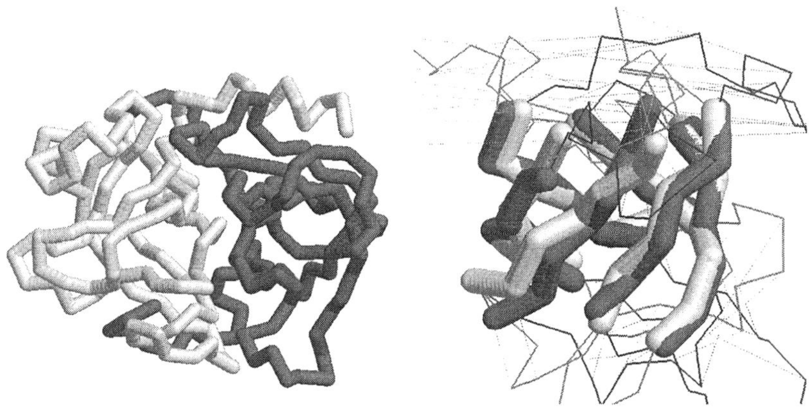

(*a*) (*b*)

Figure 5.8. Trypsin domains. (*a*) The domains in trypsin (PDB: `1sgt`) are shaded dark and light. (*b*) The domains are superposed with equivalent β-strands drawn thickly.

5.2.5.4 Excluding badly broken β-sheets

The search for the best granularity level provides an opportunity to check the integrity of the β-sheet structure, allowing this to be controlled without affecting operation of the basic method. If a β-sheet was found to be broken by separating the domains, in either the native or the smooth structure, then that solution was not accepted and the search continued, as described above.

As the current (or any) calculation of β-sheet structure can never be completely reliable, some tolerance is desirable in the strictness with which the integrity of β-sheets are maintained. It was estimated that this should relate to the size of the protein, and a rough and reasonably generous level was set at the square root of the length of the protein. A small protein of 100 residues can therefore tolerate an error of 10. (It would cost 12 to split a 4×4 sheet in two.) Whereas a large protein of 400 can accept double this error.

5.2.6 Performance and examples

The current method, in its basic form, is one of the simplest that has been applied to the problem of domain definition yet its predictions were quite acceptable for roughly 90% of proteins and deviate seriously from the accepted definitions only where the integrity of β-sheets are not preserved. Since the basic method has no inherent knowledge of protein structure, special treatment of these higher-order structures cannot be expected, but on the other hand, its generality allows it to work on just α-carbon coordinates, without any precalculation of hydrogen

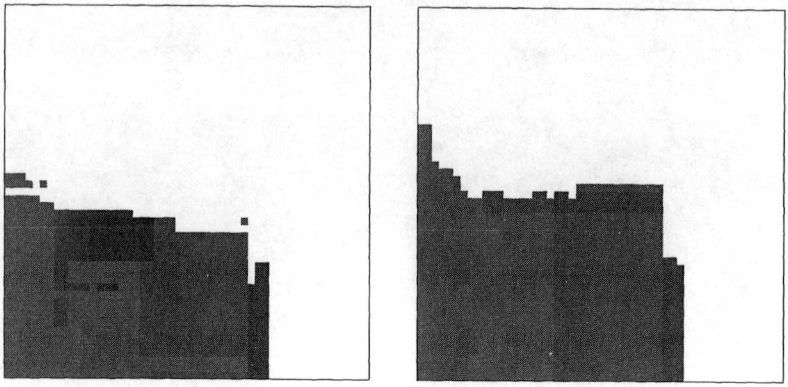

(*a*) native (*b*) smooth

Figure 5.9. Trypsin domains with varying cut-off distances. The number of domains predicted for the trypsin `1sgt` with varying r (equation (5.4)) and h (equation (5.4)) in the range 5–10 Å (along X) and 10–20 Å (along Y), respectively. The number of domains are shown in different shades of grey. A single domain is shown in white. The degree to which any β structure has been disrupted is shown by superposed shades of black (darker is greater error). The results are plotted using (*a*) the native structure and (*b*) the smoothed structure (see figure 5.8).

bonding or solvent surface areas, and to have considerable tolerance to chain breaks (and other unexpected features in the PDB files).

5.2.6.1 Incorporating β-bias

To conform to expectations of unbroken β-sheets, the H-bond network interactions within the β-structure were calculated in which a β-bias was applied only to the starting conditions (the labelling) or as filters to the results (rejecting proteins with badly broken sheets) leaving the computational core untouched. The combination of the basic method with the β-bias and variable granularity, developed on the small subset of structures, gave good results when tested on the much larger data-set. For the easier parsing problems, agreement was almost perfect (to within a residue or two) and for the majority of the more problematic proteins, the current method gave acceptable results that in many instances were an improvement over the recorded definition (Taylor 1999c). Some protein families remained difficult, including the trypsins, pepsins, lactate dehydrogenase and a few TIM-barrel structures.

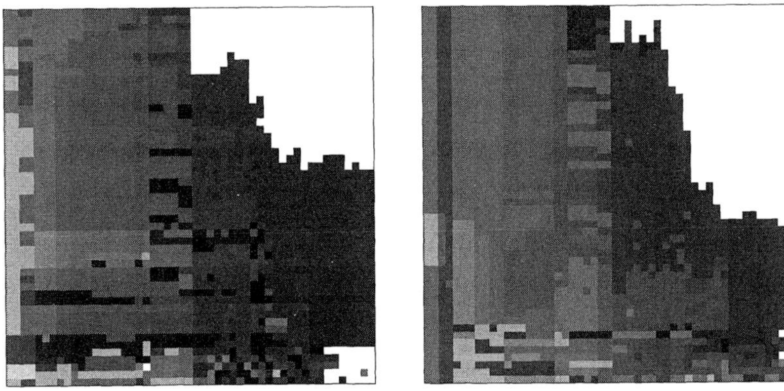

(*a*) native (*b*) smooth

Figure 5.10. Pepsin domains with varying cut-off distances. The number of domains predicted for the pepsin `4pep` with varying r and h. Regions of multiple domains are shown coloured and a single domain is shown in white (with darkness representing greater β error). See legend to figure 5.9 for details.

One of these structures, however, holds a special place in the history of the concept of a domain (McLachlan 1979a) and could not easily be ignored (with a clear conscience). Like the problem of black-body radiation in late 19th century physics, this small anomaly led to a re-examination of the approach to domain definition. The protein considered by McLachlan was chymotrypsin but in the discussion below, the related protein trypsin (PDB: `1sgt`) will be referred to and used as an example. This protein contains two β-barrels both $N = 6, S = 6$ (section 4.2) and is thought to be the product of a gene duplication (McLachlan 1979a). (Figure 5.8). It was clear that the tightly packed trypsin domains are intuitively viewed with a finer level of granularity. As the level of domain granularity can be easily influenced in the current method, it was considered whether there might be some way in which an internally derived value might be found for this property that was optimal for each individual protein (similar to the critical point in some physical phenomena (Bruce and Wallace 1992)).

5.2.6.2 Domain phase-space

To investigate the 'trypsin problem', the phase space of domain definition was plotted for the two parameters: r, the interaction radius (equation (5.1)), and h, the distance for the declaration of a β-sheet interaction (section 5.1.1). The phase space was plotted both for the native protein and the smoothed chain which gives

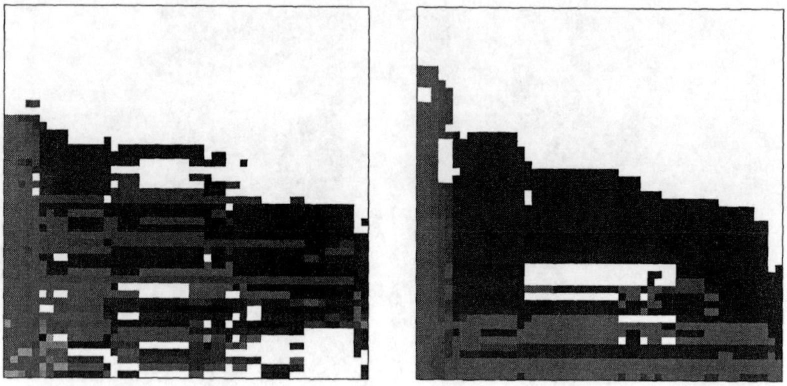

(*a*) native (*b*) smooth

Figure 5.11. Narbonin domains with varying cut-off distances. The number of domains predicted for the pepsin `1nar` with varying r and h. Multiple domain regions are coloured and single domains are white (darkness represents β error). See legend to figure 5.9 for details.

an idea of how likely it is for the two models to find agreement in the number of domains.

For the trypsin example (PDB code: `1sgt`), both models gave a large area of phase space corresponding to a two-domain solution (figure 5.9). As would be expected, with large values of d or large values of h, a single domain was found as this allowed weak links between the two sheets to be followed or for the basic Ising-like model to converge on a single domain.

For comparison, a different double domain protease was examined. This was pepsin (PDB code: `4pep`). With this protein, the phase space was much more complex with up to four domains being defined in scattered regions of the space. More worrying, however, for the native/smooth chain method of finding agreement was that for the native structure, the dominant solution was for three domains whereas for the smooth chain the dominant solution was for two domains (figure 5.10).

A third example considered was the TIM barrel protein narbonin (PDB code: `1nar`). This is generally considered as one domain but because of the stagger between the strands, some of the links across the barrel are weak, giving a tendency for two domains to be found. Like the trypsin example, the phase space was divided into single and double domains (figure 5.11). However, the double domain phase was broken and scattered—reflecting the tentative nature of the definition. Unlike trypsin, the double domain phase generally had a high

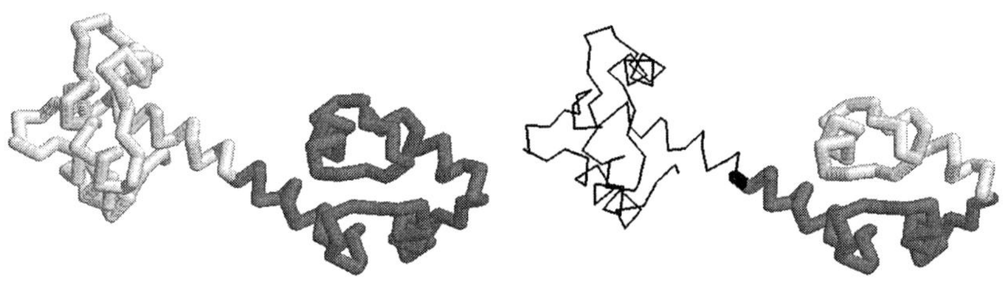

(*a*) (*b*)

Figure 5.12. Examples from stochastic domain definitions. The small all-α protein calmodulin (PDB: `1osa`) gave different domain definitions under stochastic perturbation. (*a*) The dominant two-domain solution and (*b*) an example of a minor domain split within one of the dominant domains.

error-score for breaking hydrogen bonds in the sheet, restricting the acceptable double domain solutions to an area with low h where the intact sheet had not been found in the first place.

5.2.7 Simultaneous definition on multiple structures

The flexibility of the labelling system, allows the labels to be taken, not simply as a residue position in a single structure but as a position in a multiple sequence alignment. Rather than suffer the distortions inherent in defining domains on an averaged multiple structure, or taking the averaged domain definitions after individual domain definition, the current method can allow simultaneous (interacting) domain definition across all the structures. This was achieved using the the basic method on each structure individually but with the labels derived from a multiple structure alignment. Between each iteration of equation (5.1), the individually evolving labels can be synchronized by taking an average over the label values at each position across the structures that are represented at that point in the multiple alignment.

The proteins discussed above were taken as test examples for the multiple structure extension of the method to see if pairwise combinations of related structures would help resolve their domain definition. For the pepsins, the widely scattered single structure definitions converged on a three-domain solution with a small β-linker domain being defined between the two commonly accepted domains. The trypsins showed some signs of movement towards a double domain but the lactate dehydrogenases remained with one domain, even with the recruitment of the distantly related malate dehydrogenases. It is probable that combinations of more structures, or simply more remote relatives may help these difficult definitions.

5.2.8 Probabilistic definitions

It is clear from the above examples, that the pragmatic solution of finding agreement between two different representations of the structure (native/smooth) has its limitations. Problems were encountered above, where it was found that solutions involving fewer domains will be preferred as this introduces less scope for differences. In the extreme, the agreement between single domains is perfect and, if encountered, will be accepted as best. This situation is not unexpected as the approach described here is essentially a clustering problem and, in general, the choice of the 'right' number of clusters is an open problem.

Before leaving the problem, the original inspiration of the method as an Ising model can be re-examined. The method diverged from a classic Ising model not only in its large variety of states but also by neglecting the stochastic (random or thermal) component of the model. This can be easily reintroduced by adding some 'thermal' noise to each state assignment and letting the system evolve. In doing this, unstable domain assignments will be perturbed and be seen as transient assignments while stable assignments will remain a feature in each solution.

To avoid the added complication of the β-sheet parameter, this approach was tested on a small all-α type protein (calmodulin: `1osa`) which has two very clear domains each comprising two less distinct domains (calcium binding motifs called EF-hands) (figure 5.12). It is less easy with this approach to view the results but this has been done by labelling each domain with a letter in each assignment (run with a different random seed) and sorting results alphabetically. This was done for three values of the radius r giving a range of solutions from two very stable domains to a value that gave highly variable multiple domain assignments (figure 5.13). One of the triple domain assignments from the latter set of results is shown in figure 5.12(*b*) where the C-terminal domain has been split into its two component EF-hand motifs. This is a common but not exclusive assignment and examples were seen in which the subdomain had been split in the orthogonal direction with the two opposing pairs of helices forming a subdomain (this can equally occur in either main domain). An alternate triple domain assignment was to declare the long linker-helix as a separate domain leaving two more compact globular domains at either end of the dumb-bell. In larger sets of results all possible combinations of these options will appear with differing frequencies.

This approach potentially opens a path to a method for a more absolute assignment of domains in which the frequency of all combinations of domains are analysed. The choice still remains as to what the best degree of randomization should be but following the application of Ising models in more conventional physical systems, it may be possible to identify a level of randomization that corresponds to the critical temperature of the system at which point all sizes of domains are observed.

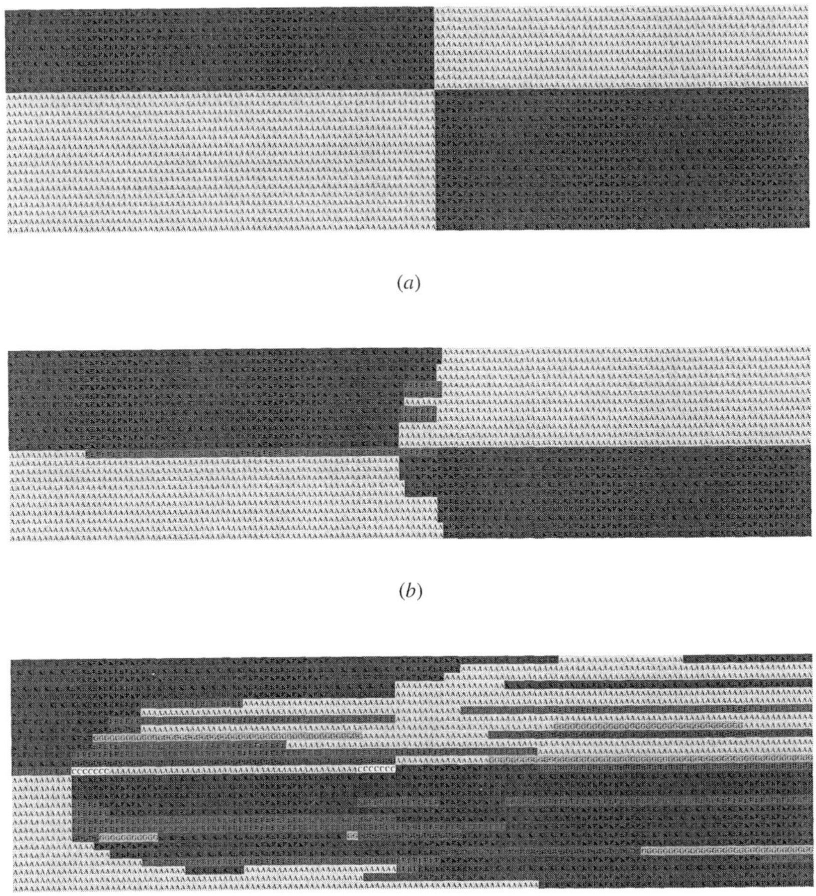

Figure 5.13. Probabilistic domain assignment. The small all-α protein calmodulin (PDB: `1osa`) gave different domain definitions under a range of stochastic perturbation (*a*)–(*c*) progressing from low 'temperature' to high 'temperature'. In each line of each panel, residues in a domain are assigned the same letter but to make the domains easier to visualize, they have also been given a different shade and the lines within a panel are sorted alphabetically. In (*a*), only one solution is obtained (figure 5.12(*a*)). In (*b*) some boundary variation is seen. In (*c*), many subdomains are defined.

5.2.9 Reparsing domains

A more direct approach to defining domains and subdomains is to take a hierarchic approach and progressively subdivide the structure. This approach does not proscribe a stopping point but gives a natural way to analyse the structure of a protein from its intact form through domains and subdomains down to individual secondary structure elements. The analysis can be made using the domain assignment method described above but many other approaches have also been employed in the same way.

The domains defined by the basic method can be automatically represented to the method to check if they could be further divided. This was done by linking the broken ends of the chain in the excised fragments and treating them as a new 'intact' protein. These reconnections were necessary, since a protein with chain breaks would behave differently in the Ising model from an equivalent connected chain (they also make the excised domains much easier to visualize). The connecting loops were 'grown' recursively from the broken ends in the direction of the centroid of the deleted segment until they came within bonding range.

When the domain assignments are good, then the resulting units are almost indistinguishable from intact proteins, with short links between the severed ends of the chains that led to the excised domain. Less optimal assignment more typically have long overreaching connecting loops. This aspect could, of course, be incorporated into the score to decide the domain breaks but will not be pursued here.

As with the problem of granularity discussed above, the problem of when to stop dividing does not have a trivial solution. Except for some general notions of domain compactness and lengths of connecting loops, there are few geometric properties that can be taken as absolute measures. Like the situation with the definition of β-sheets, we have a continuum of solutions and the point at which to draw the line can only really be found through the consideration of the known structures. If the partition of a large protein results in domains that have been seen elsewhere as intact single domain proteins, then some confidence of an evolutionary/biological nature can be placed in the result. To do this requires an unbiased way to compare protein structures and this will be considered in the next chapter.

Chapter 6

Protein structure comparison

6.1 Overview of comparison methods

The vast variety of protein sequence and structure found in the current databases could not have been anticipated by a polymer chemist looking only at bonds and forces. Indeed, the best effort from pure stereochemical considerations was made by Linus Pauling who predicted the α-helix from first principles before any protein structures were known. However, this did not prepare people for the sight of the first structures, which were much more irregular than expected[1]. The most important things we know about proteins have come therefore, not from theory, but from observation and the comparison of sequences and structures.

Equivalent proteins from related species usually have similar structures and sequences and a comparative analysis of these can tell us about residue substitutions and how the structure adapts to accommodate them. However, if one is interested in the stability and versatility of protein structure under greater degrees of sequence variation—in other words, how far can a structure be pushed by evolution, then it is necessary to compare the most distantly related proteins. This has driven those who develop methods to compare protein structures to continually 'push-back' the range of comparison methods with the hope of discovering further and perhaps more fundamental similarities.

Methods of protein structure comparison have been based on a great variety of approaches aimed at different aspects of structure: folds, fragments or secondary structures. (For reviews, see (Brown *et al* 1996, Eidhammer *et al* 2000).) With improved computer power, some of these methods have even been applied to the comparison of the complete protein structure databank, giving an automated analysis of what had previously been the monopoly of a few experts. To assess the results of such analyses (which will be considered in section 7.1) it is necessary to know how the various methods work as many behave quite

[1] From comments reported by Francis Crick in 'What Mad Pursuit' (Sloan Foundation Science Press 1990), it appears that Max Perutz had anticipated a much more regular array of packed helices.

differently. Such knowledge is equally vital when choosing a method for a particular comparison problem.

6.1.1 Structure representations and degrees of difficulty

Structural similarity is of interest at many levels, from the fine detail of backbone and side chain conformation at the residue level, through the coarse similarity of folds at the tertiary structure level, to a simple count of secondary structures. Similarities may also be locally confined or extend globally over whole structural domains and even involve more than two structures. These issues are reflected in the methods that will be discussed below: spanning comparisons of almost identical structures through to highly dissimilar ones.

The simplest applications are concerned with studies on a single protein. Examples include studies of conformational change between states of the same protein (including multiple NMR structure solutions), and the comparison of mutant forms of a protein where the structures being compared usually have very similar structure at all levels of detail and negligible or no insertions and deletions of sequence (indels).

Applications of intermediate difficulty include comparison of closely related proteins to analyse evolutionary divergence, inference of weak sequence homologies on structural grounds, characterization of conserved structural features such as functional sites within families. Conversely, structure comparison may help in the analysis of similar folds that apparently result from evolutionary convergence (Orengo *et al* 1993a). Sometimes the requirement is to screen a specified structural fragment (motif) against a database of protein structures, searching for strong matches. In these examples, the structures of interest are relatively similar, so that indels present a limited problem.

The most difficult and general structure comparison applications arise in the classification of the known protein structures into different fold families. This rationalises the organization of the structure databank, and may indicate hitherto unsuspected structural similarities, evolutionary relationships, or constraints on folding. Comparison methods must be able to deal with structural similarity at all levels of detail, must handle indels of arbitrary length and position in the respective structures, and must identify structural similarities even when these form a relatively small proportion of the structures being compared.

This diversity of applications is addressed by a corresponding variety of automatic or semi-automatic comparison methods, some suitable for comparing highly similar structures at a specific level of detail or *element size* (residue, backbone fragment, secondary structure, etc), while other more general methods may operate at several element sizes or may be applicable to more remote comparisons.

The common aims of each method are to compute some quantitative measure of similarity, and often to generate a structurally derived alignment of one protein sequence against the other(s). The set of element equivalences so defined may be used to drive a rigid body superposition to facilitate visual comparison, either as an intrinsic part of the method, or as a separate step (Rippmann and Taylor 1991).

The ability of proteins to lose or gain sequence elements over evolutionary time (relative insertions or deletions: jointly referred to as *indels*) has led many methods of structure comparison to follow the simple model of evolutionary change which is used in sequence alignment methods. This assumes that the only processes at work are substitution of amino acids (or rather the underlying nucleotides) and their deletion or insertion. More complex operations such as reversals, translocation and duplication events are 'forbidden'. This model further assumes that these processes are uniformly applied along the sequence length and are the same for all proteins. In addition, most alignment methods implicitly assume that the substitutions[2] in one place do not affect substitutions elsewhere. From our knowledge of protein structure this latter assumption is clearly untrue (one part of a structure can influence any other part) but, despite this limitation, the sequence alignment model provides a good starting point.

6.1.2 The `DALI` method

Holm and Sander devised a two stage algorithm, implemented as a computer program called `DALI`, which uses simulated annealing to build an alignment of equivalent hexapeptide backbone fragments between two proteins Holm *et al* (1992), Holm and Sander (1993b), Holm and Sander (1993a). The approach is equivalent to aligning collapsed distance matrices of the proteins from which insertions and deletions have been excised. In the first stage, hexapeptide contact maps are matched and similarity scores generated by comparing all distances within the hexapeptides. An 'elastic' score proportional to the relative differences between distances is used, making the method more tolerant to distortions in longer range distances. Hexapeptides whose contact maps match above some threshold are stored in lists of fragment equivalences. To reduce the amount of information considered, only hexapeptide pairs having similar backbone conformations are compared. Similarly, although residues occur in a number of overlapping contact maps, the map with the closest contacts to any other segment is selected for a given residue.

In the second stage, an optimization strategy using simulated annealing explores different concatenations of the fragment pairs. Similarity is assessed by comparing all distances between aligned substructures. Each step consists of addition, replacement, or deletion of residue equivalences, in units of hexapeptides and, since hexapeptides can overlap, each step can result in the addition of between one and six residues. In the next iteration step, the alignment is expanded by adding substructures that overlap with those already equivalenced. Once all candidate fragment pairs have been tested, the alignment is processed to remove fragments with negative contributions to the overall similarity score.

[2] Substitutions are realized as mismatches in an alignment of two sequences.

An advantage of the approach is that the alignment need not be constrained by fragment sequentiality, so that fragments can be equivalenced in a different order along the sequences. The method has been used to compare representatives from all the non-homologous fold families in the PDB (Holm and Sander 1994b, Holm and Sander 1997, Holm and Sander 1998). (see section 7.1 for further details).

6.1.3 Geometric hashing approach

Geometric searching techniques are used in small molecule applications to match a query structure against a database of molecules and Lesk (1979) has described a geometric searching method suitable for proteins or other macromolecules. This computes a sorted list of interatomic distances in the query structure and associates with each atom a bit-string where the ith bit is set if the atom has a neighbour at the ith position in the distance list. The bit-list is thus a discrete signature for that atom. Similar bit-strings in terms of the same distance list are constructed for each database structure in turn and compared with those of the query structure to derive tentative atom equivalences. The number of comparisons may be reduced by only considering 'like' atoms by some property, e.g. atom type.

The final, and computationally demanding, step samples all combinations of matched atoms for each database structure to find the best equivalence set by superposition onto the query structure. Brint *et al* (1989), also working with Cα interatomic distances, describe an optimization of Lesk's algorithm, which speeds up the method by replacing the combinatorial sampling step with a backtracking tree search.

Since the query specifies the substructure to be matched, these methods are unsuitable for the general problem of identifying unknown common substructure. In contrast, an application of the computer vision technique termed geometric hashing is suitable for database searches using defined patterns or to discover unknown similarities (Nussinov and Wolfson 1991, Fischer *et al* 1992, Bachar *et al* 1993). The method is demonstrated using Cα atoms, although any atoms can be discriminated on type, or other properties.

A triple of (nonlinear) atoms in a protein defines a 3D reference frame and, in general, all such triangles are computed and the side lengths are hashed to compute an address in a hash table, at which the protein identity and three atom coordinates are stored. The hash table is populated in this way for all proteins. Once compiled, it can be used for any comparison, and new proteins can be inserted without recomputing existing entries. A simplified outline of the algorithm is shown in figure 6.1.

Matching a query protein against the database proceeds by looking up the query protein triangles in the hash table. Each match found constitutes a vote for the triangle entries stored at that address. High scoring matches represent reference frames common to both proteins, and the rigid body transformation

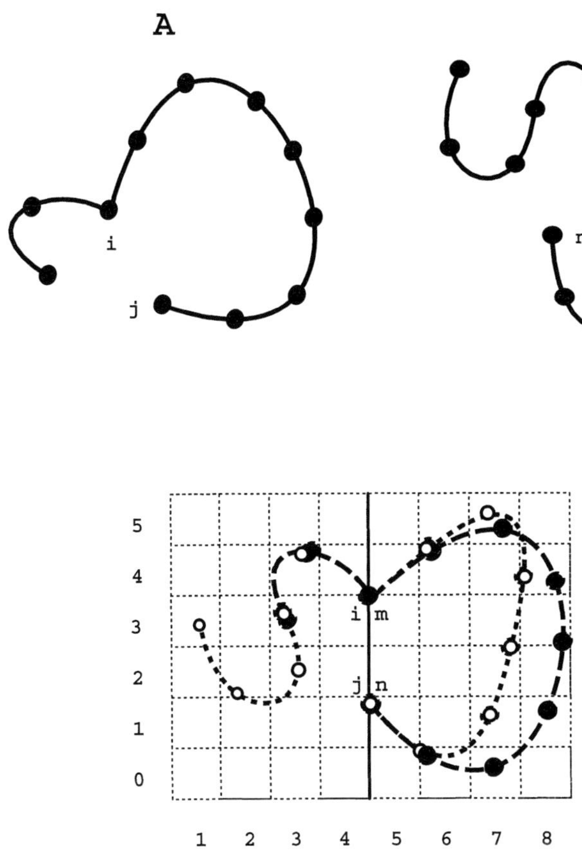

Figure 6.1. Two protein structures ***A*** and ***B*** are shown schematically. Two pairs of positions (i, j in ***A*** and m, n in ***B***) are selected. Both structures are centred on the origin of a grid (below) at i and m and orientated by placing a second atom in each structure (j and n) on the vertical axis which is (coincidentally) the terminal atom of each structure. (In three dimensions, three atoms are required to define a unique orientation.) Atoms in both structures (open and filled circles) are assigned an identifier that is unique to the cell in which they lie (the *hash* key). For simplicity, this is shown as the concatenation of two letters associated with the ordinate with the abscissa (XY). For example; atoms in structure ***B*** are assigned identifiers AD, BC, CC, CD, etc. The number of common identifiers between the structures provides a score of similarity. In this example these are CD, CE, FE, GF, HE and FA (not counting i, j and m, n) giving a score of six. The process is repeated for all pairs of pairs, or in three dimensions, all triples of triples and the results pooled.

required to superpose triangles is an estimate of the overall superposition for the molecule, while the participating atoms are an estimate of the desired atom equivalences. Many matched reference frames correspond to essentially the same transformation, and these are merged to produce a larger set of atom equivalences. These then prime a series of superposition and assignment steps to further refine and extend the equivalence list.

The same basic method has been applied to comparing protein surfaces at ligand binding sites (Fischer *et al* 1993, Fischer *et al* 1994).

6.1.4 Using structural superposition

Barton and Sternberg (1988) describe a specialized application of dynamic programming to finding residue equivalences. Given an initial superposition based on the cores of two closely related proteins, the `LOPAL` program determines residue equivalences between variable loop regions, which may differ in the number of residues as well as spatially. Each such region is represented by a distance matrix holding all Cα distances from one loop to the other and dynamic programming is used to find the best global alignment, effectively aligning the Cα atoms by their 3D coordinates. This approach was later developed into a more general method that used sequence alignment to establish an initial correspondence (Russell and Barton 1992).

In the later development of this approach, the program `STAMP` (Russell and Barton 1992), used multiple pairwise sequence alignments to construct a binary tree ordered by sequence similarity. Structures were then superposed using a conventional pairwise algorithm in the order dictated by the tree starting with the most similar pairs at the leaves and terminating at the root, using averaged atomic coordinates when merging more than two structures at an internal branch. At each merge, α-carbon equivalences are assigned using modified spatial and orientation probabilities (as in Rossmann and Argos (1976)). A matrix of probabilities for every possible α-carbon equivalence is computed, using probabilities averaged over all possible pairs of structures being merged. The best path through this matrix is assessed using a local dynamic programming step Smith and Waterman (1981) to select the most likely sequential Cα equivalences. Again like some of the older methods, cycles of equivalence assignment followed by superposition are applied until the equivalence list is stable, and the process repeats for the next merge (May and Johnson 1994, May and Johnson 1995, May 1996).

At the secondary structure level, Murthy (1984) describes a fast, two stage, superposition method, in which helices and strands are represented by their axial vectors. The first stage derives from Rossmann and Argos (1976) in which rotational space is systematically sampled and at each step a matrix of angular orientation scores for the secondary structure vectors between each protein is produced. Each cell indexed by a pair of secondary elements from the two proteins is assigned a weighted score that is maximal for parallel vectors.

Dynamic programming is then used to determine the best alignment and overall similarity score for each matrix, these being ranked and the highest selected as identifying the secondary structural equivalences. In the second stage, these equivalences are then used in another rotational search to achieve superposition by minimizing the differences between vectors linking all pairs of elements in one protein and equivalent vectors in the second. The score is modified depending on how well vectors between equivalent secondary elements can be superposed. Finally, these modified scores are plotted as a function of the three Eulerian angles giving a contour map where strong structural similarities are revealed as peaks.

6.1.5 The `SSAP` program

The `SSAP` program (Taylor and Orengo 1989b), and its derivatives (Taylor and Orengo 1989a, Orengo and Taylor 1990, Orengo and Taylor 1993, Taylor *et al* 1994a), (see Orengo and Taylor (1996) for a review) uses a double dynamic programming algorithm to manipulate two tiers of matrices[3]. A high-level matrix is used to accumulate alignment paths from a series of lower level matrices which are used to compare relationship sets of each possible pair of residues. The principal relationship employed uses a local structural environment about each residue comprising a simple reference frame defined by the geometry of the C_α atom. Every other residue is defined in this frame by a set of interatomic vectors (figure 6.2). Residue equivalences given by the resulting structural alignments are used directly to produce a weighted superposition by the algorithm of McLachlan (1979a) using the alignment score at each equivalent position (Rippmann and Taylor 1991).

Other relationships examined include interatomic distances, H-bond energies, virtual H-bonds extending through sheets, and disulfide bridges, while features include residue accessibility, secondary structure assignment, backbone angles, solvent accessible area and sequence similarity (Taylor and Orengo 1989a). Multiple features and relationships are scored using a weighted polynomial scoring function, with choice of features, relationships, and weights under user control.

The full double dynamic programming algorithm is computationally demanding, but Orengo and Taylor (1990) show that only a small subset of lower level comparisons is necessary—an aspect that has been exploited in later developments (Taylor 1999b) (more fully described below). A local alignment version using a modified Smith and Waterman (1981) algorithm (Orengo and Taylor 1993), and a multiple alignment version (Taylor *et al* 1994a) (using the progressive multiple sequence alignment algorithm of Taylor (1988)) were developed also. The latter generates a concensus structure by averaging vectors between equivalenced residues. Information gathered on structural variability

[3] This older implementation of the algorithm will not be described in detail here since the current iterated algorithm will be described in the following section.

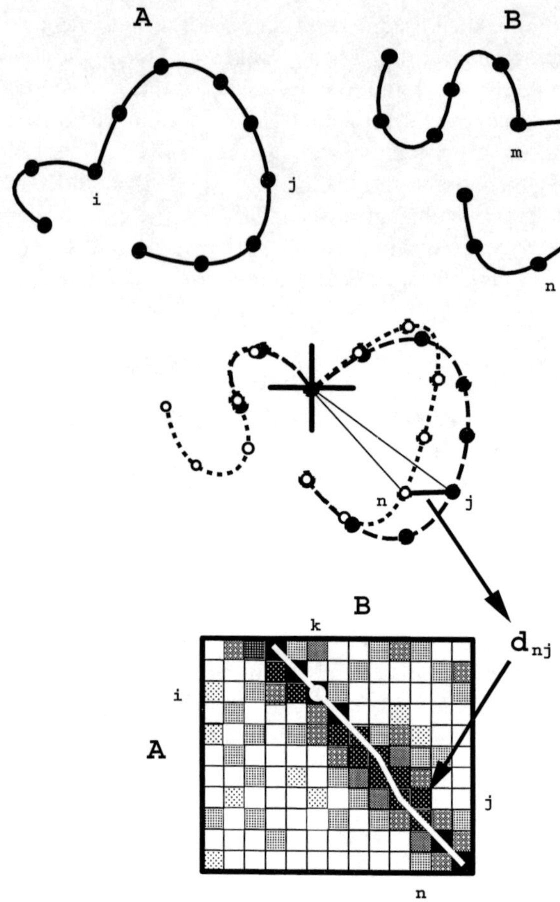

Figure 6.2. Two protein structures ***A*** and ***B*** are shown schematically. A pair of positions (i in ***A*** and m in ***B***) is selected. Both structures are centered on i and m and orientated by a local measure (indicated by the large cross). In this superposition the relationship between all pairs of atoms (e.g. n and j) is quantified, either as a simple distance (d_{nj}) or by some more complex function. All pair values are stored in a matrix and an alignment (white trace) found. The arbitrary choice of equating i and m is circumvented by repeating the process for all possible i, m superpositions and pooling the results. In the SSAP algorithm a final alignment is extracted from the summed results by a second dynamic programming step.

of individual vectors and environments can be used, for example, to weight structurally conserved positions as more structures are added to the alignment.

SSAP can also align secondary structure elements (SSEs) (Orengo *et al* 1992) using secondary structure features (hydrophobicity, length, surface area) and relationships (buried area, overlap, tilt and interaxial angles, interelement vectors). The resulting alignment of SSEs can then be used to constrain a subsequent residue level alignment. The method is fast and allows sorting of the structure databank into unique fold families (Orengo *et al* 1993a).

6.1.6 Iterated double dynamic programming

The `SAPit` program (for structure alignment program with iteration) described in this section was derived from the related `SSAP` program (Taylor and Orengo 1989b, Taylor and Orengo 1989a) (section 6.1.5) and is largely a simplification of its predecessor but is based on a refined iterative algorithm. The method is described fully here as some of its results are used below in section 6.2 and the program is used frequently through this work. The core comparison algorithm underlying both `SAPit` and `SSAP` has also been used in some sequence/structure comparison methods (Jones *et al* 1992, Taylor 1997b))

6.1.6.1 Double dynamic programming

The computational difficulty in structure comparison programs that use dynamic programming arises through trying to obtain a measure of similarity between two sets of internal relationships in different proteins. To compare the internal relationship of, say, residue i in protein $\boldsymbol{A}$ with a residue m in $\boldsymbol{B}$ relies on matching the individual relationships (such as $\{i, j\}$ in $\boldsymbol{A}$ with $\{m, n\}$ in $\boldsymbol{B}$) (figure 6.2). If this is known (even for one such i, m pair) then the comparison problem would be solved before the first step was taken! To break this circularity, the following computational device was used: given the assumption that two residues (one from each of the two proteins) are equivalent, then how similar can their relationships (or structural environments) be made to appear while still retaining topological equivalence?

This aspect of the calculation is described in figure 6.2 in which the score matrix is referred to as the *low-level* matrix (R). The scores along the best path through this matrix are then summed into a 'master' matrix (S), referred to as the *high-level* matrix. After all residue pairs have been considered and their path-scores summed in S, the best path is now found through S giving a best-of-the-best (or consensus) result. Representing the application of the dynamic programming algorithm as a matrix transform function $\mathcal{Z}$ that sets all matrix elements to zero except those that lie along the best path, then the full algorithm can be summarized as:

$$\text{path} = \mathcal{Z}(S) = \mathcal{Z}\left(\sum_i \sum_j \mathcal{Z}(R_{ij})\right) \tag{6.1}$$

where the sums are over all residues (i) in one protein with all residues (j) in the other.

The basic alignment (dynamic programming) algorithm is thus applied at two distinct levels: a low-level to find the best score given that residue i is equivalent to j, and at a high-level to select which of all possible i, j pairs form the best alignment. This double level (combined with the basic algorithm) gave rise to the name '*double dynamic programming*' (DDP).

6.1.6.2 *Selection and iteration*

The DDP algorithm described above, requires a computation time proportional to the fourth power of the sequence length (for two proteins of equal length) as it performs an alignment for all residue pairs. To circumvent this severe requirement, some simple heuristics were devised based on the principle that comparing the environment of all residue pairs is not necessary. By considering local structure and environment, many residue (indeed most) pairs can be neglected. This selection is based on secondary structure state (one would not normally want to compare an α-helix with a β-strand) and burial (those with a similar secondary structure and degree of burial are selected) but a component based on the amino acid identity can also be used, giving any sequence similarity a chance to contribute.

An iterated algorithm was implemented previously (Orengo and Taylor 1990), using heuristics on the first cycle to make a selection of a large number of potentially similar residue pairs. In the reformulated algorithm, a small selection (typically 20–30) pairs are selected initially and gradually increased over several iterations. This initial sparse sampling can, however, be unrepresentative of the truly equivalent pairs and to avoid this problem, continuity through the early sparse cycles was maintained by using the initial rough similarity score matrix (referred to as the *bias* matrix) as a base for incremental revision (figure 6.3). As the cycles progress, the selection of pairs becomes increasingly determined by the dominant alignment, approaching (or attaining) by the final cycle, a self-consistent state in which the alignment has been calculated predominantly (or completely) from pairs of residues that lie on the alignment.

6.1.6.3 *Sampling alternate alignments*

A useful 'spin-off' from the iterated DDP approach is to augment, or bias, the evolving selection of pairings (referred to as the current selection). This can be done using external information such as sequence or structural patterns (Jonassen *et al* 1999), or by adding random displacements to the scores on which the selections are based. This latter approach introduces some of the aspects of the stochastic methods such as the `COMPARER` program (Šali and Blundell 1990) and is equivalent to sampling the population of high scoring alternate alignments.

Knowledge of the distribution of subalignments gives an idea of how unique the highest scoring alignment is (and indeed, whether this best alignment was found by the 'one-shot' algorithm). If the best alignment is unique (few similar scoring alternatives) then it can be treated with confidence whereas if there are a lot of equally scoring alternatives then care must be taken in interpreting it

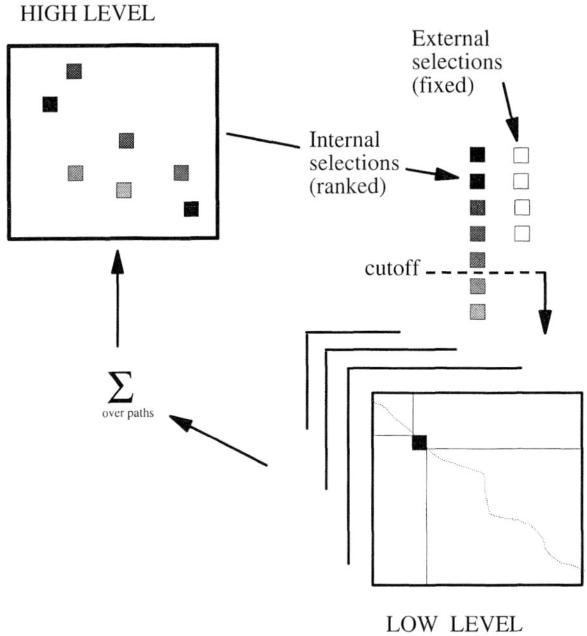

Figure 6.3. Outline of the iterated double dynamic programming algorithm. Values from the *HIGH-LEVEL* score matrix are ranked and a prespecified number (represented by the broken *cut-off* line) are passed to the *low-level* for evaluation. These are joined by a fixed number of externally specified pair selections. The resulting alignment paths are summed back into the *high-level* score matrix and, after normalization of the values, a new selection is made. Five cycles of iteration are typical.

in detail. Ways in which this can help also in assessing the significance of an alignment score are discussed in section 6.2.

6.1.7 Secondary structure graph matching

The comparison of proteins at the secondary structure level developed from some early attempts (Kuntz *et al* 1976) through the more complex 'meta-matrix' analysis of Richards and Kundrot (1988) to the automated POSSUM method (Mitchell *et al* 1989, Artymiuk *et al* 1990) which compares the geometric relationships between α-helices and β-strands abstracted as axial vectors (stick models). In this method, proteins are represented as fully connected graphs whose nodes are secondary structure elements and whose edges are pairwise closest approach and mid-point distances and torsion angle. A standard subgraph isomorphism algorithm detects subgraphs in each protein in the database equivalent to that of the query structure, having the same types of nodes with similar valued edges within user specified distance/angle tolerances.

Ordering of secondary structure elements in the query is under user control. There is no alignment score and a detailed residue level alignment is produced by conventional superposition. Like the geometric searching techniques (Lesk 1979, Brint *et al* 1989) to which it is related, the method is unsuitable for the general problem of identifying unspecified common substructure (i.e. discovery of unspecified common subgraphs). Nevertheless, it is appropriate for fast database searching with known motifs to identify candidates for more refined comparison.

The subgraph discovery problem is addressed by `PROTEP` (Artymiuk *et al* 1992b, Artymiuk *et al* 1992a, Grindley *et al* 1993), which uses an established maximal common subgraph algorithm to compare the same secondary structure types and relationships. This identifies maximal fully connected subgraphs or cliques that are shared between structures. As with `POSSUM`, the method allows fast database searching, but does not give a residue level alignment or superposition.

Subbarao and Haneef (1991) also represent protein structures as partially connected graphs whose nodes and edges are Cα atoms and interatomic distances, respectively, within some user specified cut-off. Two graphs are compared to identify the maximal common subgraph corresponding to structurally similar regions using a standard algorithm. The set of Cα atom equivalences mapped by the subgraph is used to drive a conventional superposition (external feature) from which a new set of Cα equivalences within a 3 Å limit is used produced to drive a final superposition.

6.1.8 Stick-figure comparisons

6.1.8.1 Angle and distance matching

Connected secondary structure stick-figures (of the type described in section 3.2) might be compared directly to each other using some of the structure comparison methods (described above)—for example; the program `SAPit` could take these data directly. However, if the connectivity (fold) of the ideal forms is not specified then such a direct comparison would require testing every possible fold over the ideal form. Even for small proteins (ten segments) the number of combinations are large and quickly become excessive with larger proteins. To avoid this, the stick-figures can be further reduced into a matrix of pairwise segment interactions. As in other similar comparison methods, such as those based on graph matching methods (Artymiuk *et al* 1990) (section 6.1.7) these were characterized by their distance and angle. The former was taken as the closest approach of the two line segments while the latter was the unsigned dihedral angle. These two measures are independent of line direction and so eliminate the difference between parallel and antiparallel interactions. Some interactions will be more important than others but this can be quantified by the extent of their line segment overlap. This was defined by a measure that summed a series of finely spaced lines as shown in figure 3.7 (in section 3.2.1.1).

6.1.8.2 Finding the best match

In the SAPit program, consecutive triples of points are taken in each structure and the similarity of the remaining points compared in the coordinate frame defined by each triple. This assessment was made on the basis of point separation and relative orientation and the best matching pairs found by dynamic programming (Taylor and Orengo 1989b, Taylor 1999b). (See sections 6.1.5 and 6.1.6.) The current problem can be approached in a similar way but with each triple being selected on the basis of local structural similarity with points not necessarily adjacent in the sequence. Similarly, the dynamic programming algorithm cannot be used as it assumes that the equivalent points will be in linear order. Instead the 'stable marriage algorithm' (Sedgewick 1990) was used to reconcile the matrix of conflicting preferences into a one-to-one pairwise assignment (Taylor 2002b). This algorithm is of a type called 'bipartite' graph algorithms and is less constrained than clique-based algorithms (Bron and Kerbosch 1973, Ullmann 1976) as used by the methods mentioned in section 6.1.7.

From the alignment of segments generated by the preceding method, it is possible to construct an ideal stick-figure with the same fold as the real protein. This reintroduces direction to the sticks and allows a direct comparison between the two structures. To make this comparison even more direct, the stick lengths of the real protein were set to the same length as their ideal counterparts (typically 10 Å), similar to that shown in figure 3.10(*a*). These equivalent stick-figures were then passed to the SAPit program for a full 3D comparison (figure 6.4).

6.1.8.3 Nested solutions and normalization

The method described above allows a (real) protein structure to be compared to each of the ideal forms (frameworks) giving a quantified measure of each comparison. The fit of a structure to a framework will not be unique, and in general, all substructures of a framework should find a better match than the full framework itself. This is illustrated by matching a group of small ($\alpha/\beta/\alpha$) type proteins against a series of nested frameworks beginning with two α-helices packed above and below a five-stranded β-sheet(designated 2-5-2). The goodness-of-fit was evaluated by the RMD deviation of the real stick-figure from the ideal stick-figure, as calculated by the SAPit program, based on the aligned segment end-points (table 6.1).

It can be seen from table 6.1 that the smaller ideal forms almost always have a better fit to the real proteins. If the best fit is defined by the smallest RMSDs, then this creates the problem that the best fit will always be the most trivial. To

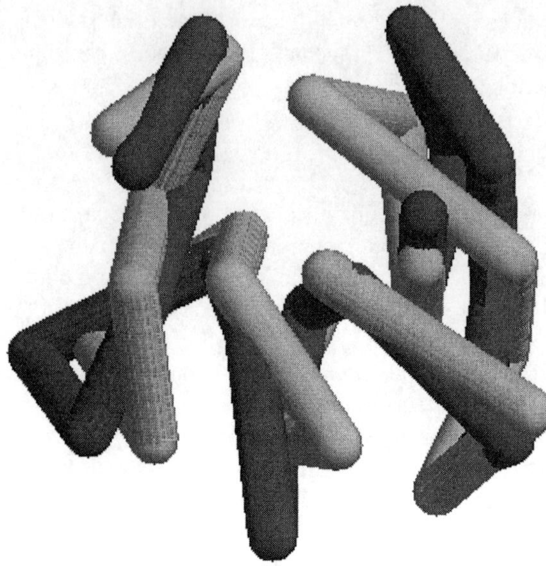

Figure 6.4. Superposed stick-figures of `3chy` and its ideal form. The stick-figure representation of 3chy (dark grey) superposed on the corresponding stick-figure of the ideal form (light grey) is shown in the same orientation as figure 3.16. The structures match with a 3.4 Å RMS deviation over all 20 matched end-points.

avoid this a normalization is required for the number of superposed points and the square root of the number of points seemed to be a good practical rule . The results of this application are shown for the small $\beta\alpha$ chemotaxis Y protein (PDB code: `3chy`) in figure 6.5. The native protein was matched against a set of ideal forms with three secondary structure layers (α-helices packed either side of a β-sheet) and a set with four layers (α-helices packed either side of a β-sandwich). These were based on the models shown in figure 3.10 (see also figure 7.2). The former class to which the native protein corresponds is clearly 'favoured' and the ideal form corresponding to the native structure which has two helices above and three helices below a five-stranded sheet (`2-5-3`) scores second highest.

This approach thus provides an automatic classification for each protein structure that is further developed in section 7.3 (Taylor 2002a) where the quality of match obtained with a small $\beta\alpha$ protein and a larger TIM-barrel protein are shown in figure 7.5.

6.2 Assessment of significance

Like sequence alignment methods, almost all of the methods discussed above will produce a match when presented with two structures—whether these structures

Table 6.1. RMS deviations from the ideal forms for a range of small β/α class proteins specified by their PDB codes. (See text for details). Each column gives the RMS deviation to the ideal form specified by its 'locomotive' class (see section 3.2.6.3) corresponding to the number of α-β-α segments in each layer. The RMS values are unweighted over all the equivalent end-points of the secondary structures, the number of which is given in parentheses at the top of each column. A dash indicates that either no solution for found by the matching program, or that which was found did not incorporate all the elements of the ideal form. Each match was examined and all were found to be a good topological match.

	2-5-2 (18)	3-5-2 (20)	3-6-2 (22)	3-6-3 (24)
`3chy`	3.305	3.260	—	—
`5nul`	4.002	4.471	—	—
`2fcr`	4.997	5.073	5.237	—
`3adk`	5.774	5.070	—	—
`1etu`	5.418	5.484	5.821	—
`5p21`	4.917	5.227	5.428	6.773
`1kev`	2.800	2.891	3.264	—

share any similarity or not. An important aspect of structure comparison is to decide when a match is significant. This is difficult as we have seen that, beyond close similarity, there is no uniquely correct structural alignment of two proteins and different alignments are achieved depending on which biological properties and relations are emphasized in the comparison (Taylor and Orengo 1989b, Godzik 1996, May and Johnson 1994, May 1996).

For a proper statistical assessment, the scoring found for a structure comparison must be compared against what is expected by chance. This is often implemented as what is expected by aligning random structures or fragments of non-related proteins.

6.2.1 Score distributions from known structures

Alexandrov and Go (1994) made an analysis for finding the significance of similar pairs of proteins using their program `SARF`. For a fixed length L, they picked up all fragment pairs of this length in two unrelated structures, and found the value R_L such that only 1% of pairs have smaller RMSD. Similarly, Russell (1998), made an analysis using distance RMSD, related to his method for detecting side chain patterns. Random pairs of structures with different folds were chosen, and random patterns of two to six patterns were derived.

Alexandrov and Fischer (1996) and Holm and Sander (1993b) used a Z-value statistic to measure significance whereas Gibrat *et al* (Gibrat *et al* 1996), in their VAST program, compute a P-value for an alignment based on how many secondary structure elements are aligned as compared with the chance of aligning

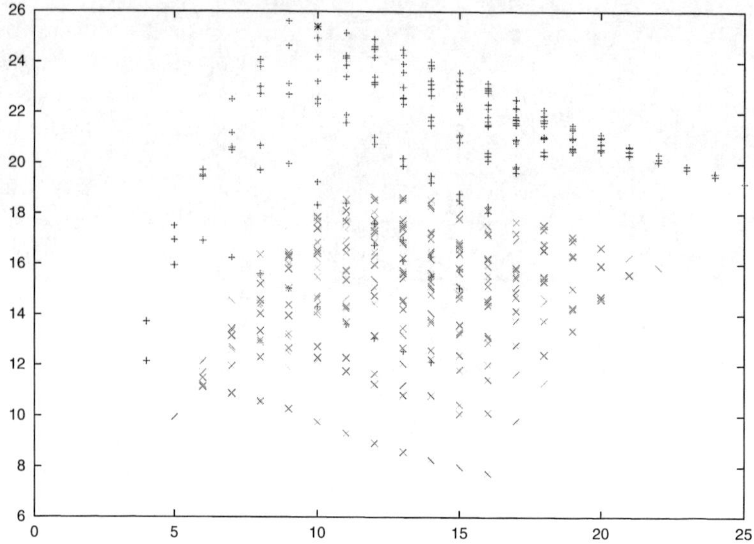

Figure 6.5. Match of che-Y against the ideal forms. The native structure of the chemotaxis-Y protein (PDB: `3chy`) was reduced to a 'stick' model and matched against the idealized stick structures (forms) representing the three-layer ($\alpha\beta\alpha$) class (pluses) and the four-layer ($\alpha\beta\beta\alpha$) class (crosses). The score of the match is plotted against the number of secondary structures in the form (N), normalized for size by $\sqrt{N}$. The largest form to be matched had six helices above and below a 13-stranded sheet (`6-13-6`, $N = 25$) but in practice such forms that are clearly much bigger than the native probe can be avoided. The form corresponding to the native structure (`2-5-3`) is marked as an asterisk. The smaller form that scores slightly higher is the `2-5-2` form, corresponding to the core of the protein.

elements randomly. Levitt and Gerstein (1998), made a comparison of the scoring of their iterated dynamic programming/superposition program to RMSD. The P-value of a score S for fixed N (number of matched residues) can be found by fitting to an extreme value distribution. The same statistics can be developed for use of RMSD and both can then be compared by a method of Brenner *et al* (1998) in which the E-values of each structure pair giving 1% false-positives was taken.

6.2.2 Random structural models

In sequence comparison, the generation of a set of random models is easily achieved by generating random sequences either as a Markov process or through shuffling the native sequences. However, no such simple method can be used for structures and the best random model against which comparison scores should be compared depends on the degree to which the inherent non-random features of protein structure in general should be considered significant (Taylor 1997c). Random chains can be generated and compared (McLachlan 1984) but the best random models would be those generated with secondary structures. Ideally, these models should be calculated for each comparison to match the length of the native comparison and the secondary structure composition and distance geometry methods can be used from this (Aszódi and Taylor 1994b). However, these models are complex to generate and cannot be 'tailor-made' for each individual comparison without excessive computation.

Models involving symmetry operations on the protein (chain reversal and reflection) can be used in situations where the comparison method restricts its calculation to the α-carbon atoms of the protein since the arrangement of the other main chain atoms is directional (Taylor 1997c, Maiorov and Crippen 1994). Considering just α-carbons, the conformations of local structural features (such as secondary structure and their chirality of connection) in the reversed chain model is indistinguishable from a forward running 'native' chain[4]. This principle of reversal applies equally at the level of the sequence and has been used previously to provide a random model for sequence pattern matching (Taylor 1986b, Taylor 1998). In both sequence and structural data the reversed chain model preserves the length and composition of the protein, including directionally symmetric correlations associated with secondary structure, while additionally in the reversed structural model, the bulk properties of packing density and inertial axes are also preserved. The latter are difficult to maintain in randomly generated structures (Aszódi and Taylor 1994a).

The reflected chain is clearly not an ideal model for proteins as they contain both large- and small-scale chiral features which will change hands under reflection. However, the use of greatly simplified lattice models avoids this problem and based on this analysis, Maiorov and Crippen (1994) proposed a definition of the significance of the RMSD in which they take two conformers to be intrinsically similar if their RMSD is smaller than that when one of them is mirror inverted.

6.2.3 Randomized alignment models

In general, the closer the random model is to preserving the properties of the native proteins, the more difficult it becomes to generate plausible alternatives. This problem is particularly acute for the reversed chain random model discussed above since, for any given protein, there is only one. This problem can be partially circumvented, however, at the stage of calculating the alignment. At this point the alignment with each random model can be expanded into a population of variants by introducing 'noise' into the score matrix and repeating the calculation of the alignment path from each noisy matrix. This generates a family of near-optimal subalignments and the spread of scores for this population can provide a measure

[4] The chirality of the torsion angle formed by successive α-carbon atoms is not affected by the chain direction (see figure 2.5).

of the stability or uniqueness of the answer. An advantage of this approach is that it can be applied not only to structures belonging to the set of randomized models but also to the native structure itself and the two resulting score distributions can be tested statistically to see if they are distinct.

If there is sufficient 'noise' introduced into the alignment method, and the population is large enough, then almost all reasonable alignments for a pair of proteins can be sampled. Plotting these solutions by their number of aligned positions against RMSD revealed a 'cloud' of points which was diffuse at high RMSD but had a sharp boundary on its lower edge. This edge represents the limit, for a given number of aligned positions, below which a smaller RMSD cannot be found. As judged by the sharp edge to the distribution, this limit is not restricted by the method of comparing the proteins and so provides an absolute standard against which other methods can be compared. For a few protein pairs, the results of other methods (gathered by Godzik (1996)) were plotted and compared to these lines. Most of these results were found to lie above the line, indicating that the optimal solution in terms of minimum RMSD had not been attained. Only a few results lay on the line and these mostly involved a smaller number of equivalent positions. It should be noted that assessing methods by the use of the RMSD value is sometimes unfair since for many of the methods, their aim is not to minimize the RMSD value.

6.2.4 Scoring and biological significance

When a structure is compared to every other structure (or to a representative selection), then scores will result ranging from the clear relationships of homologous proteins to a large number of poor scores for obviously unrelated pairs. Between these extremes lies a 'twilight' zone within which it is very difficult to assess the significance of the score. This problem is exacerbated because many proteins contain similar substructures, such as secondary and super-secondary structures and the problem is to decide when a similarity is just a consequence of being protein-like and when it indicates a more specific relationship between the two proteins.

Because of its common currency, most considerations of this problem have focused on the significance of the RMSD measure based on the comparison of proteins or protein fragments of equal length (see above). Others, such as the DALI method, have adopted a similar approach based on the scores achieved over matches of protein fragments (Holm and Sander 1993b). Both these approaches require that the selected fragments are unrelated to the proteins being assessed, however, this raises the problem of what criterion can be used to make this distinction and, in principle, it should not be a weaker method than that used for the current comparison. It is not acceptable, either, to consider completely unrelated proteins since, to take an extreme example, if the two proteins being compared contained only α-helices and the clearly unrelated control set contained

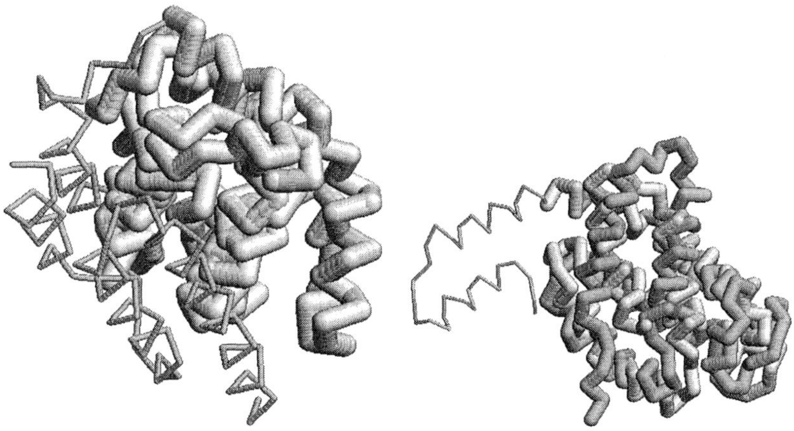

(*a*) Colicin (*b*) Phycocyanin

Figure 6.6. The globin fold in colicin-A and phycocyanin. The two structures are drawn to show their backbone as linked α-carbons with the region corresponding to the globin fold drawn more thickly. (*a*) Colicin [`1colA`], which has extra helices towards the carboxy terminus. The core region matched 97 residues with an RMS deviation of 3.2 Å. (*a*) Phycocyanin [`1cpcA`], which has two extra helices on the amino terminus. Both structures were compared against the hemoglobin structure `1lhlb` (sea cucumber). The core region matched 85 residues with an RMS deviation of 5.4 Å.

only β-structure, then the two α proteins would appear more related than they should do.

An alternate approach to this problem is to use the reversed structure (as described above). When this is matched against the structure databank a similar range of scores should result—since the reversed structure has exactly the same length, overall shape and secondary structure content as the native probe. What will be lost is any specific overall similarity to proteins that are homologous to the native probe. In addition, if the probe structure is a particularly simple fold (such as four α-helices) then the reversed structure will also embody this property so a specific match will need to capture more than a few matched helices to gain significantly over the background of scores derived from the reversed structure.

6.2.5 Examples

6.2.5.1 Distant globin similarities

A globin fold is also found in the plant phycocyanin proteins which have the same core fold of six helices with two 'extra' ones preceding this core (Pastore

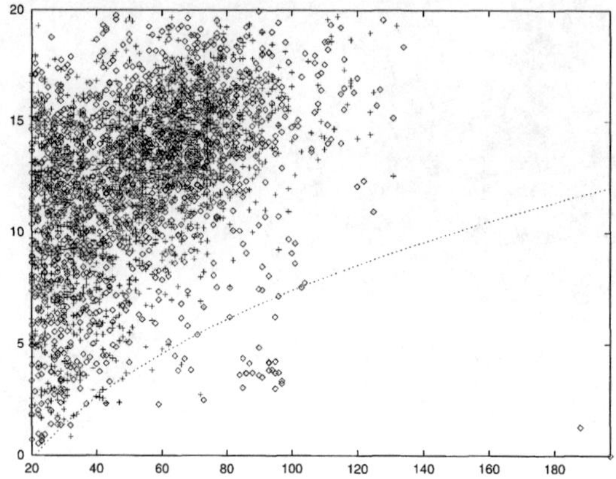

(*a*) Colicin

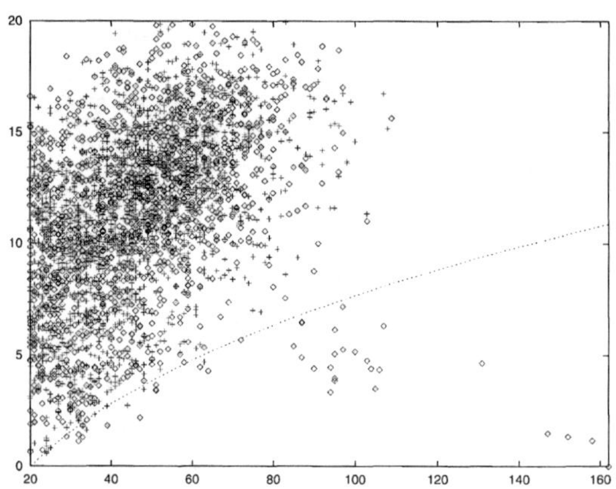

(*b*) Phycocyanin

Figure 6.7. Structural searches with colicin and phycocyanin. RMS deviation is plotted against number of residues aligned (diamonds). The structures match themselves (point on the lower right corner) and any homologues (clustered in the lower right corner). The probe structure was then reversed and rescanned (crosses). From these results a line (dashed) was fitted that excludes 99% of the reversed matches. (*a*) Colicin [`1colA`], has only one homologue and the cluster of matches around 2–4 Å RMS and 80–100 residues include globins and phycocyanins. (*b*) Phycocyanin [`1cpcA`], has a few homologues and matches the globins in a cluster around 4–6 Å and 80-110 residues. Colicin was not found at a significant level.

and Lesk 1990). These proteins have no significant sequence similarity and only a vague relationship: both bind cofactors, however, the phycocyanins are electron transport proteins specific to the photosynthetic complex and do not bind oxygen as do all the globins (figure 6.6(*b*)).

The globin fold has also been found in the bacterial toxin protein colicin-A (figure 6.6(*a*)). As with the phycocyanins, this is larger than the globins, but in this relationship, the equivalent fold must be extracted from an otherwise well-packed bundle of eight helices (Holm and Sander 1993a, Orengo and Taylor 1993). Here no amount of imagination can lead to a plausible functional or evolutionary link with the globins (or the phycocyanins).

Progressing in the other direction, many small folds can exhibit similarity with part of the globin fold—in the extreme, this might involve matching a single α-helix. A relationship of a small protein with the globins that has been considered significant was noted in the DNA-binding domain of a bacteriophage repressor protein Subbiah *et al* (1993). The authors suggested a possible evolutionary relationship here since the DNA- and hæm-binding sites are located in similar parts of the structure.

6.2.5.2 Assessment against chain reversal model

The three globin fold similarities were re-examined using the reversed chain model to generate a background score. A globin fold was scanned against a non-redundant selection of the protein structure databank (PDB)[5] and found all globins and all phycocyanins as significant matches. Similarly, a phycocyanin probe found all phycocyanins and globins.

The globin/colicin-A relationship was tested using the reversed colicin structure as a control and was found to retain its significance (figure 6.7(*a*)) with the globin finding colicin and colicin finding the globins. However, while colicin could find a phycocyanin, a phycocyanin probe did not find colicin at a significant level (figure 6.7(*b*)).

When tested in a similar way, the relationship between the bacteriophage repressor and the globins was not found to be significant when using a globin as a probe, however, with the repressor as a probe a number of globins were found to lie just on the border-line of significance.

[5] This selection was made by choosing one representative for all sequences that share greater than 50% sequence identity. The member taken to represent each family fulfilled a variety of criteria but generally had the best resolution and the lowest average B-value (an indicator of refinement quality). Details can be found in Taylor (1997b).

Chapter 7

Classification and fold spaces

7.1 Protein structure classification

Ever since they were first seen, protein folds have remained a source of fascination, and even wonder. The different folds adopted by proteins become more enigmatic when it is realized that they bear little relationship to the function of the protein. Proteins with different folds can perform the same function (for example; catalyse the same reaction) while proteins with the same fold can perform different functions and, beyond clear evolutionary relationships (homology), it is not possible to predict a function from a fold with any degree of certainty.

One of the greatest mysteries in molecular biology is why proteins adopt the folds that are observed: are they simply the result of prehistorical accident or do they represent a partial or complete manifestation of the folds allowed by 'laws' that we do not yet know (Denton and Marshall 2001)? Until the first extraterrestrial genome is determined (and some would exclude Mars from this), we cannot directly answer the first part but some idea of the second option might be gained through analysis of the known folds. Luckily, there are now several experimental programmes aimed at elucidating as many protein structures as possible and if there is some underlying order, we will soon have our best chance to see what it is.

Some of the methods described in the previous section on structure comparison have been applied further to try and define a 'fold space' for proteins. In this section we investigate some of these approaches and ask whether they can be extrapolated into these more tentative relationships between very distantly related proteins and indeed, whether the idea of a protein fold-space is a practical (or even possible) target.

7.1.1 Practical approaches to classification

In the protein data bank (PDB) (http://www.rcsb.org/pdb) the number of structures doubles almost every 18 months. Despite this flood of data, increasingly efficient and robust methods for protein 3D structure comparison have made it feasible to perform all against all comparisons of all known 3D structures (for a review, see Holm and Sander (1994b) and Orengo (1994) and also the previous chapter for details of the methods). These exhaustive comparisons reveal that proteins can share a common fold despite lacking any 'significant' sequence identity (section 6.2.5) and, furthermore, proteins with the same fold may have different functions. Their main aim, however, is to try and bring some order into the description of protein structure by imposing a classification.

The three most popular classifications, all of which are accessible via the World-wide Web (WWW) are:

(i) SCOP: (`http://scop.mrc-lmb.cam.ac.uk/scop/`)
a 'structural classification of proteins' database (Murzin *et al* 1995, Hubbard *et al* 1997) which is essentially a manual classification.

(ii) CATH: (`http://www.biochem.ucl.ac.uk/bsm/cath/`)
(Orengo *et al* 1997) which is constructed using both manual and automated approaches.

(iii) FSSP: (`http://www2.ebi.ac.uk/dali/fssp/`)
(Holm and Sander 1997, Holm and Sander 1998) which is built in a totally automated fashion using the program `Dali` (section 6.1.2).

(iv) HOMSTRAD: (`http://tom.cam.ac.uk/homstrad/`)
(Mizuguchi *et al* 1998) which is built in a semi-automated way. This database is related to CAMPASS (Sowdhamini *et al* 1998) which includes distant similarities.

All three classifications use a hierarchical data structure with a nested set of partitions grouping similar proteins.

7.1.1.1 Automated approaches to classification

Given an approach to define a topological equivalence between a pair of 3D structures we need a measure to describe their extent of similarity or distance (a metric). Most metrics specify the pairwise DISsimilarity: for example, the most common dissimilarity measure is the root-mean-square deviation (RMSD) after rigid-body superposition[1].

Unfortunately, as we have seen in the previous chapter, unlike amino acid sequence alignment, the problem of 3D structure alignment is not trivial. Although sequence alignment using dynamic programming guarantees the optimal solution (mathematically but perhaps not biologically), the comparison of 3D coordinate data is not as well defined as the comparison of 1D strings of amino acids. This gives even more scope for the measures produced automatically to differ, as an alignment between 3D structures depends on the nature of the

[1] It is important to specify over which atoms the RMSD is calculated. In the current discussion it can be assumed that only the main chain α-carbon atoms are considered but any different choice obviously affects the result.

objective function. For instance, intermolecular distances might be minimized in a rigid-body superposition (e.g. (May and Johnson 1994, May and Johnson 1995)), or they might be compared in a pairwise manner, as in the `SSAP` (Taylor and Orengo 1989b) and `Dali` (Holm and Sander 1993b) programs.

Another consideration for structure alignment, is the balance between the number of topological equivalences and the attendant RMSD (May 1996) in this case, the goal is to maximize the number of equivalences while simultaneously minimize the associated RMSD. The question arises then as to how to identify the alignment with the most meaningful compromise between the two factors (May 1996, Taylor 1999b).

7.1.2 Organization of the classifications

7.1.2.1 The unit of classification

Despite the differing philosophies behind the three classifications, (SCOP, CATH and FSSP) there is consensus on the unit of classification: the protein domain (section 1.2.7). There are several algorithms for domain identification from coordinates (Taylor 1999c, Holm and Sander 1994a, Swindells 1995a) but even a structure-based definition is non-trivial. For instance, there are often extensive interfaces between domains leading to ambiguity about the appropriate level of granularity for domain definition. Another complication lies in the fact that domains can comprise sequential (continuous domains) and non-sequential (discontinuous domains) parts of the polypeptide chain (figure 1.7). Continuous domains are easier to identify than discontinuous ones (Jones *et al* (1998)). Not surprisingly, differences in domain assignment have been shown to be an important factor between the classification schemes (Hadley and Jones 1995) although other, less-structural, criteria are involved such as folding (independently folding units) or function (functional units).

7.1.2.2 Hierarchical organization

Although all three major classifications agree on a hierarchical paradigm, they differ in the detailed organization. For example, the top level of the hierarchy in SCOP and CATH is protein class. However, SCOP and CATH differ in the number of classes used. While SCOP uses the original four classes of Levitt and Chothia (1976), CATH merges the α/β and $\alpha+\beta$ classes into a single one.

CATH has a unique level within its hierarchy: architecture. Architecture is the overall shape of a domain as defined by the packing of the secondary structure elements but ignoring their connectivity. The current release of CATH (version 1.6 June 1999) consists of 35 architectures which have been assigned by eye. (A more systematic approach will be outlined in section 7.3.)

All three classifications agree on a fold level. The fold of a protein describes its architecture together with its topological connections. However, there is a difference in how folds are assigned. For instance, it is done automatically in

CATH on the basis of structure similarity score derived by SSAP (Taylor and Orengo 1989b). However, in SCOP, fold definition is done by eye.

Although proteins are grouped into families and superfamilies, once again the operational definition of these terms can vary. Families comprise proteins believed to be homologous i.e. those related by divergent evolution from a common ancestor. Clear evolutionary relationship is usually assigned on the basis of significant sequence identity. Here there are differences: SCOP uses a threshold of $\geq$30% sequence identity while CATH uses $\geq$35%. Of course, in those cases where family membership is assigned on the basis of common fold and function, in the absence of significant sequence identity (e.g. as with the globin examples discussed in section 6.2.5), then a problem remains in the definition of a common fold. Superfamilies comprise proteins deemed to share a probable common evolutionary origin on the basis of a common fold and often function but in the absence of significant sequence identity. (A detailed comparison of SCOP, CATH and FSSP is described in Hadley and Jones (1995).)

7.1.3 Analysis of the classifications

7.1.3.1 Number of folds

Not surprisingly, there has been much speculation as to the total number of protein folds in nature. One, often quoted, estimate is that there are 1000 folds (Chothia 1992). However, a recent calculation puts the figure at around 2000 (Govindarajan *et al* 1999). In fact, the only area of agreement within the community is that the number of protein folds in nature is finite! Whatever the actual answer, we need to consider the question of how many folds remain to be seen. Of course, this is not just an academic question given the investment required for structural genomics. Clearly, classification helps to define sparsely populated regions of fold space and so can help to direct protein 3D structure determination.

From an evolutionary perspective, one of the most interesting aspects of these extensive analyses is to determine the common cores of ancient proteins that are found across a wide range of phyla, and often in all living organisms (Green *et al* 1993). These ***a***ncient ***c***onserved ***r***egions (ACRs) provide an indication of the a minimal set of proteins required to support basic life functions. The number of such proteins may be as low as 900 and a representative of 600 of these may already be found in the current databases. This is comparable to a similar estimate of 1000 for the number of families based on the analysis of recurrence in a recent influx of sequence data (Chothia 1992) (see below). However, this correspondence may be coincidental (Green 1994).

7.1.3.2 Super-folds

With proteins gathered and classified into some sort of order, one of the more obvious observations was that some folds are much more common than others. Even before any systematic count of folds was made, it was apparent that folds such as the TIM barrel (figure 1.5) seemed to recur with surprising frequency (Chothia 1988, Orengo *et al* 1994). This structure was first seen in triosephosphate isomerase (TIM) (Banner *et al* 1975), an enzyme in the key metabolic pathway glycolysis. In fact, not only are all the glycolytic enzymes α/β structures but also the last enzyme of the pathway, pyruvate kinase, contains another TIM barrel domain. Approximately 10% of all known enzyme 3D structures have a TIM barrel fold despite having different amino acid sequences and different functions (for a recent review, see Reardon and Farber (1995)). Along with a few other folds, the TIM barrel has been termed a 'superfold'(Orengo *et al* 1994): defined as a fold common to at least three non-homologous proteins (i.e. with no significant sequence identity).

Classification has also made it possible to explore global relationships between protein 3D structure and function. For example, originally Nishikawa and Ooi (1974), and more recently, Martin *et al* (1998) have shown that most enzymes have α/β folds. It is also possible to identify densely populated regions of fold space—referred to as 'attractors' by Holm and Sander (1998).

7.1.3.3 Future prospects

Since we do not yet have a complete library of protein folds, any classification can only be a snapshot of a dynamic situation and this means that the classifications need constant updating. This emphasizes an important difference between the three classifications: FSSP, because its construction is entirely automated and so is always up to date; however, SCOP and CATH need considerable human input and so are behind the latest release of the structure data. More fundamentally, as we have seen, there is an unacceptable level of disagreement about the usage of certain terms and what is important in a classification. It is to be expected then that even when a complete set of protein folds is available there will be many discrepancies between classifications.

In Rutherford's division of science, protein fold classification currently bears a greater similarity to stamp collecting than to physics! In many ways, it represents little more than fact accumulation and sorting. Indeed, one might wonder whether attempts to classify protein folds are simply a reflection of an innate human desire to impose order and certainty on an otherwise unconnected collection of folds? What we lack at the moment is a general physical theory to synthesize the current data. This might come from a better understanding of how a 1D amino acid sequence specifies a particular 3D structure (the 'protein folding problem') but at the moment we can do little more than catalogue each new protein 3D structure and hope, as occurred in natural history of the mid-nineteenth century, for the arrival of a new Darwin!

While we await this 'second coming', interesting pathways can be explored. Those that look most promising involve the definition of a 'fold space': providing a metric or a symbolic connection spanning the Universe of protein folds. These ideas will be considered in the following section.

7.2 Protein fold spaces

In the previous section we have seen that the most immediate approach to search for patterns in the world of folds is to use a computer program to compare all proteins to each other and make links between the most similar pairs in decreasing order of similarity until all proteins are linked. This results in a dendrogram (or tree) which gives an accurate portrayal of the relationships between homologues but becomes almost arbitrary for proteins that are not clearly related. For example; using these methods, it is not meaningful to compare a protein constructed from α-helices with one based on β-structure. In this section we investigate alternative classification approaches, concentrating on those that have the capacity to provide a systematic classification over all proteins.

7.2.1 Distance geometry projection

Given a measure between protein structures (or from the review of methods in the previous sections, any number of different measures of similarity or difference), a natural progression is to try and visualize the relationship between all the structures. One of the most immediate approaches to this is to use a multidimensional scaling or distance geometry (DG) method as described in section 2.2.

This approach was initially attempted by Orengo *et al* (1993b) using a measure of similarity calculated by the `SSAP` program (section 6.1.5). The projection in the plane of the two largest eigenvalues produced a clover-leaf projection of three clusters converging towards the centre (figure 7.1). Not unexpectedly, these three lobes corresponded to the three major classes of structure: all-α, all-β and α/β. Within these lobes it was difficult to discern any fine structure but when different subclasses were projected separately, these were found to segregate into their distinct fold types. A similar contemporary projection was made by Holm and Sander (1993b) (using a statistical method related to distance geometry) and more recently by Hou *et al* (2002). The remarkable feature of the latter is that even after 10 years, the shape of this space has remained unchanged.

In the full plot of proteins, the reason for the loss of fine detail is clear from the analysis of DG in the preceding section (section 2.2) where the equivalent problem was encountered in trying to reconstruct protein chains. In simple DG, individual pairwise distances cannot be weighted. This means that the close links between homologous structures cannot be preserved in the face of a large number of random distances derived from the comparison of non-homologous structures. If we assume that the homologous structures are singly-linked, then the random background of distances will outnumber these by a factor of N (the number of proteins).

Given this situation, it is surprising that there is any meaningful structure to be seen in the points projected into the plane of the two (or space of the three)

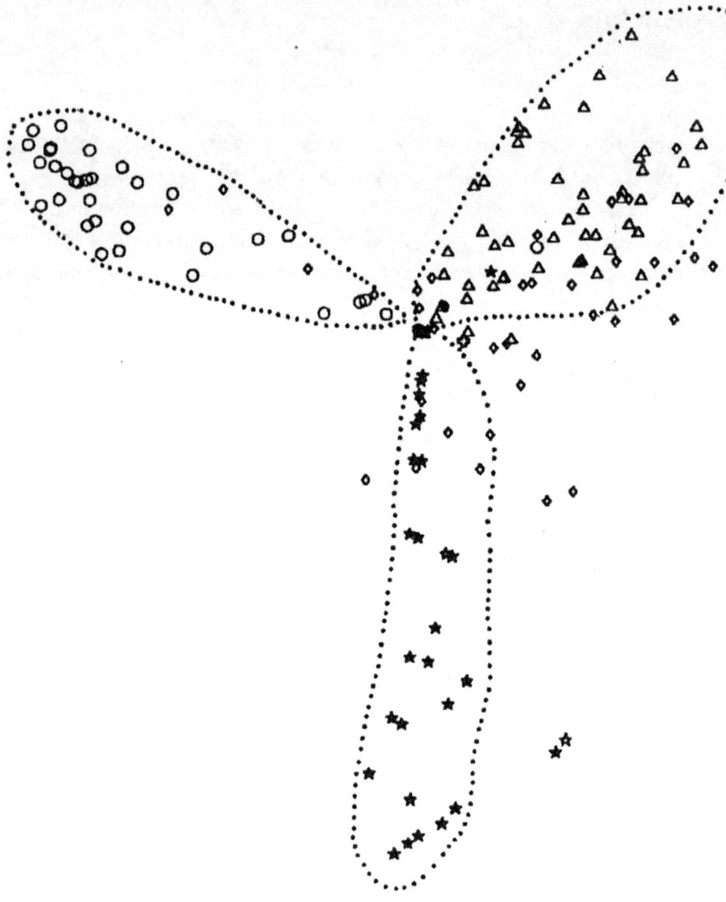

Figure 7.1. Fold space visualized by projection. The RMS distances between a set of proteins was projected into two dimensions. The main classes of proteins can be seen in the lobe-shaped clusters: all-β (circles), all-α (stars), alternating β/α (triangles) mixed with other β/α (diamonds).

largest eigenvalues but the observed trilobed structure is exactly what would be expected, where the few local similarities are being preserved while 'trying as best they can' to avoid any other proteins to which they are not linked (similar to electron orbitals). The large number of 'random' relationships masks the visualization of the useful fine structure and all that remains is a plot of secondary structure composition with clusters of the more common folds.

This problem can be overcome, to some extent, by focusing on fragments of proteins and quantifying relationships by their largest shared fragment (Harrison *et al* 2002). In this analysis, the clusters become populated by 'gregarious' proteins that share a fragment. In the end, however, the fragment can become a trivial piece of protein substructure (such as a single α-helix) and this metric also breaks down for distant relationships.

7.2.2 Simplified fold space

An alternative approach is to simplify folds. This is commonly taken to the level of secondary structure elements, but need not stop there. By repeatedly smoothing the backbone chain, folds can be reduced to a straight line (unless they contain a knot) (Taylor 2000a) so at some point, all proteins (whether α or β) become identical. This is equivalent to the statement that, in strict topological terms, all proteins are just pieces of string. If we now compare all smoothings of all proteins to each other, we have a continuous metric that is defined over all pairs of proteins. This measure is still not ideal: the trivial problem that all proteins are most similar when they are straight lines must be avoided but, more fundamentally, the smoothed traces can differ depending on domain structure. This leads to the problem of finding the largest common fragment.

The computational approaches based on fragments and smoothing, in turn: avoid the domain problem but neglect the fold problem, then tackle the fold problem but neglect the domain problem. While some clever combination may achieve both simultaneously, these difficulties suggest that a meaningful fold space based on a simple comparison metric is not possible across all proteins. A less automated route to a protein fold space, that has been pursued over many years, is to start with simple folds and 'grow' them, through the application of rules of structure addition, into the folds we know (Efimov 1997). This is a problematic approach since for it to succeed, it is necessary both to know what these transformation rules are and also to have a clear definition of what distinguishes one fold from another.

The definition of a fold has recently been formalized over quite a large part of the 'universe of proteins' in terms of layers of secondary structures (Taylor 2002a) which can be used to 'digitize' the path of the chain and so encode a fold uniquely (section 3.2.6.3). This helps codify one part of the Efimov approach but before we return to the current problem, this classification scheme will be described more fully in the next section.

7.3 A 'periodic table' for protein structures

7.3.1 Classification using ideal stick forms

The stick comparison method described in section 6.1.8 can be used to find matches of a protein (reduced to a stick-figure) with a collection of ideal forms (figure 6.5). This opens the possibility for its use as a classification tool. Given a series of ideal forms, it is necessary only to present these in order of size and select the largest solution. Unlike the visual analysis of 'topology cartoons', this approach is completely automatic and is focused on the well-packed core elements of the structure (which are not always obvious in topology cartoons). Finding solutions based on the core also means that two proteins can be compared even though they do not have the same overall fold. This can be done by looking back at their match to smaller ideal forms and if a common solution is found then this can be taken as a measure of relatedness. The further development of this approach into something resembling a 'periodic table' of proteins will be described more fully in this section[2].

The method outlined here is based on a set of idealized structures that are compared with all known structures. The domain definition problem is less directly solved but as the ideal structures are all of domain size, then the best match can define (or bias) the definition of the domains. This approach shifts the classification from a clustering problem to that of finding the best set of ideal structures that can account for as much protein structure as possible. As the ideal structures will be generated from rules applied to basic forms, this can be viewed as finding a minimal basis set of generating forms (figure 7.2).

These Forms were derived from the models described in section 3.2 in which the hydrogen bonded links across a β-sheet impose a layer structure onto the arrangement of secondary structures in a protein domain (Chothia and Finkelstein 1990, Finkelstein and Ptitsyn 1987). These layers can consist of either α-structure (packed α-helices) or β-structure (hydrogen bonded β-strands). There are seldom more than four layers in any one domain and each layer tends to be exclusively composed of one of these two types of secondary structure. The spacing between the axes of packed α-helices is typically 10 Å, as is the spacing between β-sheets and between helices and sheets (Cohen *et al* 1980, Cohen *et al* 1982), while the spacing between the hydrogen bonded strands in a sheet is close to 5 Å. This makes 10 Å a convenient unit with which to 'digitize' protein structure. (See section 3.2 for further details.)

7.3.2 Structure layers become valance shells

As we have seen in section 4.1, β-sheets normally have a twist and the whole structure follows this twist, resulting in a staggered arrangement for the secondary structure elements in the outer layers. Sheets can also curl[3] and incorporate a stagger between adjacent strands and combinations of these 'distortions' can result in the sheet forming a complete hydrogen bonded cylinder or barrel (Murzin *et al* 1994a). While all these parameters (twist, curl, stagger) might be varied with different numbers of layers (of different composition), as a simple beginning, the

[2] This section is reproduced in part from Taylor (2002a) with the kind permission of Nature Publishing.

[3] The term '*curl*' was used specifically in section 4.1 to refer only to the bending of a β-sheet along the hydrogen bonding direction (with '*bend*' being reserved for the chain direction). In this section, *curl* is used more loosely to indicate the combined effect of the two transformations.

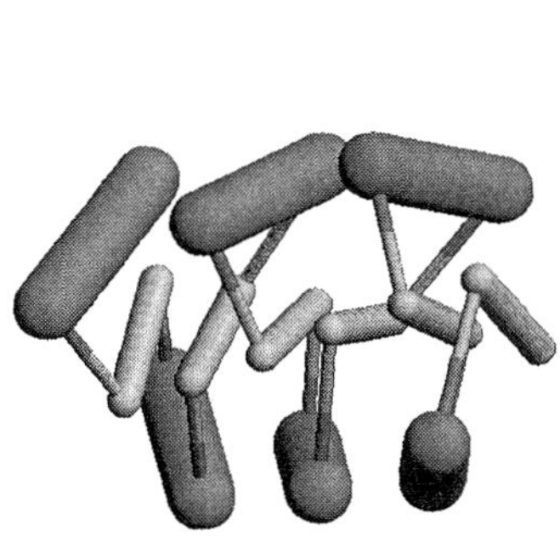

(a) $\alpha\beta\alpha$ layers

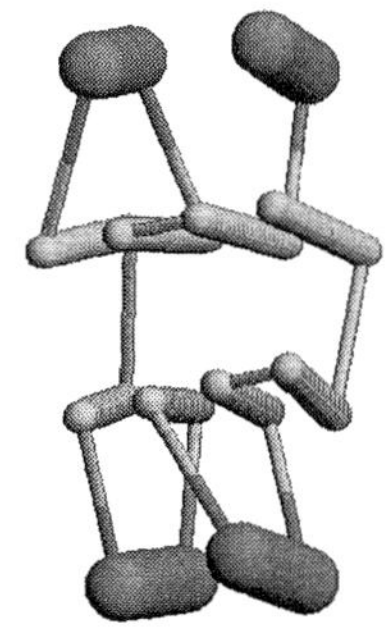

(b) $\alpha\beta\beta\alpha$ layers

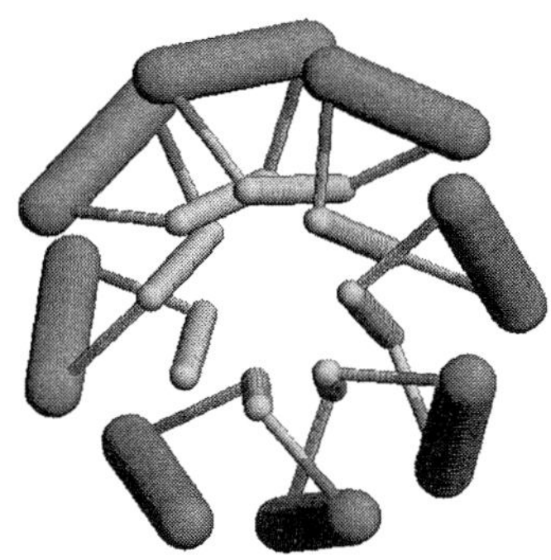

(c) $\alpha\beta$ barrel

Figure 7.2. Stick-figure representations of the basic forms. Each of the basic generating forms are represented by 'stick' models in which α-helices are darker and drawn thicker than the β-strands. (a) $\alpha\beta\alpha$ layers. Six strands are shown but the sheet can extend indefinitely. (b) $\alpha\beta\beta\alpha$ layers. As in (a), the sheets can be extended (removal of the α layers leaves the common β-'sandwich'). (c) Eight-fold $\alpha\beta$ barrel. Similar barrels with 5–9 strands were constructed (see section 3.2.4 for construction details). By deleting helices and strands from these models, almost all known globular protein domains of β and $\beta\alpha$ types can be generated.

more limited layer combinations shown in figure 7.3 were considered and the curl and stagger covaried to give either topologically flat sheets or cylindrical sheets (barrels), with the curled sheets being represented by partial barrels (figure 7.3).

The resulting arrangement is not dissimilar to the periodic table of elements. In this loose analogy, the layers are equivalent to valance shells that become progressively filled with electrons (secondary structure elements): first, the inner

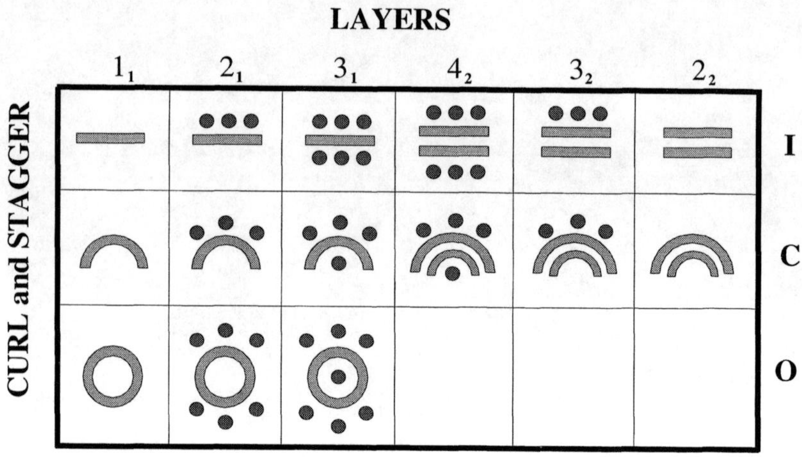

Figure 7.3. Simplified layer structure of proteins. Layers of secondary structure (β = grey α = black) are combined to make globular protein domains. The β-sheets are represented as bars and circles, as they would appear when viewed looking along their component strands. Each sheet has a left-handed twist between the strands (not depicted) onto which can be added curl and stagger. All possible deletions are made for each basic structure: those for O2 are shown in figure 7.4.

β layer (S orbitals) followed by the outer α layers (P orbitals) then repeating with a second β layer (D orbitals). Extending the model to incorporate the permutations arising from additional α layers would even be reminiscent of the interjection of the rare earth series. (Even delocalized helices in the outer shell might be imagined acting like metallic electrons.)

In figure 7.3 it is imagined that the β-sheets can progressively 'deform' from a topologically flat sheet into a cylinder (or barrel). The two end-points and one intermediate stage are represented by the rows in the figure and indexed as I (flat), C (curled) and O (barrel). For each of these, up to four layers of secondary structure are shown. For simplicity, not all possible layer combinations have been represented, in particular, those with adjacent α-layers have been omitted as the boundary between these is not well defined. Most biologically important structures can be generated from three of those represented above: referred to as the basis set. (See figure 7.2 for 3D 'stick'-figures of the basis set.) Using the ICO index plus layer number: $I3_1$ can generate $I2_1$ and $I1_1$ (by the deletion of helices), while $O2_1$ can generate $O1_1$, and with the removal of strands from the barrel, also $C2_1$ and $C1_1$. Similarly, $I4_2$ can produce an $\alpha\beta\beta$ and the common $\beta\beta$ layer structures. Of those remaining, only $C3_1$ is biologically important and will be reconsidered later.

As there is no strict energetic difference between different protein structures, the filling of the layers with secondary structures was allowed more freedom than their electronic counterparts and, for each of the basic forms described above, any combination of secondary structures can be removed. For example; the eight stranded barrel with eight helices around it (type O2) can give rise to thirty variations shown in figure 7.4. A similar combinatorial enumeration of variants was made for barrels of five up to nine strands and flat sheets from 3 up to 13 strands. The intermediate curled strands (row C in figure 7.3) were made by successive deletion of all but three strands from each of the barrels (and the helices similarly permuted). The application of symmetry considerations greatly reduced the number of possible combinations but for some larger structures it was necessary to impose a limit. Variations on the large 'flat' layers were ranked by compactness and (for alphabetic reasons) just the 26 most compact combinations were considered for matching. Variations on the larger barrels were also limited by allowing only one break in the sheet layer and similarly restricted to 26 variants on any given combination of secondary structures. In total, 12 640 variations were generated.

7.3.3 Matching against all stick forms

In order to match all ideal structures against all known protein structures, each native structure was firstly reduced to linear segments (Taylor 2001). For each comparison, a fast bipartite graph matching algorithm was used as a pre-filter and also to 'prime' a more exhaustive double dynamic programming comparison algorithm (Taylor 2000c). Each comparison began by pairing-up line-segments ('sticks') irrespective of their length or direction but when a good fit was found, the lengths of the native protein 'sticks' were set to 10 Å and the connectivity of the ideal sticks set to match the native protein. This allowed the two matching stick-structures to be directly superposed in 3D using a conventional comparison program and the quality of the match was taken as the root-mean-square deviation (RMSd) between two sets of stick end-points. From examination of the solutions, anything less than 5 Å RMSD between two structures is a good fit while between 5 and 6 Å RMSd is acceptable. An example of two solutions in these ranges is shown in the figure 7.5.

7.3.3.1 Matching to the PDB

The method was initially tested on the single domains defined to have distinct folds in the CATH database (Orengo *et al* 1997). From these, the all-α domains were excluded as were integral membrane proteins. Although ideal structures can be described for these classes (Murzin and Finkelstein 1988, Taylor *et al* 1994b), a comprehensive nomenclature has not been devised (see section 3.2.6.2 for a full discussion of this problem). Some small domains that do not have packed secondary structure were also excluded, leaving a sample of 418 domains. Of these domains, only 34 had no acceptable fit to an instance of the ideal forms, and most of these were small with fragmentary beta structure.

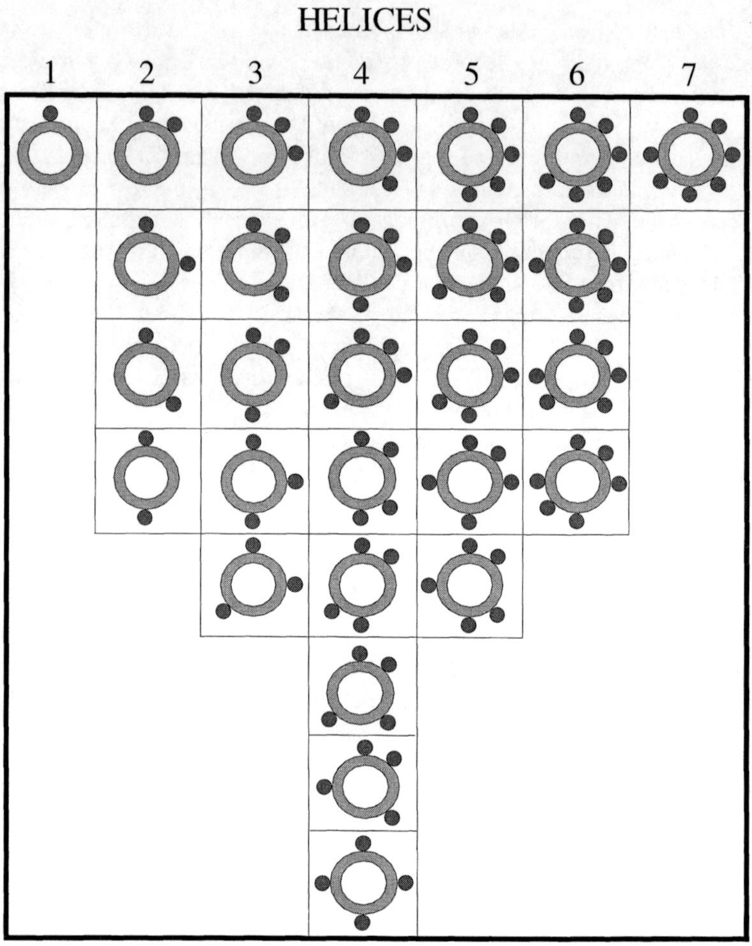

Figure 7.4. β-barrel helix packing variations. The basic barrel form (O2 in figure 7.3) with eight strands and eight helices can generate 30 packing variations through the deletion of helices (28 depicted above plus the two not shown with eight and zero helices).

Of the remaining 384 domains 85% of the matches accounted for over 50% of the structure with more than 70% of the structure being matched in over half of the domains. Partially matched domains comprised large β-propellor structures (figure 10.10) and viral coat proteins that have long 'unstructured' loops. The remainder was typically composed of α-helices which were either packed in a distinct subdomain or suggested an additional α-layer.

Rather than adapt the current model to account for these elements, it was clear that it would be necessary firstly to reassess the definition of domains. This

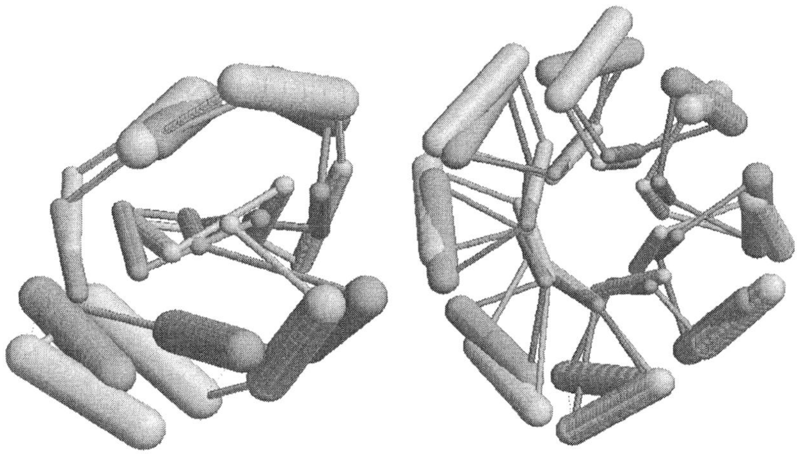

(*a*) poor fit (*b*) good fit

Figure 7.5. Range of match quality. The stick representation of native structures (light grey) were matched to their ideal forms (dark grey) and the quality of match assessed by the RMSD over equivalent stick end-points. Matches over 6 Å were rejected and part (*a*) shows the match of a small protein (PDB code: `1a3aA`) that approached this limit (5.6 Å over 20 points) while part (*b*) shows a large protein (PDB code: `1a49B`) that matched well (3 Å over 32 points). α-helices are drawn thicker than β-strands (as in figure 3.10).

was done using an automatic (Ising-like) method (section 5.2 and Taylor (1999c)) in which it was possible to bias the matched portion of a protein to remain distinct from the rest. Any remaining material was then presented again for matching to the ideal structures and the process repeated. The full chains from a non-redundant set of 2230 protein structures were then processed by this algorithm, resulting in an improved coverage by the ideal forms which, on average, now accounted for 80% of each protein, with 70% of the structure being matched in 75% of the domains (previously 50%).

7.3.3.2 *Table of forms*

Each form can be indexed by the number of secondary structures in each layer using the 'locomotive nomenclature' (see section 3.2.4 for details) and from the index of its matching form, each domain can be allocated a grid reference and plotted in space. Any step in these tables of forms represents the addition or removal of a secondary structure while one of the dimensions in the barrel grid represents the opening of the barrel (for a fixed number of strands). Using

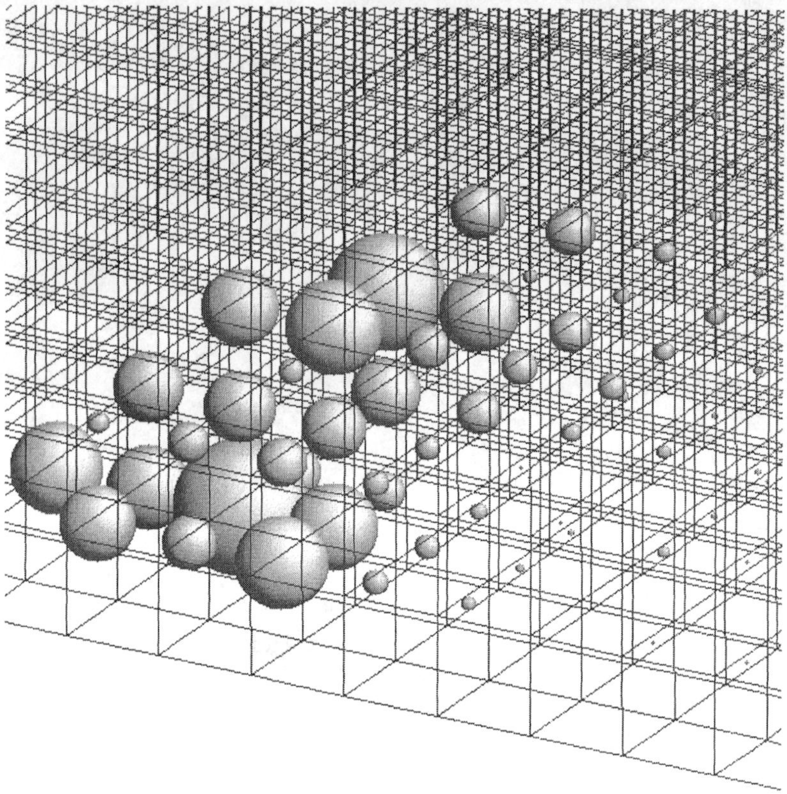

Figure 7.6. Structure tables for flat sheets. The number of helices and strands in each secondary structure layer of the ideal forms is represented as a grid: *a*, $\alpha\beta\alpha$ and *b*, $\alpha\beta\beta\alpha$. Grid cells contain a sphere if their corresponding ideal structure constitutes the best match to a native protein domain. This has a radius to indicate the number of proteins that match. The most populated cell has the form `0-4-2` (α-β-α) with 48 members, closely followed by the `2-5-3` form (42 members). The former includes 25 different topologies while the latter includes only 18 (being dominated by the common flavodoxin fold).

this representation, the full range of known protein structure can be visualized (figure 7.6).

Besides the all-α class of protein, the only structures not 'captured' by the set of three basic forms used above were β proteins that contain internal repetition. These included not only the series of propellor proteins (figure 10.4) but also those with a triangular arrangement of structure, including β-trefoils, β-prisms and β-helices (figure 10.3) (Chothia and Murzin 1993). Clearly triangles do not map well onto layers (or cylinders) and the only solution for these may be to generate a 'triangular' version of figure 7.3, or because of their detectable internal sequence repeats, to treat them as exceptions. For the moment, the latter route will be followed but a series of propellor forms will be added to the basis set, along with the neglected $C3_1$ form from figure 7.3 (which includes the β-grasp structures).

7.3.4 Reintroducing topology

This classification of structure has not explicitly considered topology but has concentrated on the prerequisite of defining the secondary and tertiary links. From this base, the previously difficult issue of topology becomes almost trivial since two proteins matching the same ideal form will either have the same connection of their secondary structure elements (SSEs) or will not (see section 3.2.6.3 for nomenclature details).

7.3.4.1 Uniqueness of folds

Topology strings were written for each protein in each grid cell and sorted for uniqueness so giving the number of folds for each form. This value was then plotted along with the number of unique secondary structure strings for the three-layer forms (figure 7.7). For example; a typical topology string for a flavodoxin-like protein (the chemotaxis Y protein `3chy` fitting form `2-5-3`) is:

```
+B0.-A0.+B-1.-C0.+B1.-C1.+B2.-C2.+B3.-A1
```

where A, B and C are the three layers prefixed by their relative orientation to the first strand in the sheet and suffixed by their position relative to the first element in each layer. The corresponding string of secondary structures is: $\beta\alpha\beta\alpha\beta\alpha\beta\alpha\beta\alpha$, which (by the binomial distribution) can have 252 possible arrangements. In figure 7.7, the number of different (unique) folds observed for this form is plotted against the number of different secondary structure orders observed. When these are equal, each SSE order corresponds to a unique fold and when they diverge, many folds are seen for a single SSE order (for some the number of possible arrangements is also plotted).

This simple analysis shows that for the smallest forms, all linear arrangements of secondary structure have been observed and there is typically twice as many folds as secondary structure arrangements. With over ten secondary structures (around 35 in the ranked forms in figure 7.7), this balance changes and almost every secondary structure arrangement corresponds to a unique fold. This suggests that with these larger $\beta\alpha$ proteins, sequence-structure matching (threading) methods (Jones *et al* 1992) need only concentrate on the correct prediction of secondary structure to find an unknown fold and not on their 3D interaction.

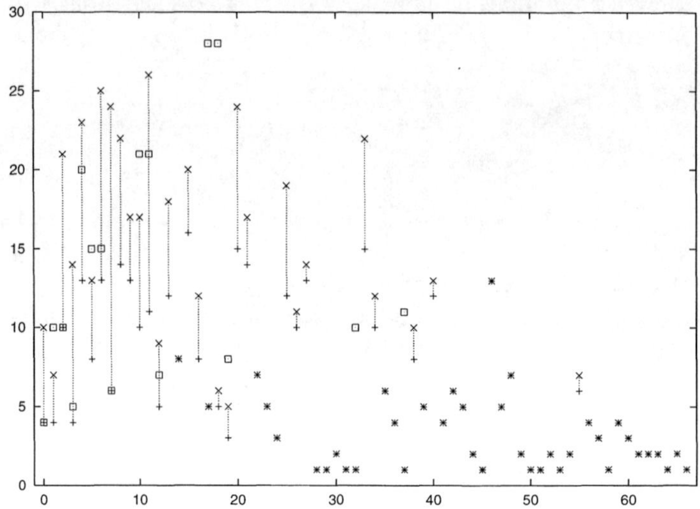

Figure 7.7. Fold numbers for ranked forms. The three-layer forms (figure 7.6) were ranked by the number of secondary structures they contain (X-axis) (Those of equal size were subordered by their number of β then α elements.) Against this is plotted the number of different secondary structure arrangements seen for each form (+) and the number of different folds (×). The number of observed arrangements and the number of folds are linked by a dashed line and when they are equal, appear as an asterisk (∗). For reference, the number of possible arrangements is plotted as a box (□) but this only appears for some of the smaller forms as, for most, it is a very large number.

7.3.4.2 Reduction to a common core

If a structure matches, say, a `2-5-3` (α-β-α) form then it will also match any substructure at least as well (`2-5-2`, `1-5-3`, `2-4-2` ...) and two proteins which have a different topology when matched against the `2-5-3` form might have the same core `2-4-2` topology. The question of whether two proteins have the same fold now becomes relative and should be posed as: 'What is the largest ideal form under which two proteins have the same fold?'. While the largest form with common topology will be of greatest interest, it is also informative to view all the submatches in the format of the grid-tables used in figure 7.5. Such a representation allows not only pairs, but also whole groups of proteins to be analysed and it is simple to determine both visually and automatically if they share a common core (figure 7.8).

For the larger proteins (e.g. figure 7.8(*c*)), the matched subforms are not unlike a plot of the stable isotopes of an element. In the latter, if the balance between protons and neutrons becomes too imbalanced, the isotope is unstable. So in proteins, if the balance of secondary structure elements between the layers becomes too imbalanced, then the structure is not likely to exist. In such a plot, each step through the grid from one protein to another (via submatches) represents a deletion or addition of a secondary structure element. This allows a path to be defined between any two structures in the same table (no matter how dissimilar) and a minimum path length to be computed. (This aspect will be returned to in section 7.4).

7.3.5 Expanding the classification tables

It is apparent from their arrangement in figure 7.3 that all the forms are interconnected. For example: there is little difference between a 'flat' sheet of five strands and five 'curled' strands taken from a nine-strand barrel, both with two helices on one side (`0-5-2` on `2-5.9` = 3 Å RMSD). Similarly, it takes only the deletion of a single strand in a 'broken' barrel to make a β-sandwich (`3-8.9` on `1-3+4-2` = 3.7 Å RMSD). The pathways of structural change discussed in the main text above are therefore not limited to one grid table (such as figure 7.6) but can link proteins across the full range of structure outlined in figure 7.3.

Each form also relates to figure 7.3 through the progressive filling of the layers with secondary structures (the periodic table analogy). While there is no strict rules, the forms can be arranged to minimize differences between adjacent entries. A possible arrangement is shown in figure 7.9 for the top row (I) of figure 7.3. (The cyclic order, however, has been changed to place the break between the `A-B+B-A` and `A-B-A` forms as this transition corresponds to a greater structural jump.) The minimum sheet size considered is three strands and the number of helices on a sheet cannot exceed half the number of strands in the sheet.

The decomposition of structures into their basic forms has opened a different approach to the classification and analysis of protein structure that is both flexible and automatic. Unlike clustering methods, it does not require the comparison of all structures to each other and can therefore be incrementally up-dated as new structures are determined. The major result from this approach is that it allows the difficult problem of protein topology to be rigorously addressed. The fold of any individual protein can be specified as that of the largest matching form, while for pairs or groups of proteins, the fold is that of the largest common form. This implies that the topology of a protein can only be defined under the specification of a given ideal form which, in turn, means that proteins do not have a unique intrinsic topology or a unique position in any classification based on topology.

In the following sections, we will explore the application of this approach to the definition of topological transitions between proteins and to the classification of rough model proteins that are poorly distinguished by an RMSd measure.

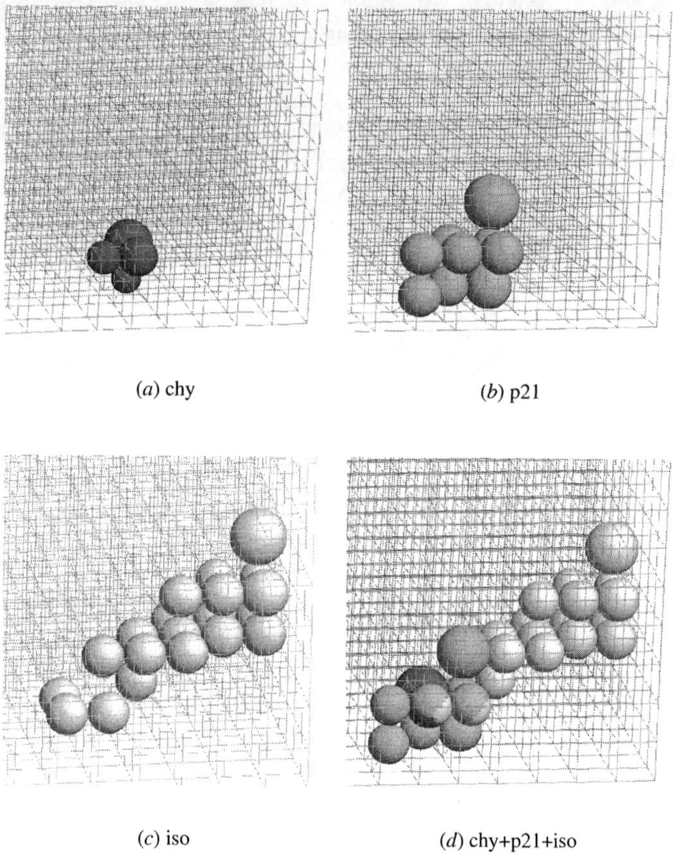

Figure 7.8. A common core from multiple matches. Three $\beta\alpha$ proteins were matched to the three-layer forms and all matches displayed on a grid as in figure 7.6 except that the size of the sphere here represents the strength of the match (not the number of proteins matched). The proteins (PDB codes) were (*a*) the small Che-Y protein (`3chy`), (*b*) Ras p12 (`5p21`) and (*c*) a large isocitrate dehydrogenase (`1iso`). In frame (*d*) the three sets of matches (*a*) (dark grey), (*b*) (light grey) and (*c*) (white) are replotted. A common match can be seen near the bottom left as a sphere divided into three shades. This represents a common core for these proteins with form: `2-4-2` and the same topology.

7.4 'Evolutionary' steps in fold space[4]

From the pioneering work of Ptitsyn and colleagues (Ptitsyn and Finkelstein 1980, Finkelstein and Ptitsyn 1987), it has been shown that the 3D arrangement of

[4] This section is reproduced in part from Johannissen and Taylor (2004) with the kind permission of Oxford University Press.

4(2) A-B+B-A	3(2) B+B-A	2(2) B+B	1(1) B	2(1) B-A	3(1) A-B-A
1-3+3-1	0-3+3-1	0-3+3-0	0-3-0	0-3-1	1-3-1
1-3+4-1	0-3+4-1	0-3+4-0	0-4-0	0-4-1	1-4-1
1-3+4-2	0-3+4-2			0-4-2	1-4-2
1-3+5-1	0-3+5-1	0-3+5-0	0-5-0	0-5-1	1-5-1
1-3+5-2	0-3+5-2			0-5-2	1-5-2
1-3+6-1	0-3+6-1	0-3+6-0	0-6-0	0-6-1	1-6-1
1-3+6-2	0-3+6-2			0-6-2	1-6-2
1-3+6-3	0-3+6-3			0-6-3	1-6-3
1-4+6-1	0-4+6-1	0-4+6-0			
1-4+6-2	0-4+6-2				
1-4+6-3	0-4+6-3				
2-4+6-1					
2-4+6-2					
2-4+6-3					
1-5+6-1	0-5+6-1	0-5+6-0			
1-5+6-2	0-5+6-2				
1-5+6-3	0-5+6-3				
2-5+6-1					
2-5+6-2					
2-5+6-3					
1-6+6-1	0-6+6-1	0-6+6-0			
1-6+6-2	0-6+6-2				
1-6+6-3	0-6+6-3				
2-6+6-1					
2-6+6-2					2-6-2
2-6+6-3					2-6-3
3-6+6-3					3-6-3
+	+	+	+	+	+

Figure 7.9. Ideal forms arranged as a periodic table. Each of the ideal forms derived from the 'flat' templates in figure 7.3 (row I) are arranged to minimize the change in secondary structure arrangement from one adjacent form to another. The gaps introduced while one layer exhausts its different combinations are not unlike those seen in the periodic table of elements as electrons fill successive orbitals.

helices and strands in larger proteins can be obtained from the stepwise addition of secondary structure elements (SSEs) to basic structural motifs (Efimov 1993, Efimov 1997). Whether this accretion of SSEs reflects either a possible folding pathway for the protein or an evolutionary history is debatable but irrespective of any of these rationalizations, it provides a valid approach for the classification of protein folds.

Using this approach, the protein folds become organized into a phylogenetic tree (which may include 'missing-links'). Unlike the methods discussed in section 7.1 that clustered by similarity, a fold tree can be arbitrarily deep, so relating the most dissimilar folds. A disadvantage, however, is that the construction of the trees is a manual operation that embodies an implicit set of assumptions and rules that are only to varying degrees stated explicitly.

7.4.1 Matching ideal forms

The organization of known structure based on their ideal forms embodies many of the principles discussed above for the alternative approaches: for example; the identification of the largest common form shared between two proteins corresponds to their largest isomorphic subgraph in the graph-based methods (section 6.1.7), while steps within the table are equivalent to the addition of secondary structures onto a core structure. The 'growth' of a large structure from a simple core can then be viewed as a pathway through the table of forms (ToF) (figure 7.8) and a distance metric between structures can be established as the shortest path (or edit distance) between the structures.

In this section, the connections between the Evolutionary approach to protein fold analysis and pathways through the table of forms are explored. The aim of this investigation is to develop both a metric of structure similarity based on an edit–distance and also to establish in a rigorous way, the set of rules that can be used to constrain steps through the ToF.

This section concentrates on the three-layer $\alpha\beta\alpha$ structures which provide a rich collection of structures that have been extensively analysed by Efimov.

7.4.2 Largest common fold

7.4.2.1 Topology strings

As we have seen in section 3.2.6.3 and the previous section, the encoding of protein architectures as layers of secondary structure allows the fold of the chain to be described as a series of steps between the layers. In the three-layer $\alpha\beta\alpha$ architecture, each layer can be designated by the letters A, B and C (respectively) which together with a number encoding relative position in the layer (signed by orientation) specifies the fold of the chain (see the examples shown in figure 3.17).

For our current task of measuring the similarity of two folds, this simple encoding embodies all the necessary information. However, as noted in section 3.2.8.1, the scheme is sensitive to the orientation of the first element and if two structures are identical except for the addition of an extra strand on the amino terminus of one, then the complete string will be changed. The simple solution to this problem, developed later, is to 'flip' the orientation of the strings and compare both flipped and original forms with each other.

7.4.2.2 Dynamic programming solution

Given two topology strings, the problem of finding the largest common fold can be solved using a variation of the dynamic programming algorithm commonly used to align sequences. In this initial consideration of the problem, we will assume that the assignment of layers and the orientation of the first strand is the same in both strings.

Given the topology strings corresponding to the two proteins shown in figure 3.17:

```
+B0.-A0.+B-2.-C0.+B-1.-C1.+B1.-C2.+B2
```

and

```
+B0.-A0.+B-1.-C0.+B2.-C1.+B3.-A1.+B1
```

a matrix of scores is calculated by accumulating a positive score (+3) for a match in SSE position and orientation relative in the last SSE in the same layer and a negative score (-1) for a mismatch. Taking the two example topology strings (figure 7.10), the first strands match and score three, as do the following helices (giving six), the next pair of strands have the same orientation but differ in their displacement from the first strand, so they mismatch, reducing the score to five. Following this algorithm to completion identifies the highest scoring pathway through the matrix which corresponds to the elements that match in the common core (see figure 7.10 for further details).

7.4.2.3 Substructure matching

The basic algorithm described above is limited by the requirement that SSE matches are dependent on their orientation and displacement relative in the first instance of an SSE in the same layer and so cannot be used to find substring matches (corresponding to local alignments in sequence matching). However, following the use of dynamic programming in 3D structure matching (double dynamic programming; section 6.1.6.1 and figure 6.2) (Taylor and Orengo 1989b), the common core can be calculated for each 'enforced' matching of all pairs of SSEs (of like type). It may now be necessary to 'rotate' the subfragments into the same orientation but this can be easily done typographically: a 180° rotation in the plane of the topological diagram requires only the interchange of A↔C characters combined with the negation of all position values, while a flip in up/down orientation requires the negation of the orientation sign combined with either of the preceding transformations. As the orientation is set by the forced match of the initial elements in the substrings, only one alternative string needs to be tested and the better match of the two retained as the solution.

7.4.2.4 Additional constraints

To produce biologically meaningful results, the algorithm was modified so that solutions that contained a gap in a β-sheet were discarded. This implies that the

(a) SCORE MATRIX

protein A

protein B	+B0	-A0	+B-2	-C0	+B-1	-C1	+B1	-C2	+B2
+B0	*3	0	1	0	1	0	1	0	1
-A0	0	*6	3	0	0	0	0	0	0
+B-1	1	3	5	4	*7	4	1	0	1
-C0	0	0	4	7	4	*10	7	8	5
+B2	1	0	3	4	6	7	9	8	11
-C1	0	0	2	5	5	8	8	*11	8
+B3	1	0	1	2	4	5	7	8	10
-A1	0	3	0	0	3	2	6	5	9
+B1	1	0	5	2	2	1	9	6	8

(b) ALIGNMENT

```
  protein A     +B0 -A0 +B-2 -C0 +B-1 -C1 +B1  -  -C2 +B2
                 |   |   |    |   |    |   |   |   |
summed score     3   6   5    4   7    10  9   8   11
                 |   |   |    |   |    |   |   |   |
  protein B     +B0 -A0  -    -  +B-1 -C0  -  +B2 -C1 -...
```

Figure 7.10. Example topology string alignment. The basic algorithm to align topology strings is illustrated for two proteins 'A' and 'B'. (*a*) Their strings are arranged to form a score matrix which is filled with values beginning at the top left. Matching SSEs score 3 and mismatches score -1. The scores are accumulated towards the lower right with each new score taking the highest score from the adjacent cells of the submatrix to its top left (of which it forms the bottom right corner). After a match ($*$) only a diagonal transition is allowed. A bias of 1 is added for matching β-strands when they have different relative positions in the sheet and a penalty of -3 is given to cross-layer mismatches. Since the match is dependent on the relative position to the first match, each initial pairing of SSEs is made in turn for both interchanges of the A and C α-layers see text for details. (*b*) The resulting alignment.

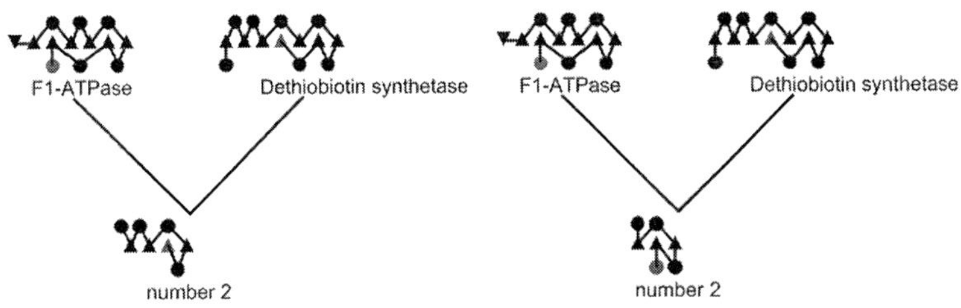

(*a*) global (*b*) local

Figure 7.11. Local *versus* global algorithm. The difference in results between the two alternative algorithms is illustrated. (*a*) The global algorithm allows the deletion/insertion of SSEs at both edges of a domain, while (*b*) the local algorithm allows deletion/insertion only at the termini of the chain. This corresponds to the longest common substring in the topology strings (as defined by structure, not character matching). Deleted portions in both structures are shown in grey. A triangle represents a β-strand while a circle represents an α-helix.

algorithm will not skip over core β-strands within a substructure but these are still able to be discarded from the edge of a sheet as mismatches in the normal way.

The solution to a single comparison is not necessarily unique and it can be seen from the example in figure 7.10 that there are two equally high scoring paths (one substitutes a final helix in the match for a strand). In this situation, a decision was made to take the solution with most β-strands.

An option was provided to report the largest common substructure that does not contain any internal mismatches (corresponding to a strict subdiagonal in the score matrix). This constrains solutions to be derived only by deletions from the termini of the chains. By analogy with sequence alignment, the two variants are referred to as 'global' and 'local', with the latter being the more restrictive. The different behaviour of this option is shown in figure 7.11. It was considered that the option that allows general deletion of structure was more biologically realistic and the results presented below were all generated with this option.

7.4.3 Trees of structures

Proteins can be clustered according to the common substructures that they share and the largest common substructure was chosen as the next node in the tree. This new node is added to the pool of structures, and the two larger structures are removed. The common substructure of two proteins is then represented at their joining node as an 'ancestral' structure, unless this also corresponds with one of the structures (i.e. the smaller structure is contained within the larger), in which case the smaller structure itself occupies the 'ancestral' node.

The resulting tree (figure 7.12) was compared to a simplified representation of the tree constructed by Efimov (1997) (figure 7.13). The majority of the tree structure is isomorphic with that of Efimov, including the relationship of the thiolase (1) and adenylate kinase (3) branches. A minor rearrangement occurs among the structures 12 (PDC), 13 (HGPRT) and 14 (POx). Using our automatic method, these join at a common node whereas Efimov links 12 to 14. while this may seem like a trivial change, the connection between 12 and 14 involves the 'deletion' of an internal (buried) β-strand from the sheet (with the subsequent closure and reformation of an intact sheet). This is an operation that was specifically disallowed by our algorithms but has been allowed on this occasion by Efimov who allows some flexibility in the application of his rules. Specifically, he states: *'a structure obtained in the preceding step is maintained [but] it can be slightly modified'* (Efimov 1997).

The largest rearrangement, however, occurs between the two branches carrying Subtilisin (10) and PFK (5). Efimov links Subtilisin (10) by a long branch of seven 'ancestral' structures back to a point where there is a common node with our automatically generated tree. Similarly, DBS (9) is linked back through five 'ancestral' structures to PFK (5). By contrast, these two structures are linked by a common node ('number 17') in our construction (figure 7.12). This association involves the loss of two helices and an edge strand from DBS to recreate the unusual left handed $\beta\alpha\beta$ unit found at the N-termini edge of the Subtilisin sheet. Although this transition can be accomplished in three steps it is not an obvious route to take as it involves inserting/deleting a helix that lies between two existing helices.

While the automatically generated trees are in broad agreement with the Efimov tree, there are details, involving the desirability of making particular insertions/deletions in core positions that need to be further examined. If it is assumed that Efimov provides a 'gold-standard' for the relationships between structures, it would now be possible to modify the algorithm (adding further constraints or relaxing existing ones) to optimize the behaviour of the algorithm on its ability to regenerate the trees of Efimov. Rather than pursue this route, a more fundamental investigation would involve an attempt to derive the underlying constraints from the data. Using the automated approach described here, it will be possible to calculate trees of structures rapidly for each formulation of the rules, so allowing these to be varied until the most parsimonious tree is obtained.

The primary application of this approach is to impose a minimal hierarchic description on the relationship of protein folds in a way that clearly states the assumptions and 'rules' that have been applied. While the resulting classification is, in itself of value, it is also interesting to speculate whether the relationships might have resulted from a corresponding series of evolutionary events. If this were so then the 'ancestral' nodes that have no equivalent known structure might correspond to relic structures that are yet to be discovered. Alternatively, it might be postulated that the resulting order reflects the constraints of similar folding pathways—with the accretion of secondary structure imitating the assembly of the

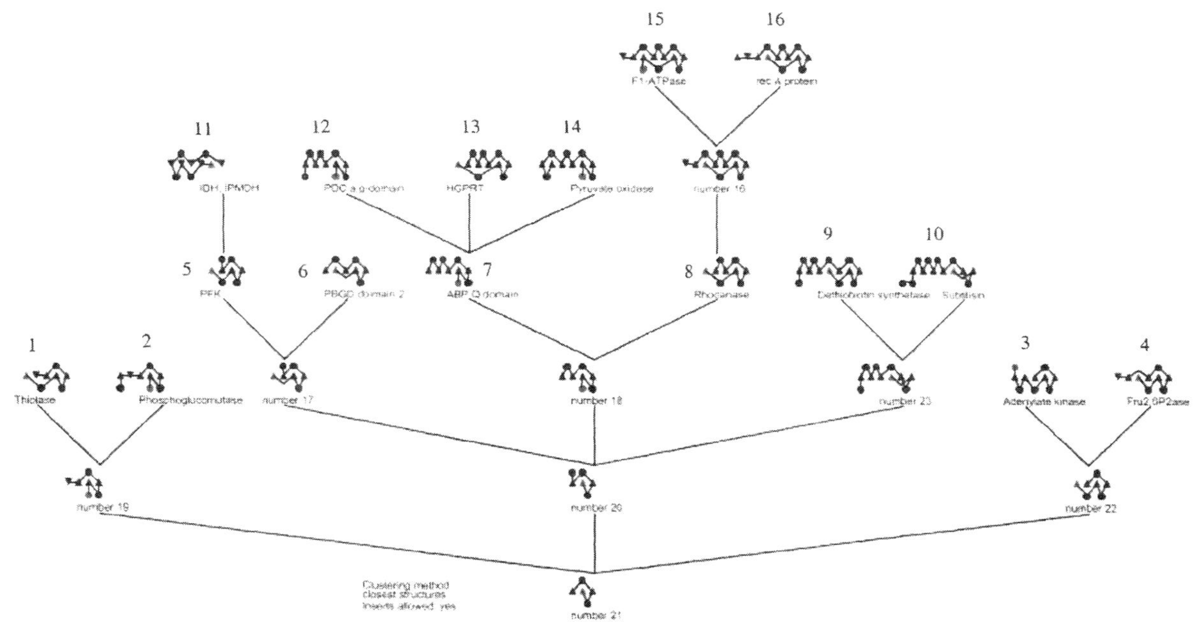

Figure 7.12. LCS clustering—distance-based algorithm. The dendrogram shows the clustering of the 16 protein structures using the refined clustering algorithm operating in the distance of each structure from their their largest common substructure (LCS) with reclustering after each pair of structures have been replaced with their LCS. 'Ancestral' or branching node structures that do not correspond to a known structure are given an arbitrary number. (As these are automatically generated, they do not correspond to those in figure 7.12.)

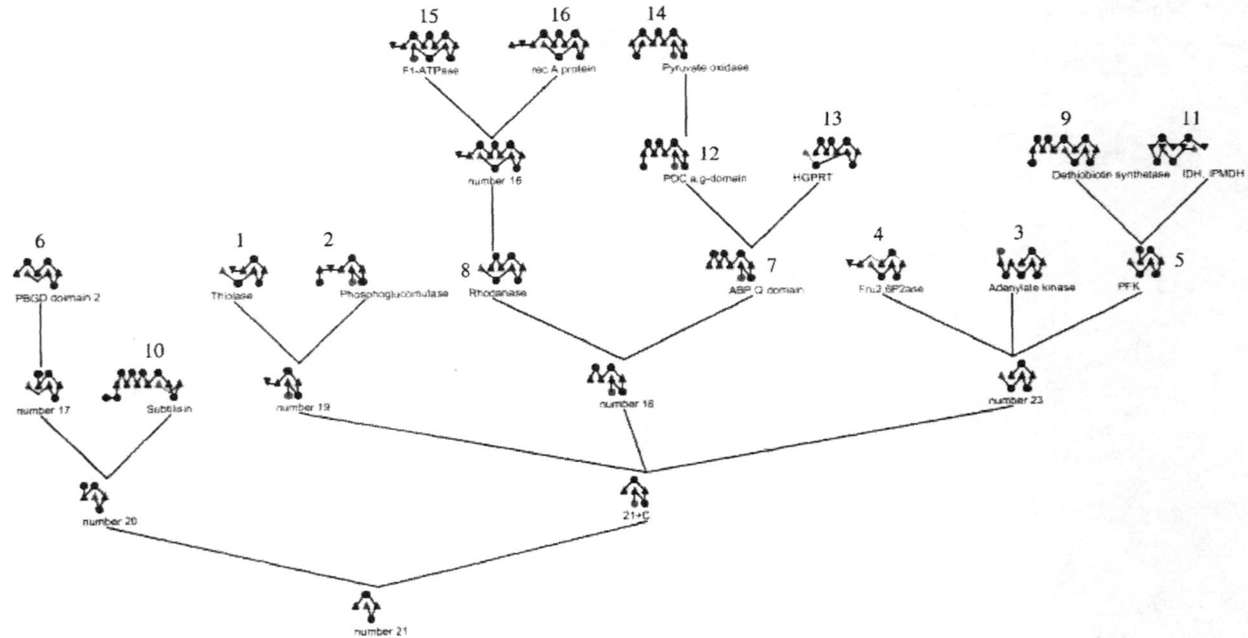

Figure 7.13. Efimov tree. The major part of the β/α tree constructed by Efimov (1997 table 5) is shown in simplified form in which each of the 'ancestral' structures that do not constitute a branching node have been removed.

protein structure as it folds (as originally postulated by Ptitsyn and colleagues). These two options may even be simultaneously true: in the way in which ontogeny recapitulates phylogeny in embryonic development, so the protein folding pathway might equally recapitulate its evolutionary history.

For the further application of the method to a large set of protein structures, the problem of secondary structure definition will need to be addressed. For this, a method that is insensitive to the fine details of hydrogen bonding is required such as the line segment based approach (Taylor 2001) (See chapter 3). More importantly, for the method to become fully automatic, it will also be necessary to have a robust method to define protein topology. As discussed above, the matching of ideal forms can be used for this (Taylor 2002a) but this introduces the added complexity that there is not always a unique match of a form to a structure. To overcome this problem, the current method has the potential to be extended to consider alternative topology strings derived from multiple forms. Thus, rather than force the choice of a 'best' match for each protein, the matches that give rise to the best path can be selected.

Rather than have an expert pick a 'best' pathway, an alternative is to consider all possible pathways. This is similar to the calculation of quantum phenomena that are (ideally) made over all possible intermediate states. The physical cost of each step would need to be assessed (for example; it should be easier to delete a β-strand on the edge of a sheet compared to the middle). Rather than complex energy functions, these costs could be weights (or penalties, like the gap penalty in sequence alignment) that could be optimized to produce the shortest sum of paths needed to connect all proteins. Such an approach combines the best of both the alternatives considered above. Superfluous domains or subdomains would be removed through a series of deletions to focus on a core structure, while the fold of the protein would be retained as the basis of the metric.

7.4.4 Links and islands in fold space

What would the resulting foldspace be like if these calculations could be made? There are branches in this fold space such as the all-α and all-β classes, that can be connected only by deleting all α-helices and re-growing all β-strands. However, these proteins were probably never related by evolution anyway, emerging independently from the precellular soup. The structure trees within a class is in reasonable agreement with the trees of Efimov but the interesting parts are the intermediate structures along the pathways that have no equivalents among the known structures. These missing folds might soon be revealed by the structural genomics programmes, or perhaps they (and many others) have become extinct? It is also possible that some folds might have arisen spontaneously, say, through a frame shift or the mutation of a STOP codon (Taylor 1997a). In the current scheme, these would appear as fold islands with a long connection (of many steps) to other parts. Other fold islands might have arisen through more complex genetic events, the most common of which would be full or partial gene duplication combined with deletion.

To incorporate these more complex transitions would complicate any model to the extent that the available protein data would be insufficient to discriminate between possibilities. We can, of course, wait for the various structural genomics programmes to complete but, from the genome sequence, we know that the data from these is finite and when reduced to a non-redundant set (with no homologies), may be quite limited in the number of folds.

Kendrew concluded his Nobel lecture with the vision that, with two globin structures, 'we have merely sighted the shore of a vast continent waiting to be explored'. Now, we must wonder whether this continent will be big enough to allow any firm conclusions to be drawn on the origin and evolution of protein folds. So even after all the structural genomics programs come to an end, we may still need to send out some space probes.

PART 3

TOPOLOGY

'Beauty is truth, truth beauty,'—that is all Ye know on earth, and all ye need to know.

John Keats, from '*Ode on a recian Urn*'.

Chapter 8

Folds, tangles and knots

In the previous chapters there has been much discussion of protein topology. Although it was declared early on that the term 'topology' was used somewhat loosely, it is now time to address the problem with a little more rigour. This will allow us to describe new ideas and tools that can then be applied to the further analysis of protein folds and fold spaces.

8.1 Topology and knots

8.1.1 Introduction

Although topology is a highly complex and abstract branch of mathematics, its roots can be traced back to simple practical problems. Knot theory, in particular, started as a subfield of applied mathematics. The first scientific application of knot theory was Gauss' work on computing the inductance of a system of linked circular wires, and Listing, who was a student of Gauss, coined the term *topology*. Since then, topological considerations have often played a role in theoretical problems in physics. For example; when studying the hydrodynamics of perfect fluids, Helmholtz proved that a vortex tube (a solid torus in the flow), once created, would persist in the flow forever. While his theorem illustrates the beauty and usefulness of topology in capturing the invariances in physical problems, they probably also induced Rutherford to postulate that knotted vortices in the æther might explain the different elements. Although, not supported by experiment, this intriguing theory lived long enough to give a major boost to knot theory.

As we have seen often in the preceding sections, the word 'topology' is applied to the description of the various features in the structural hierarchy within protein molecules, from the connection patterns between secondary structure elements to the overall fold of the protein. In this section, however, we discriminate between the 'true' topological features of proteins in the strict mathematical sense (such as intrinsic chain topology, the presence of knots and links) and the qualitative (and ill-defined) concept of the spatial arrangement of chain segments which we shall call the *fold* of the chain. This is important since in the absence of intrachain cross-links all polypeptide chains share the same intrinsic topology, namely that of the straight line segment, and are therefore indistinguishable from each other in the strict topological sense.

Before turning to proteins, we will briefly review the terminology and application of topological ideas in chemistry, giving a more general background from which applications to proteins might arise.

8.1.2 Chemical topology

Geometric considerations have been playing an increasingly important role in chemistry since van't Hoff postulated the tetrahedral geometry of carbon atoms in organic compounds. In fact, the development of organic chemistry provided a seemingly limitless variety of molecular shapes, the understanding of which would not be possible without the tools of topology.

Molecular structures may be regarded as graphs, where the atoms are the vertices of the graph and the edges correspond to the bonds between the atoms. Chemical graph topology has proved very useful in formalizing the hitherto qualitative concepts of 'molecular similarity' and 'molecular shape'. Similarity of structures can be characterized through subgraph isomorphism matching, a technique which enables the identification of common structural motifs within molecules (see section 6.1.7 for application to protein structure comparison). The shapes of molecules can be described by various topological invariants, i.e. mappings which assign (real) numbers to graphs. Topological invariants have been used for automatic compound cataloguing and retrieval, for predicting physicochemical properties and in quantitative structure-activity relationship (QSAR) studies.

Despite the variety of organic compounds, the overwhelming majority of them can be described by simple acyclic graphs (trees) or graphs containing a few cycles. Knots and links have not been observed and their synthesis proved difficult. The first interlocked organic molecules were synthesized as late as 1960 by Wasserman, who named them catenanes, from the Latin word *catena* (chain). These compounds contained a novel type of chemical 'bond', the *topological bond*, since they were held together by the topological arrangement of their constituent atoms, rather than by direct interatomic interactions.

8.1.3 Polymer topology

Natural and synthetic polymer molecules introduce an additional layer of complexity of structure which brings us closer to potential applications to protein structure. When studying the topological properties of polymers, it is often convenient to distinguish between the *intrinsic* topology and the *spatial embedding* of the structure. The intrinsic topology of the molecule is determined by the (covalent) connectivity graph of the constituent atoms, whereas the spatial embedding corresponds to the conformation of the molecule as described by the coordinates of the atoms. For example, all circular polymers have the intrinsic topology of a closed circle, but the spatial embedding of an unknotted circle is different from that of a knotted one. Conformational changes which do not require the making and/or breaking of chemical bonds are considered topologically equivalent, in line with the conventional definition of topological transformations which allow continuous deformations but no 'cut-and-paste' operations.

The intrinsic topologies of polymers can be divided into a small number of major structural classes which will be discussed below. It must be noted, however, that the topology of a given molecule depends on the definition of the underlying molecular graph. In the following, we will investigate polymers at 'low resolution', by constructing molecular graphs where the nodes correspond to monomers and the arcs to bonds between monomers, thus ignoring the details of the arrangement of atoms within monomers. In some biopolymers, weaker interactions such as H-bonds often play a crucial role in structure formation; therefore, a distinction shall be made between covalent and non-covalent topologies.

8.1.3.1 Bond direction

In some polymers, including proteins, it is possible to assign a direction to the bonds linking the monomers. For example, in polypeptides the—NH_2 groups of the amino acid monomers form bonds with the—COOH groups and therefore each peptide bond has an amino $\rightarrow$ carboxy direction (N $\rightarrow$ C for short). Such polymers can be represented by directed graphs in which the arcs have 'polarities'.

8.1.3.2 Linear polymers

The spatial embedding of all linear polymers are topologically equivalent since even the most tangled conformations can be transformed into a straight line by pulling the chain at one end until the whole string 'flows' out smoothly. This theoretical assertion sometimes seems to contradict sharply with the practical experience concerning 'knots' on ropes and tangled telephone cords, as well as folded polypeptide chains, which at first sight do not resemble straight lines at all.

The apparent inadequacy of the topological approach to describe these situations (which are directly related to the application to protein structure) can be rationalized by observing that topology concerns itself with the existence of transformations which do not change abstract properties: while the nature of the physical forces determining the conformation of a protein or a telephone cord influences the probability with which these transformations occur. However, as we shall see below, polymers with linear covalent connectivities often exhibit more complex intrinsic topologies when weaker intermonomer interactions are taken into account, thus enabling the construction of non-trivial topological models.

8.1.3.3 Branching polymers

The connectivity graphs of branching polymers are trees, i.e. acyclic graphs in which there exists only one path between any two nodes. Branching polymers can also be directed if the linear branches are made up by 'head-to-tail' polymerization. At branching points, the monomers should be at least trifunctional, which is the most common case. Similarly to linear polymers, branched polymers cannot have knots or links. Natural branched polymers can be found among polysaccharides, the properties of which can be manipulated by controlling the degree of branching during synthesis.

8.1.3.4 Circular polymers

Circular polymers, which have the intrinsic topology of a closed loop, are particularly interesting because they can be embedded into space as knots. Also, two or more loops can be linked, giving rise to an additional topological variety. The most important circular polymers can be found among nucleic acids. In particular, the study of topological transformations of double-stranded circular DNA molecules initiated the development of the whole field of biochemical topology (Cozzarelli and Wang 1990).

8.1.4 True topology of proteins

8.1.4.1 Disulfide bridges

The sulfhydryl groups in the cysteine side chains can form disulfide bridges in an oxidative reaction. As opposed to peptide bonds, the disulfide bridges are symmetrical and therefore the covalent connectivity graph of a polypeptide with disulfide bonds can be represented by a partially directed graph. The closure of disulfide bonds creates cycles in the connectivity graph and can generate complex embedding topologies. Although the majority of such bonds form simple local connections in the sequence (Thornton 1981) the possibility of interesting topologies has been a topic of study and speculation since the earliest days of structural work on proteins (Kauzmann 1959, Sela and Lifson 1959).

Crippen (1974, 1975) analysed the chances of finding a knotted topology in protein chains that had been cross-linked by disulfide bridges. He simulated protein folds of different lengths as a random self-avoiding walk on a cubic lattice and then counted the knots formed. This was done in a largely automated method using an approach similar to Reidemeister moves (Adams 1994) to reduce the complexity of the 2D projection. The chance of a knot being formed was low, at around 3% for a protein of length 128 residues but none were seen in the few multiple disulfide linked structures known at the time (Crippen 1974). This work was further extended through simulations that incorporated the sequence (cysteine positions) of the known proteins but these more realistic simulations again suggested that proteins appeared to be 'avoiding' knotted topologies. Probably, it was speculated, for entropic reasons (Crippen 1975).

On a more symbolic level, Klapper and Klapper (1980) analysed the chance of obtaining a non-planar graph in the disulfide bonded protein chain. This is a graph that cannot be drawn in 2D and is the minimal requirement for what would be considered a knotted configuration (although the Klappers used the less restrictive term of 'loop penetration'). The chance of obtaining a non-planar graph clearly increased with the number of disulfides and again the results suggested a greater chance of non-planar topologies than was later found in known protein structures. Their approach had the advantage that the disulfide bonding pattern can be known from chemical sequencing studies without having the full 3D atomic structure. However, while one case was substantiated by the 3D structure (scorpion neurotoxin) their prediction for a knot in colipase was not found in the 3D structure (implying an error in the chemical bond assignment). This was later analysed more fully by Mao (1993) along with the addition of another example in the light chain of the protein methylamine dehydrogenase.

The number of possible disulfide bonding arrangements in a polypeptide chain can be determined from the following formula:

$$\alpha(M, n) = {}_{M}C_{2n} P(n) = \frac{M!}{2^n n!(M - 2n)!} \qquad M \leq 2n \tag{8.1}$$

where n is the number of disulfide bonds and M is the number of cysteines in the chain (Sela and Lifson 1959). Within these patterns, Benham and Jafri (1993) defined the special cases of *symmetric patterns* and *reducible patterns*. A pattern is symmetric if its mirror image (with the backbone direction reversed) has the same disulfide connections as the original, and reducible if it gives rise to two separate non-trivial subpatterns when cut once somewhere along the backbone. The same authors also carried out a statistical survey of the structure database to assess the probabilities with which the various subpatterns occur. Symmetric and reducible patterns were observed with a much higher frequency than which was expected from theoretical studies of random disulfide bond formation (Kauzmann 1959, Crippen 1974). However, the limited size and the bias of the database did not allow for an analysis of statistical significance.

The non-trivial intrinsic covalent topologies generated by disulfide bonds may give rise to various interesting embeddings (knots and links). However, neither true knots nor links were found in database searches (Benham and Jafri 1993), indicating that non-trivial disulfide bond topologies must be extremely rare if not absent among native proteins. The absence of true links in which the loops share no common backbone segment is all the more puzzling because pseudo-links, i.e. interpenetrations of chain segments in which the loops formed by disulfide bonds share common parts of the backbone, have indeed been observed in proteins (Klapper and Klapper 1980, Kikuchi *et al* 1986, Mao 1989, Le Nguyen *et al* 1990). However, pseudolinks are topologically not equivalent to true links as can be shown by suitable continuous deformations, and their linking number is zero.

From Crippen's work, the probability of a disulfide loop participating in a true link was about 0.15. This means that well over 250 true links could be expected to occur in a database containing 2487 disjoint disulfide loops; however, none were found Benham and Jafri (1993). This absence of true links is very unlikely to have happened by chance since the proportion of reducible bond patterns[1] is larger than that was expected from probabilistic considerations. Knots were also absent from the database, although Crippen's model estimated a 4% probability for knot formation in average proteins and the probability was found to increase with the chain length. These observations suggest that some feature of protein folding works against the formation of non-trivial topologies. It is sometimes argued that loop penetration is hindered by stereochemical constraints in polypeptides; however, penetration is not a prerequisite of disulfide knot formation since these can be constructed by appropriately twisting hairpin loops and then linking them together. If protein folding occurs in a hierarchical fashion, with small local regions of the chain folding first and then these regions packing together, coupled with disulfide bond formation at the early stages (and consequently restricted to happen within the local folding units), then the relative abundance of reducible disulfide patterns and the scarcity of knots and true links could be explained. However, neither the current theoretical knowledge nor the available experimental information is sufficient to decide the correctness of this assumption.

8.1.4.2 Other cross-links

There is a very wide variety of post-translational modifications made to proteins and many of these introduce cross-links, either through direct enzymatic modification of the protein itself, or through the binding of metals and other cofactors (see Kyte (1995) for details). Many of these modifications link two sites on the protein and so open the possibility for the creation of linked loops and knots. A wide variety of these have been analysed by Liang and Mislow (1994a, b, 1995) .

8.1.5 Pseudo-topology of proteins

Without covalent cross-linking, the formal topological analysis of proteins is greatly limited. Some further progress can be made, however, if the strict covalent bonding criterion for graph connectivity is relaxed. This can be progressed in two directions: either by considering weaker bonds, such as hydrogen bonds as valid links, or more simply, by joining the two ends of the protein chain to form a circle.

[1] These can be considered a prerequisite for link formation but to be precise, the two loops that link do have to be disjoint, since there could be other loops spanning the interval between them and this arrangement could form a true link without being reducible.

8.1.5.1 Topology of 'circular' proteins

Given a piece of string, it can usually be decided by pulling the ends whether it is knotted or not. Since we hold the ends, the string plus body combination forms a closed circle and there is no danger of untying the knot as it is pulled. One way to approach the problem of defining knots in proteins is simply to join the ends (as we do when we pick up a string). This is trivial for knots where the ends of the string are remote from the knot site—but if the ends are tangled-up together with the knot then any algorithm devised to 'pick-up' the ends creates the risk that the external connections might either untie an existing knot or create a new one. Fortunately, for proteins, the ends of their chains (being charged) tend to lie on the surface of the structure (Thornton and Sibanda 1983) and so can often be joined unambiguously by a wide loop. Usually, this was done by extending the termini to 'infinity' in a direction away from the centre of mass but the closer the termini lie to the centre of the protein, then the more arbitrary this direction will become.

With the two ends of a protein chain joined, the resulting circle can then be analysed using 'proper' knot theory. This approach was originally based on representing the crossovers in a 2D projection of the protein in a matrix. For example; if each section between crossings (specifically just under-crossings) is given an index, then for each crossing, we have a pair of indices and the type of crossing (effectively, left or right handed) can be entered into a matrix. The properties of such a matrix were analysed by Alexander who found that a polynomial of the matrix captured an invariant property that corresponded to its state of knotting. This was not a unique mapping as some knots could not be distinguished, but with further refinements, the distinction of knots was improved. Further progress came largely from the work of Vaughan Jones, who recast the problem as a series of 'edit-operations' on the knot (called skein moves), that gradually reduce the knot to a trivial form. These are 'recorded' in an algebraic way and also generate an answer in the form of a polynomial. The current and most powerful refinement of this approach is referred to as the HOMFLY polynomial—after the initials of the authors who developed it. (See Adams (1994) for a more complete history.)

Unlike DNA, protein chains are very short (relative to their bulk) and the range of features cannot be expected to be very great. Rather than finding complex linked chains or different knot topologies (as in DNA), it is rare to find a protein chain that can even be considered as a knot. Until recently, the few folds that were reported to be knotted (without considering post-translational cross-links) have one end of the chain barely extending through a loop by a few residues and all of these form simple trefoil knots (Mansfield 1994, Mansfield 1997).

8.1.5.2 'Topology' of open chains[2]

A way to avoid the unsatisfactory step of projecting the termini of the protein chain to 'infinity', is to reverse the operation and shrink the rest of the protein. This can be done gradually through repeated local averaging: in a chain of length N consisting of a set of coordinate vectors a $(a_1, a_2, \ldots a_N)$ representing the α-carbon of each residue, each position a_i can be replaced by the average of itself and its two neighbours;

$$a_i^{t+1} = (a_{i-1}^t + a_i^t + a_{i+1}^t)/3 \qquad \forall i,\ 1 > i < N \tag{8.2}$$

where t marks the time step in the iteration. To avoid the chain passing through itself (an undesirable property for topological analysis), each move $(a_i^t \rightarrow a_i^{t+1})$ was checked to ensure that the two triangles formed by the points $\{a_{i-1}^t, a_i^t, a_i^{t+1}\}$ and $\{a_{i+1}^t, a_i^t, a_i^{t+1}\}$ were not intersected by any other line segment in the chain. If they were, then the new position (a_i^{t+1}) was not accepted.

Repeated application of this smoothing function eventually shifts all residues towards the line connecting the two termini—unless there is a 'knot' in the chain as this cannot be smoothed away. In theory, this simple algorithm is sufficient to detect knots in an open chain (and is equivalent to what happens in 'real-life' when we pull a string tight) but, just as in 'real' life, the resulting knots end-up very small. Indeed, in practice, the knots can become so small that the numerical accuracy of the computer is insufficient to perform the necessary topological checks and, in a numerical equivalent of quantum tunnelling, the knots become undone. This was avoided by representing each line between residues by a tube 0.5 Å in radius.

In practice, the test for collinearity was not made at the end but an equivalent test was made to every triple of consecutive points as the smoothing progressed. When three points were close to colinear (their cosine was less than −0.99) then the middle point was removed (providing the thin triangle formed by the three points was not intersected by any other line). In addition, where the outer two came very close (specifically, fell within the tube diameter) then the middle point was also removed. This not only improved execution time but led to an even simpler test for knots as any chain that can be reduced to just its two termini is not knotted. Chains with more than two residues remaining are either knots or tangles in which a group of moves have become 'gridlocked' (like 'rush-hour' traffic at an intersection). This latter condition was eased (but not completely eliminated) by making a slight reduction in the tube diameter any time the chain became stuck. Most chains of a few hundred residues are reduced to their termini in around 50 iterations. If by 500 iterations a chain was still not reduced to two points, then the resulting configuration was analysed in more detail (figure 8.2).

[2] This section is reproduced in part from Taylor (2000b) with the kind permission of Nature Publishing.

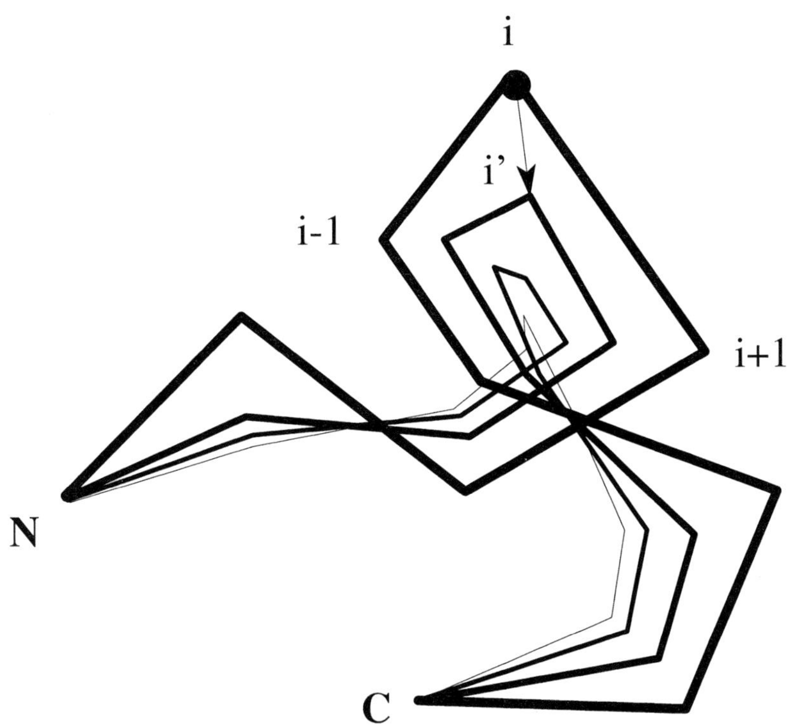

Figure 8.1. The basic chain smoothing algorithm. Protein chains are drawn schematically as lines connecting the central carbon atom in the backbone of each residue unit running from the amino (N) terminus to the carboxy (C) terminus. Beginning at the second residue, for each residue point (i) in the starting conformation, the average coordinate of i, $i-1$ and $i+1$ was taken as the new position (i') for the residue. This procedure was then repeated, and the results of this are progressively smoother chains, shown as a series of fainter lines. Note that the termini do not move. With each move, it was checked that the chains did not pass through each other. This was implemented by checking that the triangles $\{i'-1, i, i'\}$ and $\{i, i', i+1\}$ (broken lines in the figure) did not intersect any line segment $\{j'-1, j'\}$ ($j < i$) before the move point or any line $\{j, j+1\}$ ($j > i$) following.

Importantly for proteins, the algorithm is not sensitive to the direction of projection of the termini and can therefore be used to define the exact region of the chain that gives rise to the knot. This allows knots in proteins to be characterized by how deep they lie: specifically, the number of residues that must be removed from each end before they become free.

As the termini are now well separated from the knot site, they can be unambiguously joined and analysed as a 'proper' circular knot. This might be done using one of the knot-invariant polynomials (discussed above). However, the few knots encountered in proteins are so simple that they do not require any

(*a*) Unknotted (*b*) Knotted

Figure 8.2. Smoothed protein structures. Applying the smoothing algorithm described in the text (also figure 8.1) to protein structures produces a series of increasingly smoothed chains, shaded from dark to light grey in the figures. (For clarity, the native starting structure is not shown.) (*a*) Applied to a protein that has no knots (triosephosphate isomerase, [`1tph1`]) results in a straight line joining the termini. To reach this stage took 52 smoothing iterations. (*b*) Applied to the knotted protein (the carboxy-terminal domain of acetohydroxy acid isomeroreductase, [`1yveI`]), a straight line is never attained and a small knot remains deep in the core part of the protein. This is shown in isolation in figure 8.3(*a*).

sophisticated analysis and furthermore, from a theoretical perspective, not only are protein knots directional but also they have a unique break-point (between the termini) which is not taken into consideration by any of the polynomial forms. As a working tool, a simpler method was adopted to characterize these open knots based on the Dowker knot notation (Adams 1994). In this, each crossover in a 2D projection of the knot is characterized by its handedness. Beginning at the amino terminus, recording the handedness of successive crossovers as 1 or 0 generates a binary number which can be used as a reasonably unique descriptor for simple knots. To minimize the effects of projection, each knot was rotated around the axis defined by the two termini and the smallest numeric descriptor recorded.

Applying this method to a non-redundant selection of protein structures revealed a surprisingly large number of knots (Taylor 2000c). A few of these proved to be unresolved tangles, including slip knots, (both discussed further below) and some others were caused by breaks in the chain creating an unnatural short-cut. The former were all eliminated by running the program with a smaller 'tube' diameter but the latter could only be removed through visual inspection. Of the seven remaining structures, five were right-handed trefoil knots including related carbonic anhydrase structures (`1zncA 1kopA 1hcb 1dmxA`) and the protein S-adenoyslmethionine synthetase (`1fugA`) both of which had been identified previously. In addition three novel knots were found including a left-handed trefoil in ubiquitin (`1cmxA`) and two figure-of-eight knots (or Flemish knots) in a viral core protein (`2btvB`) and acetohydroxy acid isomeroreductase (`1yveI`) (figure 8.3(*a*)). These latter two are of particular interest as they include an additional crossover above the trefoil and are therefore less likely to be formed by a wandering chain during folding. This was confirmed by simulation of random and semi-random compact protein-like chains in which the trefoil was by far the most common knot type (section 8.1.5.1). The location of the two figure-of-eight knots was determined by a series of deletions from both termini of the protein chain. This revealed that the knot in `2btvB` was barely tied and was held by just the last eight residues. By contrast, the knot in `1yveI`, which is in the carboxy terminal domain of the protein, remained until 70 residues were deleted from the carboxy terminus and 245 residues (including a complete domain) were removed from the amino terminus (figure 8.3(*b*)).

8.1.5.3 A figure-of-eight knot in `1YVE`

Clear knots in the protein chain are rather rare and it is always of interest to examine closely those that are identified. Sometimes these must be treated with caution as knots almost always involve loops, which with their greater exposure to solvent are more mobile than other parts of the protein chain and hence less well resolved. This can lead to errors in chain tracing and the erroneous creation of a knot. It is always better if there is more than one independent solution of the structure (preferably at as high a resolution as possible). This was the situation for the most deeply buried knot identified (Taylor 2000a) in the structure of the acetohydroxy acid isomeroreductase which had been solved twice at 1.60 Å (`1QMG`) (Thomazeau *et al* 2000) and 1.65 Å (`1YVE`) (Biou *et al* 1997).

It is interesting to speculate how a structure with such a deep and complex knot might fold—as it is difficult to imagine over 100 residues being 'fed' through a loop in a reproducible way during the folding of the protein. Clues to the folding of this protein can be found in a clear internal duplication within the domain comprising 80 residue pairs with 2.0 RMS deviation (as measured by the program `SAP` (Taylor 1999b) over the α-carbon positions). If it is assumed that the two most deeply buried symmetrically equivalent helices initially pack together, then the remaining parts of each repeat can wrap around this core requiring only that the carboxy terminal segment can pass through the large loop between the repeats before this contracts (through the formation of α-helices) and finally packs onto the core. Following this path, the nature of the knot is determined by the chirality of the packing of the initial core helices. The symmetry in this arrangement

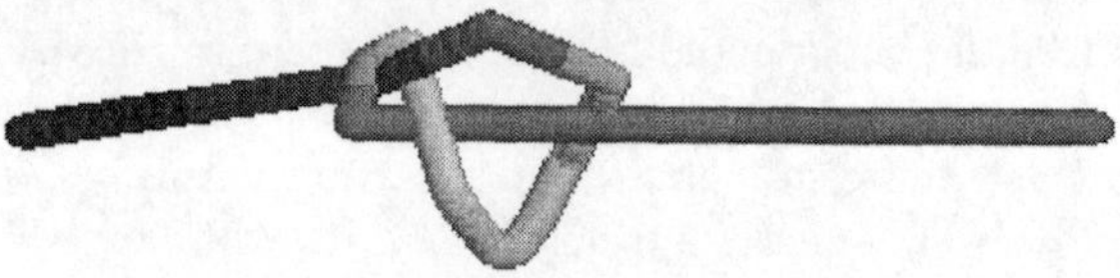

(*a*) Smoothed

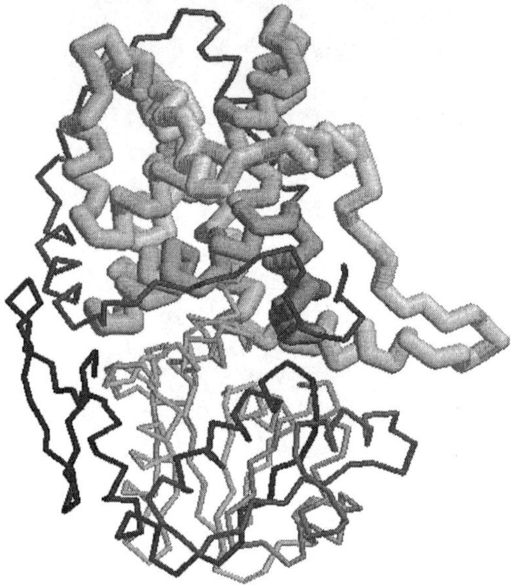

(*b*) Native

Figure 8.3. Knot in `1yveI`. (*a*) The knotted core in the smoothed representation of `1yveI` (figure 8.2(*b*)) is shown in isolation allowing the figure-of-eight knot to be seen clearly. This form was attained after 50 cycles and if continued, an irreducible core consisting of eight points was attained. (*b*) The backbone representation of the complete native protein structure with the minimal knotted region drawn thickened. This region is preceded by a complete nucleotide binding domain and followed by a long loop that wraps around the knotted domain.

suggests that the protein might have evolved from an exchange of structure or 'swap' (Bennet *et al* 1995) between two duplicated domains in which the first helix in the repeat has been transposed across the two-fold axis of symmetry so creating the knot. (See section 10.1 for a further discussion.)

The trefoil knot found in the S-adenoyslmethionine synthetase protein (1fugA) also appears to have arisen in a similar manner to the 1YVE knot through the transfer of a β-strand on the edge of a sheet from one duplicated domain to another.

8.1.5.4 A new trefoil knot

Recently, a new knot has been identified in two homologous structures of an RNA methyltransferase (Nureki *et al* 2002, Michel *et al* 2002) (PDB codes: 1IPA and 1GZ0, respectively). These have 1.9 Å root-mean-square deviation (RMSD) deviation over the 145 equivalent residues in the knotted domain. There appears to be no possibility of any segment swapping in these proteins and it seems likely that the knots have arisen by the diffusion of the 30 residue C-terminal part of the chain through its loop.

8.1.6 Topology of weak links in proteins[3]

8.1.6.1 Loop penetration

In their analysis of disulfide bonded proteins (above), Klapper and Klapper (1980) introduced the idea of ‘loop penetration’, being a less restrictive interpretation of a knotted state defined by the covalent network being non-planar. This approach was generalized by Connolly *et al* (1980) who defined cross-links to be any pair of α-carbon atoms that came within 7 Å (this includes all disulfide links). This looser definition encompassed a correspondingly wider variety of proteins and topological features which were referred to generally as ‘threaded loops’. Some folding ideas of how such features could arise were discussed.

A further generalization of this approach is to consider all distances in proteins as potential ‘cross-links’. Each link can be characterized by the number of residues that have been ‘short circuited’ by the connection and this value plotted against the two residue positions. The resulting ‘tornado plots’, while similar to the Phillips (1970) distance plots, give a good impression of the sequential packing order of the protein (Aszódi and Taylor 1993) (figure 8.13).

8.1.6.2 Hydrogen bonded pseudo-knots

A potentially knotted structure was seen in a domain of the structure of a histone lysyl methyltransferase (this is unrelated to the RNA methyltransferase above). The structure has been reported independently by five groups (Wilson *et al* 2002, Min *et al* 2002, Trievel *et al* 2002, Zhang *et al* 2002, Jacobs *et al* 2002) (PDB codes: 1H3I, 1MVH, 1MLV, 1ML9 and 1MT6, respectively). These structures revealed a multidomain protein, consisting of a catalytic domain (referred to as the SET domain) preceded by another domain (sometimes called a preSET domain).

[3] This section is reproduced in part from Taylor *et al* (2003b) with the kind permission of Elsevier.

The SET domain has a novel complex fold consisting of three interconnected β-sheets arranged around a short central 3_{10} helix.

To examine the differing topological descriptions of the SET domain, the smoothing described above (Taylor 2000a) was applied to one of the recent crystal structures: the histone methyltransferase protein SET7/9 (Wilson *et al* 2002) residues 135–343. The domain was completely reduced to a straight line by the application of the algorithm in 43 cycles of iteration, which is typical for a globular protein of this size (figure 2.2). The speed of this reduction indicates that the fold does not even contain a slip knot as these take longer to resolve than a simple fold. It was clear from this result that the SET domain does not contain a topological knot in the usual definition of the term where only the backbone topology is considered.

The 'knot' in the SET domain might be considered more than just a threaded loop as it involves hydrogen bonded cross-links. If hydrogen bonds were given the same 'status' as covalent cross-links (such as disulfide bonds) then a protein as rich in secondary structure as the SET domain would undoubtedly have a knotted topology. While this seems likely, it is not something that can be deduced with absolute certainty just by inspection. To investigate this more rigorously, the algorithm of Taylor (Taylor 2000a) was modified slightly to take account of cross-links. This was achieved with almost no change to the basic algorithm but in addition to forbidding the (virtual α-carbon) chain to pass through itself, the chain was also prevented from passing through cross-links and cross-links could not pass through other cross-links. Otherwise, the chain was iteratively reduced as before (figure 8.4). Ideally, the modified algorithm could be applied to the SET domain with all hydrogen bonds defined as cross-links. However, as some technical difficulties in the removal of redundant cross-links have not yet been resolved, the simpler approach was taken to see how few cross-links were necessary to create a knot in the domain.

Extensive trials inserting just one hydrogen bond cross-link failed to create a knot. Another pin is necessary and further trials have shown that the exact location of the two cross-links are not important, provided that they both link the relatively parallel segments between residues 231–246 and 316–330 (which are hydrogen bonded over most of their length) (figure 8.5).

From a more technical viewpoint, the loose analysis of the hydrogen bond linked 'knot' described above can be treated more exactly using methods developed over many years to analyse knots formed by disulfide (or other) cross-links (Liang and Mislow 1995). Having reduced the protein chain (plus cross-links) to a simple form, similar to figure 8.5(*b*), then the two termini must be fixed to a flat surface (to prevent any 'cheating' in which a loop is passed over a terminus). In this situation, if the rest of the chain can be placed flat on the plane without any lines crossing (including cross-links) then there is no knot. Computationally, this can be solved by representing the protein (plus links) as a graph and using the algorithm of Kuratowski to check for a planarity (Crippen 1974, Klapper and Klapper 1980).

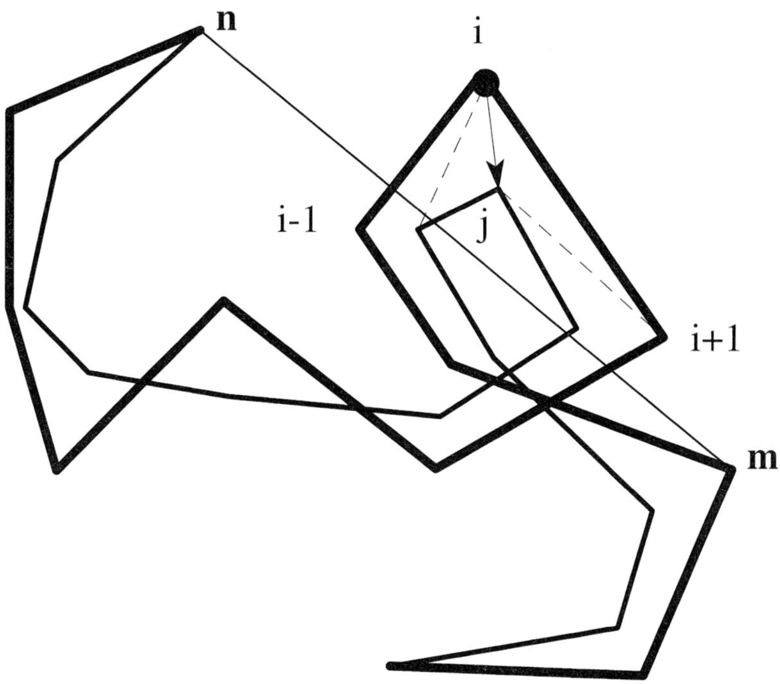

Figure 8.4. Chain smoothing with cross-links. Part of a protein chain is drawn schematically as lines connecting the α-carbon of each residue. For each residue point (i) in the starting conformation (bold line), the average coordinate of i, $i-1$ and $i+1$ was taken as the new position j (thin line). This procedure was then repeated, resulting in progressively smoother chains (the termini do not move). In the basic algorithm, the chains were forbidden to pass through each other by checking that the triangles $\{j-1,\ i,\ j\}$ and $\{i,\ j, i+1\}$ (broken lines in the figure) did not intersect any line segment $\{k-1, k\}$ ($k < i$) before the move point or any line $\{k, k+1\}$ ($k > i$) following. This has been extended in the current work to include any cross-link (shown as a fine line between the residues m and n).

8.1.7 Generalized protein knots

The analysis of the SET domain has illustrated that, in general, proteins exhibit a progression of increasingly solid cross-links from close approach (van der Waals 'bonds') through hydrogen bonds to disulfide (or other) covalent bonds. Unlike mathematical knots, the point at which a loop can be said to be a knot becomes a matter of energy, not simply logic. It would be possible to devise an algorithm to calculate the minimum energy required to undo a cross-linked knot in terms of the number of hydrogen bonds that must be broken, Van der Waals energy lost and even solvent effects. To clarify the description of knot-like structures the

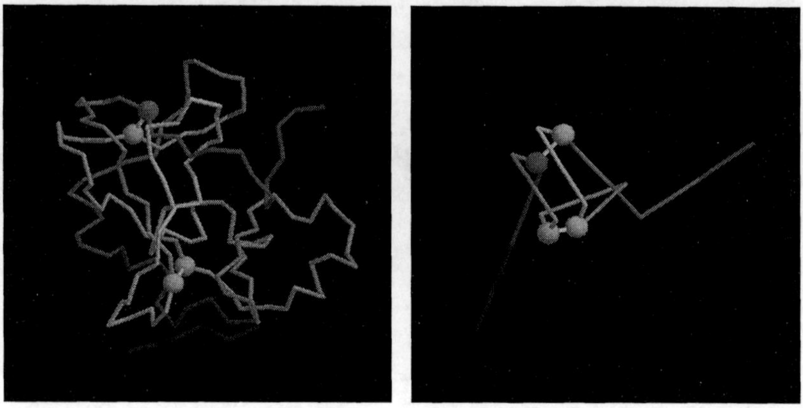

(*a*) native (*b*) reduced

Figure 8.5. Chain smoothing with cross-links. The modified algorithm of Taylor was applied with a pair of cross-linked hydrogen bonds as indicated by the linked spheres on the native structure (*a*). This reduced to the simplified form shown in (*b*). If the two termini were fixed to a surface, the structure cannot be laid flat without either the chain or the links being crossed.

term *threaded loop* should be retained for unbonded 'knots' while for the cross-linked knots, the term *covalent knot* should be used to distinguish them from the true *topological knot* in which the backbone fold alone is sufficient. This leaves the 'knot' found in the SET domain and, by analogy with the hydrogen-bonded knots seen in RNA structures, we propose that the term *pseudo-knot* is adopted for protein knots formed by hydrogen bonds.

Unlike the topological knots, there is no mystery about how covalent knots (or pseudo-knots) might form: they simply require cystine residues (or H-bonding groups) to lie close enough to become cross-linked either during or after folding. Like disulfide bonds in general, their function may be to give extra stability to the fold. This explanation raises the question of whether topological knots might perform a similar function or are they just 'harmless' tangles that have arisen accidentally? Intriguingly, both deep topological knots considered above are in the catalytic domains of their proteins, with one even running through the active site (Nureki *et al* 2002, Michel *et al* 2002). However, there is little in the overall structure of a knot that can contribute any specific function that could not just as easily be constructed by an unknotted piece of protein chain: so any advantage from their presence must derive from an indirect or entopic effect.

The distinction between topological knots and knots derived from cross-links, is not simply one of nomenclature but reflects a fundamental difference in the 'gymnastics' that the protein must perform to fold. With cross-links, it is only necessary for a loop to fold across another part of the chain in order to 'trap' it. With this event, it is irrelevant how many residues lie before or after the knot site. To form a true topological knot, however, the protein chain must form a loop and then pass one of its termini sufficiently through the loop for it to be counted as a knot. This is clearly a more demanding exercise and would be expected to occur much less frequently. Just how frequently it might happen by chance is difficult to calculate but some rough estimates will be considered in the following section.

8.1.8 Knots in random chains

From our experiences with shoelaces, we tend to think that tying a knot requires both dexterity and intent. Since proteins lack both of these qualities it was assumed for many years that there would be few, if any, knots found in protein chains—and to a large extent, this has been proved correct. It is possible to find a slightly better model for a protein chain in a string of beads, while pearls can be used, it is possible to find a cheaper source of materials in the decorations that are sometimes draped over the Christmas tree branches (Taylor and Lin 2003). To conduct an unbiased experiment: take a metre length of these (roughly 150 beads is a small protein), cover the end beads in 'Blu-Tac' and 'pour' the beads from hand to hand. When the two end beads eventually stick together, check for knots (as shown in figure 8.6). In a limited experiment (carried out by a random sample of experienced scientists at coffee time), a quarter of our 'proteins' were found to be knotted. From this result, one might expect proteins to contain a reasonable number of knots.

A likely explanation of why there are so few knots in proteins is that protein chains are not free-flowing strings of beads but are 'sticky'. Interactions within the chain are therefore predominantly local and few open loops will be formed for the termini to pass through. This means that the number (and type) of knots found in any more realistic simulation will depend critically on the nature of the interaction between elements in the chain. In the following sections, a range of semi-random models for proteins will be considered that might help in addressing this problem.

8.2 Random walks in fold space

In the previous chapter we looked at the sorts of folds seen in nature. It is clear from these studies that we have not yet seen every natural fold and despite the gallant efforts of the structural genomics programmes, it is likely that it will be some time before all protein structures are known (or even linked with a close homologue of known structure). Even when that golden age dawns, we still will not know the structures of all the proteins from organisms that are now extinct, which (as suggested by the evolutionary paths in fold space) might include lost intermediates on the pathways to the current observed folds.

Figure 8.6. Early protein topologist at work. The knotted topology of protein-like chains is being considered by an early worker in the field (*circa* 1922). So that experiments could be carried out at any convenient moment, these pioneers often kept their chains round their necks and perfected a distinctive circular randomization procedure. (Apologies to Louise Brookes.)

These speculations raise questions not just about the nature of the fold space for natural proteins but about the fold space of all possible proteins. It would, of course, be completely impossible to start constructing molecular models for every possible protein sequence but some idea of the range of possible folds can be gained by ignoring the sequence and instead asking how many distinct compact protein folds there can be. The answer to such a question depends critically on what the criteria are for distinguishing folds: too fine and they are innumerable, too course and there is just one.

From a different viewpoint, this approach depends on the degree to which we require the different features of proteins to be modelled. At one extreme, the protein structure can be represented by a self-avoiding random walk while at the other extreme, the method for generating folds should reproduce secondary structures and the typical ways in which they pack together (including all their chiral features). Collectively, these structures are referred to below as 'fake' structures to distinguish them from 'genuine' (native) structures and models that have been constructed through the application of established homology modelling methods.

8.2.1 Random walks

Random proteins have been generated many times in the past for testing purposes and a simple algorithm to do this is a self-avoiding random walk from one α-carbon to the next constrained to lie inside a sphere or ellipsoid (Cohen and Sternberg 1980a, Thornton and Sibanda 1983). An algorithm is described below (devised by Tom Flores) that is similar to this but incorporates a local, rather than a global, constraint for the chain to be confined. This was implemented by selecting the next position in a growing chain to be preferentially in contact with its predecessors. Each residue was added to the chain in a random direction that did not seriously clash with any other. For each accepted position the number of packing contacts were counted and if they fell short of a target number the position was rejected. After a good number of attempts, the target height was successively reduced until an acceptable position was found.

The density of the 'fake' structures are dependent on the number of target neighbours: if this is set to find too few contacts then the resulting structures are not sufficiently compact while if set to find too many neighbours, then the time taken increases. Trials indicated that aiming for four or more neighbours produced compact structures while aiming for only two neighbours was not sufficient. As a compromise, the models described below aimed for three contacts. The resulting structures are remarkably 'life-like' (figure 8.7), even incorporating elements reminiscent of secondary structure and, if allowed to grow large enough, 'breaking-up' into domain-like regions. (e.g. figure 8.7(*c*)).

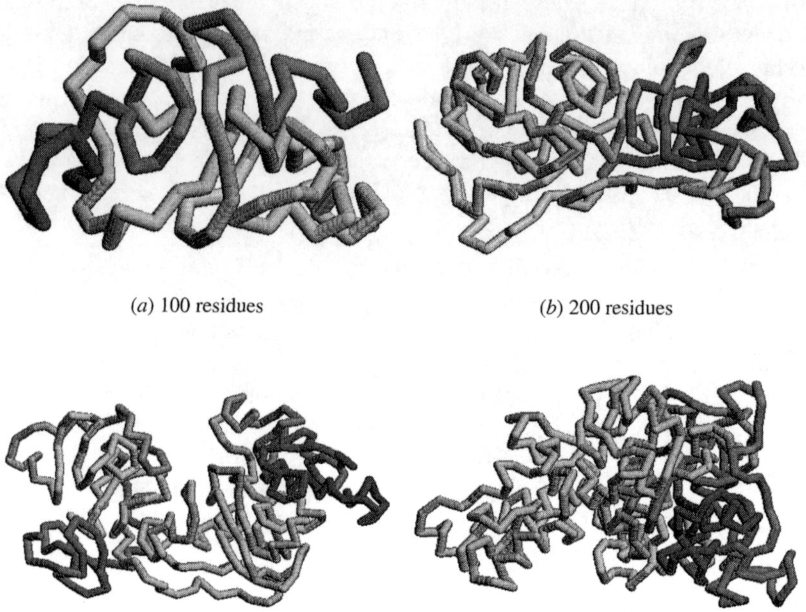

(*a*) 100 residues (*b*) 200 residues

(*c*) 300 residues (*d*) 400 residues

Figure 8.7. 'Random' protein models. (*a–d*) Typical structures produced by the random walk algorithm described in section 8.2.1 for four different chain lengths. Secondary structure and domain-like components can be seen. The figures were produced by the program RASMOL.

8.2.2 Secondary structure based fake proteins

A more protein-like 'fake' model can be based on the stick models of proteins described in section 3.2 in which each secondary structure was represented by a simple line.

8.2.2.1 Fake all-α models

To generate models for all-α proteins, for a given number of helices, all possible arrangements of helices over the polyhedral stick models of (Murzin and Finkelstein 1988) can be enumerated. The register of the sequence on the framework is set by the secondary structures with each element being placed on alternate edges of the polyhedron as the winding progresses. Computationally, this can be achieved by the application of a recursive routine which chooses a free path from each node until there is no further secondary structure units or a

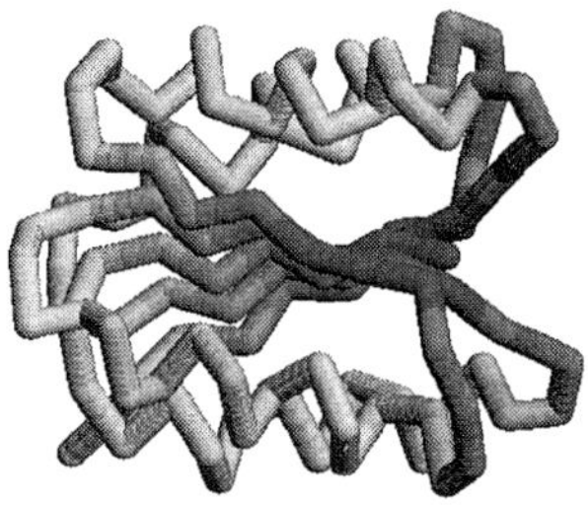

(*a*) $(\beta\alpha)_4\beta$

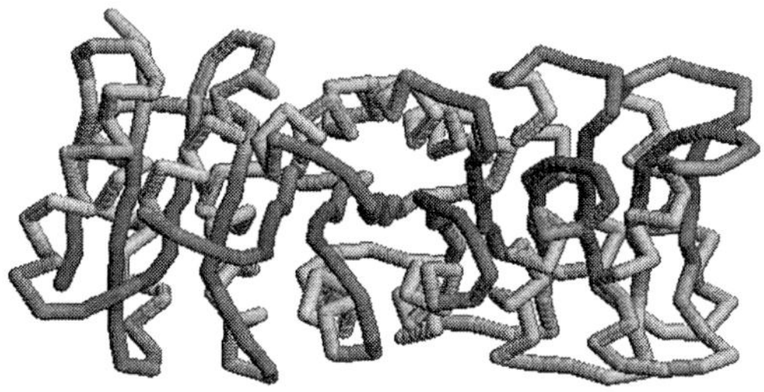

(*b*) $(\beta\alpha)_{12}\beta$

Figure 8.8. Stick-based protein models. Proteins with ideal secondary structure packing were generated ranging from small structures with a five-stranded β-sheet (a) to large structures with 13 strands in the sheet (b). The former is viewed across the sheet while the latter is viewed end-on to the sheet. This view reveals a particularly symmetric structure with each $\beta\alpha$-unit 'jumping' to the right by two strands then 'leap-frogging' back towards the left. These models have the correct chirality of secondary and super-secondary structures with no loops crossing each other. Parallel connections between sequential secondary structures were only allowed on the edge of the sheet.

dead-end is encountered. On each of these conditions the procedure 'backtracks' to the preceding node and takes an alternative path. Exhaustive application of this procedure eventually enumerates every possible path. Thus, each model has a different secondary structure packing arrangement and generally, a different fold. To create a more life-like model, using only α-carbon coordinates, loop regions were roughly modelled after the secondary structures were in place (Taylor 1993b).

8.2.2.2 Fake β/α models

Following the work of (Cohen *et al* 1982), models for the β/α class can be generated by enumeration over the stick architectures described in section 3.2 (figure 3.10). Before being realized as full 3D models, some constraints can be imposed on this class of fold including the handedness of connection of $\beta\alpha\beta$-units (figure 1.6) and the observation that loops rarely cross on the same face. Like the all-α models, these models were then also 'expanded' into α-carbon-models as described previously (Taylor 1991b). Typical examples are shown in figure 8.8. In these models, all α-helices have the correct chirality and β-sheets have the correct twist.

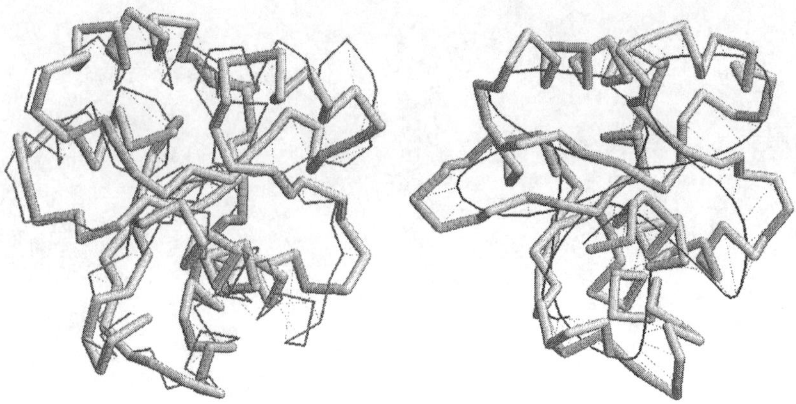

(*a*) (*b*)

Figure 8.9. 'Rambling' over native backbones. Using the `RAMBLE` program semi-random walks were generated for the small $\beta\alpha$ protein `3chy` over (*a*) the native α-carbon backbone and (*b*) the smoothed backbone. The guide path is shown as a fine line and the model backbone as a tube.

8.2.3 Off-lattice fold combinatorics

For the analysis of a fake fold space, the random walk models are too unstructured and, without regular secondary structures, are difficult to classify. By contrast, the fake folds combinatorially generated on a secondary structure lattice come ready classified but may be too regular to be truly realistic and contain externally imposed chirality constraints.

An intermediate situation would be more ideal and this can be realized by combining the two approaches as a random walk through a regular lattice. The basic random walk algorithm described above can easily be modified to implement this by adding constraints to the conditions for accepting a new residue position. Besides the original packing target, the additional constraint for the new point to lie close to a point on a secondary structure line segment can be imposed. If the line represents an α-helix, then the torsion angle with the preceding residues should also be in the range expected for a helix or the range for a strand if the guide-stick is a β-strand. A constrained random path of this nature will be referred to below as a rambling walk or 'ramble' and the program that implements the method is also called `RAMBLE`[4] (Taylor *et al* 2003a).

[4] ***R****andom-walk* ***A****b-initio* ***M****odelling* ***B****y* ***L****ocal* ***E****xtensions.*

8.2.3.1 Rambling over backbones

The method can be tested by providing the coordinates of the native protein as guide points. The resulting models (figure 8.9(*a*)) typically adhere well to the fold with an overall RMS deviation of 2–3 Å. Occasionally the path can diverge but often is recaptured. As a step towards using stick models, the smoothed backbone chain was substituted. With five cycles of smoothing (as described in section 2.1.1.1) the secondary structures are effectively straight (or bent) lines (figure 8.9(*b*)). Again the path is followed but with an increased chance of 'going off the straight and narrow'. With these models, the phase (in-out orientation) of the helices and strands is no longer directed and an added degree of divergence from the native fold is contributed by this effect.

Just as a semi-random wander can be made over guide points provided by (or derived from) the native fold, so the same can be done for guide points derived from a stick fold. Each guide point can either lie on the sticks (with the loops thrown in as parabola) or the phase of the secondary structures can be estimated from the orientation of their hydrophobic residues. This approach, pioneered by Eisenberg and colleagues (Eisenberg *et al* 1982, Eisenberg *et al* 1984, Eisenberg *et al* 1989), was used in the construction of the all-α models described above (Taylor 1993b). In the small $\beta\alpha$ proteins the α-helices are usually strongly amphipathic, giving a clear signal on how they should be orientated. The same is not so with the β-strands which tend to be hydrophobic on both sides of the sheet. Although the sequence is not an important component in the study of possible folds, it was included in the construction of the models for later use. For this, a compromise was used in which the guide points on the helices were constructed at half the true helix radius and the guide points on the strands were kept on the line segment (figure 8.10).

To estimate the accuracy of models generated from stick-figures, the stick connectivity equivalent to the flavodoxin-like fold was extracted and guide points constructed as described above. This figure only had four helices (2-5-2 form) which left the second, sometimes transient, helix to be guided only by a shallow parabola. Five hundred models were constructed based on this framework and compared using the `SAPit` program (section 6.1.6) to the native structure of the chemotaxis Y protein (PDB code: `3chy`) on which the secondary structure assignments were based. When viewed as a plot of RMS deviation against length (figure 8.11) it can be seen that the data lie well below the Maiorov and Crippen (1994) line for a random relationship. Only a few models have a poor fit indicating that their path has wandered seriously off track while the majority have an RMSd value between 6–7 Å, with the better models being around 5.5 Å. An example is shown in figure 8.11.

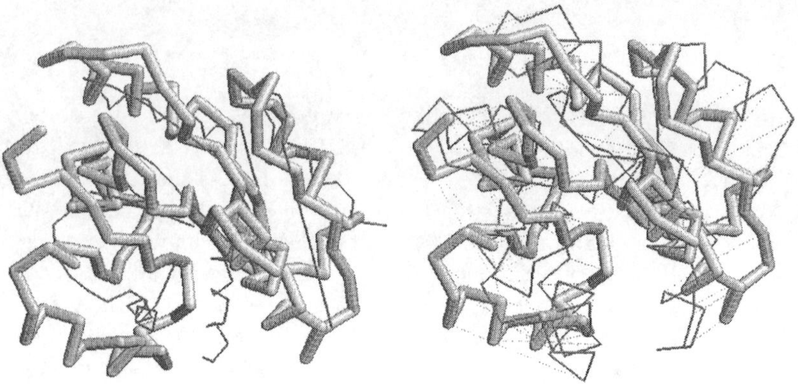

(*a*) (*b*)

Figure 8.10. 'Rambling' over stick backbones. Using the RAMBLE program semi-random walks were generated for the small $\beta\alpha$ protein 3chy over a secondary structure stick-figure with (*a*) α-helices expanded to half size and (*b*) α-helices expanded to full size. The guide path is shown as a fine line and the model backbone as a tube.

8.2.3.2 *Rambling off the beaten track*

The models generated using a stick model as a guide give rise to variations around a single representative of each fold and so give no idea of the relative probabilities with which the different folds might occur. To allow some greater freedom to the wandering walk, a probabilistic element was introduced into the fold generation. The first SSE was laid down and the second packed against it in a hairpin conformation either hydrogen bonded for a β-hairpin or packed at 10 Å for a $\beta\beta$ or $\beta\alpha$ pair. From this core, additions of SSEs were made randomly against any available edge: for α-helices an algorithm similar to that used to generate the Bernal 'lattice' (section 3.2.5.3 and figure 3.15) was used while suitable hydrogen bonds were required for $\beta\beta$ packing. The chirality of β-connections were checked and if the hand was of the wrong type, another attempt was made to find a sheet edge. If this failed, then the strand was still allow to 'grow' in a free direction and could provide a nucleus for a new sheet. (All these features were incorporated into the RAMBLE program.)

The big change with this algorithm compared to the guided path described above, is that now the starting position and direction in which the chain grows makes a difference to the resulting structure. For example; if growth always starts at the amino (N) terminus, then in the resulting folds, the first two SSEs will always be packed. (The same is also true for the constrained random walks described above but they are sufficiently random that it does not matter much). To

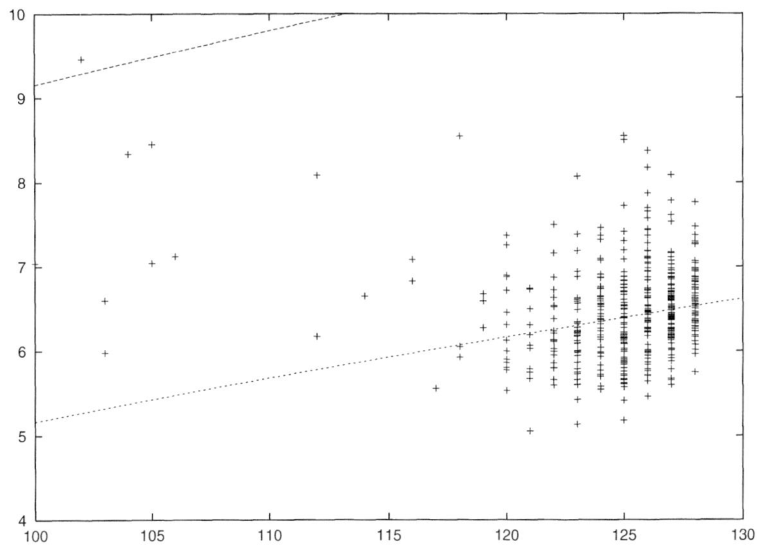

Figure 8.11. Model RMSd distribution. The models generated by the `RAMBLE` program for `3chy` were compared to the native protein using the `SAPit` program (section 6.1.6) and plotted as the RMS value (Y-axis) against the number of aligned residues (X-axis). The upper dashed line is the random estimate of Maiorov–Crippen (see figure 6.7) and the lower line is their function scaled to pass through the model distribution.

avoid this bias, the chain was allowed to grow in both directions starting at any residue position (with random switches between the termini). A collection of over 1000 folds was generated in this way, using the secondary structure definitions from `3chy`, and a sample compared as above to the native structure (figure 8.12).

Where previously, with the backbone guide points, there was a tight cluster of models with 6–7 Å deviation from the native, there is now a wide spread from 6–16 Å, extending well above the random level of significance as estimated by the Maiorov–Crippen line (figure 8.12). In this plot a subpopulation can be seen at a shorter length where a region of good RMSd can again be attained by discarding a part of the fold as an insert (typically a $\beta\alpha$-unit). In the full-length population, the best similarity is comparable to the backbone guided random walks but when the best of these was compared to the native fold it was found that it had a different topology with the two folds differing in the transposition of two adjacent strands. With all the other SSEs in roughly the right place, this local change was not sufficiently disruptive to be detected by the RMSd measure.

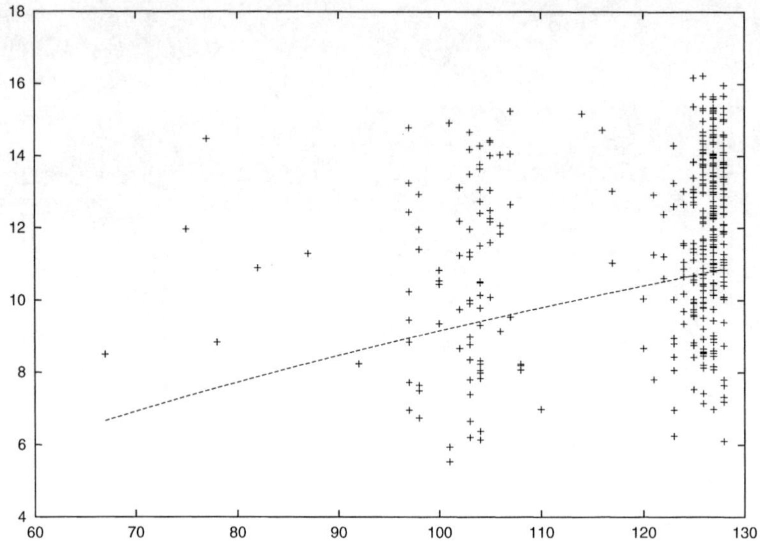

Figure 8.12. RMS distribution of off-lattice models. The models made using the secondary structure of `3chy` were compared with the native structure. Most of these do not have the same fold which is reflected in a wide RMS distribution centred around the (dashed) Maiorov–Crippen line (see figure 6.7). The subpopulation around 100 residues results from the `SAPit` program deleting a $\beta\alpha$-unit to find a better local fit.

8.2.4 Classifying topology in fake proteins

From the result of the previous section, it is clear that RMS deviation is a poor measure for topological relatedness when dealing with models at this degree of dissimilarity. A similar problem was encountered in section 7.1 where the topological relationship between distantly related native proteins was not captured well by the RMS value. A solution to the problem in that situation was to develop a 'pigeonholing' approach based on ideal forms and the same method can be used with model proteins.

Since the secondary structure definition method developed in chapter 3 can be used with α-carbon models and is robust to limited distortion in the structure, it can be applied to the rough models generated by the semi-random walk method described above. Clearly, a number of secondary structures will be missed but with sufficient models, enough will be caught and classified. As with native structures, the models can then be matched against the ideal forms and classified. This is a simpler procedure than previously as we now know that each model is (or should be) a single domain and it is reasonable to assume that the models will fall into the three-layer $\beta\alpha$ class, allowing a restricted set of forms to be matched.

After classifying the folds in this way, over a third fell into the 2-5-2 architecture (two pairs of helices either side of a five-stranded sheet). There were much fewer in the 2-5-3 class indicating that it is difficult to pack three helices together on the sheet. When sorted on their topology strings (section 3.2.8.1 and section 7.4.2.1) there were just over 300 different folds seen for the 2-5-2 class. This can be roughly compared to the 940 folds generated by combinatorial enumeration[5] Only two folds were found with the flavodoxin-like topology and both had RMDs values of 7.5 Å (over 126 residues) when compared to the native fold. Both these folds derived from starting positions for the 'random' walk near the amino terminus (positions 2 and 18) which is a region that ensures the first helix and strand are packed and that the first strand can be buried in the sheet by the addition of later strands.

8.2.5 Local *versus* global folding

The simple random walk model described above in section 8.2.1 had a directional bias in the nature of the folds it generated. By always starting at the first residue the folds in this first part of the protein will have more local contacts than the later additions (like a ball of string). Since proteins are synthesized starting from their amino terminus, it was once thought that native proteins would also exhibit this property but despite analysis, no convincing trend has ever been observed for N-terminal folding (originally proposed by David Phillips). It is now thought that the effect is masked by the action of chaperones which prevent immediate folding as the protein comes off the ribosome.

The same effect operates on the semi-random walks that incorporate secondary structure. Walks that start near the amino terminus will have more local packing in that region. Many folds, including the flavodoxin-like fold used above as an example, can be generated by accreting secondary structures around an amino terminal nucleus in a process that maintains a continual packing of the added structures without rearrangement. However, there are many folds where this condition does not hold.

The restriction to have every fold accessible by packing around the amino terminus was relaxed in the random secondary structure walks where the nucleation could begin at any point along the chain. In theory, this still confines the fold space to a subset of folds but a greater number of known folds are now included. For example, the fold of adenylate kinase is shown in figure 3.17(*a*) which is not accessible from the N-terminus but by starting at any of the SSEs on the left side (as depicted), the fold can be generated without any unconnected

[5] These values cannot be exactly compared as the combinatorial method imposed connection chirality and forbids loop crossing. Only the former was imposed in the random walk models but because of their less regular nature, the constraints were not always effective. A larger difference, however, results from the classification method which recalculates its own secondary structure definitions. This means that distorted regions will not be declared in their target conformation giving a greater variety of secondary structures and consequently a greater variety of folds.

leaps into space. The same is not true, however, for the other (imaginary) fold shown in figure 3.17(*b*). Whatever the starting point, this fold cannot be laid down without at some point finding it necessary to place an unconnected β-strand and 'backfill' the space. While these constrictions on the accessible fold space may not pose a serious practical limitation, they do contribute a bias to the sampling of folds generated by the secondary structure-based random walk approach. Not only will some folds never be seen, but those that are generated will appear with a frequency that reflects the even spread of start points along the chain. To avoid this, and indeed any directional or bias from local folding, the approach of distance geometry can be used. In the distance geometry (DG) method, there is no pathway to the final fold which 'materializes' almost fully-formed from hyperspace. This approach will be described in section 9.1 but before examining that method, the following section will consider some of the issues raised above on how to distinguish a simple fold from a complicated fold or a tangle.

8.3 Protein fold complexity

It is clear that there are different degrees of complexity associated with protein folds. In the simpler world of an open string of beads (figure 8.6) these can range from a simple fold, which would be equivalent to lowering the string onto a flat surface, through a ball of string fold where the fold is built up by accretion onto a core, to the folds described above where a jump-and-fill-in step is needed. Over this hierarchy, can be added the degrees of knottedness discussed in section 8.1.5.1, from the almost knotted (loop penetration) through H-bonded cross-links (pseudo-knots) to the true topological knots. The latter again constitute a further hierarchy of the type of knot ranked by the number of crossings (trefoil, figure-of-eight). In this section, we investigate ways of measuring this complexity.

8.3.1 Topological indices[6]

Measures of complexity have been expressed in terms of the balance of local and non-local contacts and a measure that captures this is the topological index. Topological indices are abstract mappings of graphs to (real) numbers, achieving further information compression. Topological indices have been the subject of intensive study in theoretical chemistry (for a review, see Randić (1992)). One possible application is the management of molecular structure information. For this, unique topological indices (i.e. invertible 1:1 mappings) are required which give different values even for structural isomers. The other major area for topological indices is QSAR (quantitative structure-activity relationships) where the shape of a molecule is correlated to some of its physical, chemical

[6] This section is reproduced in part from Aszódi and Taylor (1993) with the kind permission of Oxford University Press.

or biological properties (see, e.g. Klopman and Henderson (1991), Mercier *et al* (1991)).

For protein chains, a topological index measures the 'connectedness' of the folded molecule which is expressed through the notion of 'effective chain length' (Aszódi and Taylor 1993). This approach regards the folded chain as a set of loops (or cycles) held together by several non-covalent connections, thus focusing on the main topological difference between the folded structure and the unfolded chain which contains no cycles. The minimal amino acid distance for a given pair of amino acids was defined as the shortest Euclidean distance $D(X, Y)$ between all possible pairs of non-hydrogen atoms (including the main chain atoms) across the two amino acids X and Y, i.e.

$$D(X, Y) = \min_{i,j}(d_{ij}) \tag{8.3}$$

where d_{ij} is the distance between the ith atom belonging to amino acid X and the jth atom belonging to amino acid Y. This distance measure assumes low values for amino acids which are in close contact in the structure and is independent of the side chain sizes. Other amino acid distances, such as the Euclidean distance between the centres of mass of two amino acids or the distances between α-carbon atoms could have been used as well without affecting the results. The minimal amino acid distances were calculated for all possible pairs of sequentially non-adjacent amino acids in a given structure, and those pairs for which this distance was smaller than a threshold were regarded as being connected. The number of connections per amino acid was not restricted.

In a 1D string of consecutively numbered nodes, the ith node is separated from the jth node by $j - i \quad (i \leq j)$ arcs. When additional arcs (connections) are added to this string, these will provide shortcuts so that the graph topological distance (the shortest route from node i to node j measured by the number of arcs traversed) will be $d \leq j - i$. To measure connectedness in a graph, the minimal graph topological distances between all possible pairs of nodes were summed. For a 1D string with N nodes, this sum can be obtained analytically:

$$\mathrm{Conn}_{1\mathrm{D}} = \sum_{k=1}^{N-1} k(N-k) = \frac{N}{6}(N^2 - 1). \tag{8.4}$$

For general graphs, first a matrix of minimal topological distances between each pair of nodes was obtained (Sedgewick 1990), and then the elements of this matrix above the main diagonal were summed (i.e. counting each pair only once). Backbone and non-backbone connections were regarded as equivalent when calculating the connectedness index for 3D protein structures. The value of the connectedness index depends on the choice of the cut-off: if the cut-off is very large so that every amino acid is regarded as connected to all the others, then the value of the index for this maximally connected graph will obviously be $(N/2)(N - 1)$, the lowest value possible. On the other hand, it reaches its maximal value when no non-backbone connections are present, as in the case of the corresponding 1D completely unfolded string (equation (8.4)). The connectedness numbers for graphs of folded 3D structures should fall between these two limits.

To obtain a measure for the degree of foldedness, the notion of the effective chain length was introduced. This is the (usually non-integral) number of nodes in a 1D string which would give the same connectedness index as the 3D graph in question. If the index is denoted as $Conn_{3D}$ and the effective length as N_{eff}, then from the definition and from equation (8.4) the following formula is obtained:

$$\mathrm{Conn}_{3\mathrm{D}} = \frac{N_{\mathrm{eff}}}{6}(N_{\mathrm{eff}}^2 - 1) \tag{8.5}$$

which can be solved for N_{eff}.

8.3.1.1 Connectedness and effective chain length

When plotted against the chain length (the number of monomers in the chain), the effective chain length changes monotonically, in a regular fashion while the connectedness index is approximately a second-order function of the chain length over one order of magnitude (in the range of 50...500) (Aszódi and Taylor 1993). As the effective chain length is calculated from the connectedness index, it essentially conveys the same information. However, it provides an easy means of assessing the 'shrinkage' of the polypeptide chain while folding. The dependence of the effective chain length on the actual chain length is almost linear as indicated by orthogonal polynomial regression (with some small deviation for proteins with chain lengths less than 100 amino acids).

The connectedness index and the effective chain length were calculated for eight native myoglobin structures and 100 'fake' myoglobin folds generated using the methods outlined in section 8.2.2.1. The native myoglobins gave a connectedness index $\mathrm{Conn}_{3\mathrm{D}} = 5.40(\pm 0.05) \times 10^4$ and an effective chain length $N_{\mathrm{eff}} = 68.7(\pm 0.2)$. For the artificial structures, these values were $\mathrm{Conn}_{3\mathrm{D}} = 4.82(\pm 0.17) \times 10^4$ and $N_{\mathrm{eff}} = 66.2(\pm 0.8)$, respectively. Both indices were higher for the native structures, indicating that the fake proteins were more tightly packed.

8.3.1.2 Tornado plots

A convenient way of visualizing the non-backbone connections is to plot the connection spans versus the sequential positions (figure 8.13). If a connection was made between the ith and jth amino acids in the sequence, then a horizontal straight line was drawn from the point (i, s) to the point (j, s) on the graph, s being the span of the connection $(i < j, s = j - i)$. Local interactions show up on this plot as short lines at the bottom, whereas long-range interactions are indicated by long lines at the top. Connections with spans less then five were not plotted to avoid cluttering at the bottom of the graphs. Loops containing equivalent connections appear as vertical stacks of lines forming roughly equilateral triangles standing on their tips. As these structures are reminiscent of tornadoes, these plots will be referred to as 'tornado plots'. Although the plots contain the same information as contact maps, their different graphical layout facilitates appreciation of the hierarchy of protein substructures.

By their definition, non-backbone connections are indicative of closely packed regions in the protein structure, therefore tornadoes correspond to locally compact hairpin-like antiparallel substructures. Parallel strands are implicit in this representation, indicated by the presence of antiparallel motifs necessary for connecting parallel stretches of the chain. Although tornadoes often overlap, the inspection of tornado plots reveals the structural organization of the molecule in terms of domains and subdomains.

For example, the plot of an immunoglobulin domain (figure 8.13(*a*)) shows the substructure of β-hairpins while for an all-α protein like myoglobin (figure 8.13(*b*)), the helices dominate the short-range interactions at the bottom of the plot. Since helices contain no proper loops, this part of the tornado plot conveys little information. For larger-scale structures, where helical arrangements only rarely occur, the tornado plot shows the main loops and reveals a surprising twofold symmetry in the globin structure.

8.3.1.3 Connectedness indices and loop topology

From the second-order dependence of the connectedness index on the chain length, one could speculate that the connectedness index of native proteins is proportional to the square of the chain length, i.e. roughly to the total surface area of the unfolded polypeptide. This seems reasonable since in a closely packed structure the number of available interresidue contacts is clearly limited by the size of the free surface of the unfolded chain. The almost linear dependence of the effective chain length N_{eff} on the actual chain length might suggest a uniform degree of 'shrinkage' during folding.

Although the actual values of the coefficients of the interpolating polynomials do depend on the cut-off chosen, the degrees of the polynomials remain the same. This means that the nature of the relationships of the indices to the chain length is probably not influenced by the choice of the cut-off, and the indices indeed grasp a characteristic feature of native protein structures which is apparently independent of the structure type.

Based on these relationships, one can give rough estimates of the indices even for a sequence with unknown structure on the basis of the chain length only. This could, in turn, serve as a rapid guide to filter out grossly miss-folded molecules while performing tertiary structure prediction.

The tornado plot represents protein molecules as loops within loops and this simple graphical technique is sufficient to visualize the structural units of a folded polypeptide chain. Since loops are topologically the same, regardless of their

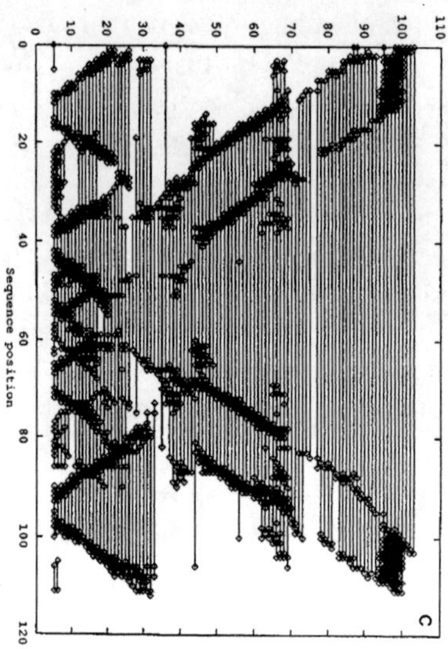

(*a*) β-protein

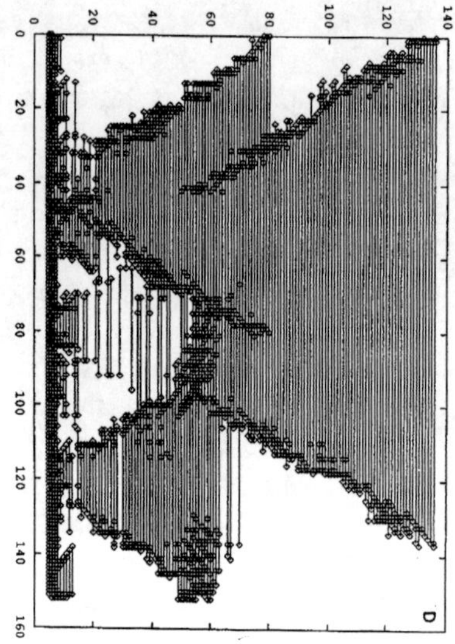

(*b*) α- protein

Figure 8.13. Tornado plots for (*a*) an all-β type protein (the immunoglobulin domain `2rhe`) and (*b*) an all-α type protein (sperm whale myoglobin, `1mbo`). Structures for these proteins can be seen in figure 1.3.

length, their hierarchy implies that the tornado plot describes the protein as an approximately self-similar object within a limited range of structural hierarchy. Self-similarity is a property of fractals (Mandelbrot 1982) (geometrical objects with non-integral dimension), and indeed, polymer molecules (including proteins) can be modelled as fractals (Li *et al* 1990, Havlin and Ben-Avraham 1982). It should be kept in mind, however, that this apparent self-similarity of the protein structure is limited to the few levels of structural hierarchy in the molecule, therefore proteins are not fractals in the strict mathematical sense.

The hierarchical organization of domains and subdomains (Crippen 1978, Rose 1979, Zehfus 1987) is also well reflected by the set of tree structures imposed upon the familiar distance maps by the loop hierarchy plots. Instead of viewing the whole forest, one could choose one of the trees of the set by dynamic programming, applying a suitable energetic criterion. This 'best tree' would then map the protein into an RNA-like secondary structure. However, this analogy cannot be taken too far as the loop topology of RNA molecules are much simpler than that of proteins. In RNA secondary structure, loops can contain each other, in which case the large loop is made up of several smaller loops branching from bifurcations. The basic topological difference is that in the 3D protein connectivity graphs, any node can have more than one non-backbone connections. Therefore, in general, the protein connectivity graphs are not planar, i.e. they cannot be projected into two dimensions without any crossover in connections. As a consequence, the topological relationships between any two loops are much more complicated than in the case of a planar graph.

8.3.2 Local and non-local packing

8.3.2.1 Contact order

The topological indices described above convey information about the compactness of a structure in a way similar to the compactness measure of a polymer chain used by Chan and Dill (1990). Their compactness is defined as $(t/t_{\max})$, where t is the number of intrachain contacts made, and $t_{\max}$ is its maximal value, the latter being determined by taking packing constraints into account. However, unlike connection topology, which is based on the graph theoretical concept of shortest paths between nodes, this simpler measure does not contain information about the main chain backbone connections already present in the chain. A similar measure was devised by Baker based on the balance between local and non-local contacts. Specifically, this is defined as the average sequence separation between residues that make contact in the native (all-atom) structure, divided by the sequence length (Plaxco *et al* 1998). This measure has the advantage that, over a limited sample of proteins, it correlated well with observed folding rates (Alm and Baker 1999).

While these measures go some way towards capturing the complexity of a fold, they do not distinguish well between the difference in complexity of some folds. For example; the TIM barrel might be considered one of the simpler folds (figure 1.5) yet it contains a close link at its termini giving a very non-local component to any Baker-like score. Using the same components (8 β/α-units) a Rossmann-like fold can be made with a chain reversal at the mid-point (figure 1.3). This should be considered a more complex fold yet its contacts are more local in nature. An approach that captures this idea of complexity was outlined in the previous section as a problem for pseudo-random walk folds (section 8.2.5) and will be elaborated in the next section.

8.3.2.2 Topological accessibility

The 'fold' of a ball of string can be reproduced only by starting at the beginning and copying each turn of the string. It would be impossible to start at the outside end and add layers of string towards the centre. If the chain must always remain locally packed, then a ball-of-string fold is accessible only from the amino terminus. A fold like the TIM barrel is accessible from every point (adding in both directions) whereas Rossmann folds are accessible only up to a point before the chain reversal (figure 1.3). If a measure of fold complexity is taken as the fraction of residues from which the fold is accessible then the TIM barrel is twice as accessible as the Rossmann fold. This measure seems to capture the intuitive sense of complexity for these folds which are poorly differentiated by other measures.

To apply this approach to real structures (as distinct from topology diagrams) requires some definitions of what constitutes a contact. In the above discussion, β-structure contacts were considered more important than α-structure contacts and in dealing with α-carbon models a 'soft' cut-off is desirable. These requirements were captured in the following exponential switch function:

$$t_{ij} = 1 - 1/(1 + \exp(r + s_{ij} - d_{ij})) \tag{8.6}$$

giving a score s for residues i and j which are separated by a distance d_{ij}. The switch point occurs when $d_{ij} = r + s_{ij}$ which is determined by a base level r and a secondary structure component s. If either residues i and j are β then $s_{ij} = 5$; unless, if either i and j are α then $s_{ij} = 10$ and if neither condition applies, then $s_{ij} = 15$. By trial, a value of $r = 3$ was found to be reasonable which means that less than 12% is contributed by $\beta\beta$ interactions over 10 Å, α interactions over 15 Å and coil interactions over 20 Å.

Given these measures, then starting at any residue, a segment can be grown by testing amino and carboxy terminal extensions on either side. If the current segment runs from residue n to c (in a chain of length N), then the best interactions are found for the extensions[7]:

$$T_{n-1} = \max\{t_{n-1}, j\} \qquad \forall\, j,\ (n \geq j \leq c) \tag{8.7}$$

and similarly for the carboxy terminal extensions:

$$T_{c+1} = \max\{t_{c+1}, j\} \qquad \forall\, j,\ (n \geq j \leq c). \tag{8.8}$$

The larger score ($\max\{T_{n-1}, T_{c+1}\}$) then determines which end is extended to form the new segment, on which the process is then repeated. With each extension, a sum is maintained as:

$$S_i = \sum \log(1 + \max\{T_{n-1}, T_{c+1}\}) \qquad \forall\, n \geq 1,\ \forall\, c \leq N. \tag{8.9}$$

(A sum was taken of the log of the values simply to keep the numbers in a reasonable range.) The sum S_i then gives a measure of how good the packing is when extensions are grown from residue i to encompass the whole chain. This value can be plotted for every residue, as shown in figure 8.14(*b*) for the small $\beta\alpha$ protein chemotaxis Y protein and the larger TIM barrel.

The S_i values can provide a measure of fold complexity but it is unclear what the best aspect might be. The obvious measure is the sum of S_i over all residues (S_{sum}). By this measure a TIM barrel would have a large sum, while a Rossmann

[7] The symbol $\forall$ means 'for all' and is equivalent to a loop in a computer program.

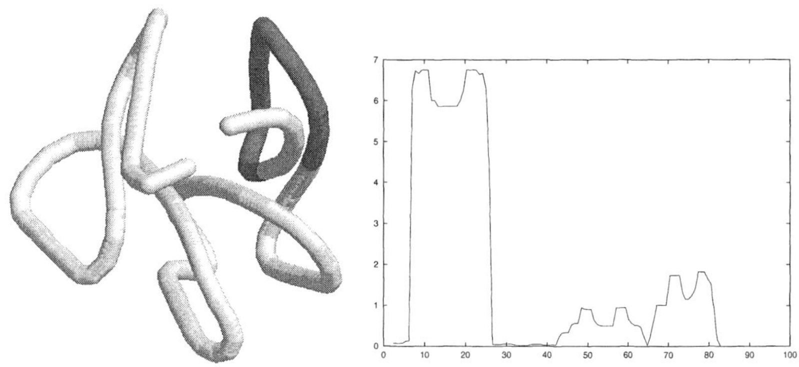

(*a*) CheY fold (*b*) CheY plot

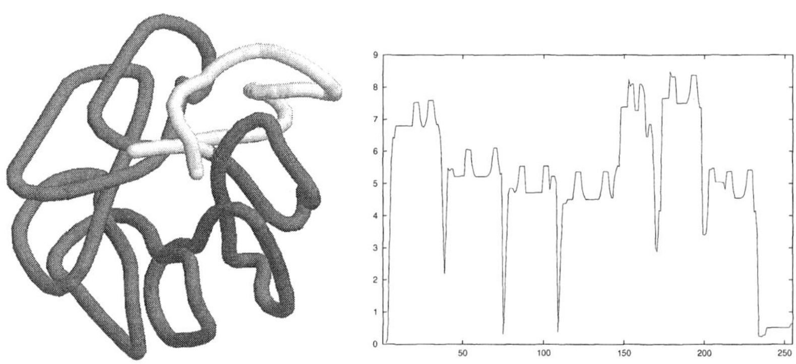

(*c*) TIM fold (*d*) TIM plot

Figure 8.14. Topological accessibility for two proteins. (*a*) The chemotaxis-Y protein (PDB: `3chy`) has a Rossman-like fold which is accessible only from the $\beta\alpha$-unit at the amino terminal segment shaded grey (to the right of the smoothed fold). The values of S_i (equation (8.9)) are plotted in part (*b*). (*c*) the TIM-barrel protein (PDB: `1egz`) which is accessible from everywhere except a small amino terminal segment and the C-terminal helix (shaded grey on the smoothed fold). The values of S_i are plotted in part (*d*). The sharp dips in this plot occur as it is difficult to initiate packing in the middle of an extended β-strand.

fold (or flavodoxin-like fold) would score less, being accessible only from the N-terminal portion. (Compare the areas below the curves in figure 8.14(*b*) and (*d*).) Alternatively, the maximum value of S_i ($S_{\max}$) might be of interest as this gives a measure of how easy it is for the structure to zip-up. For example, when assessed

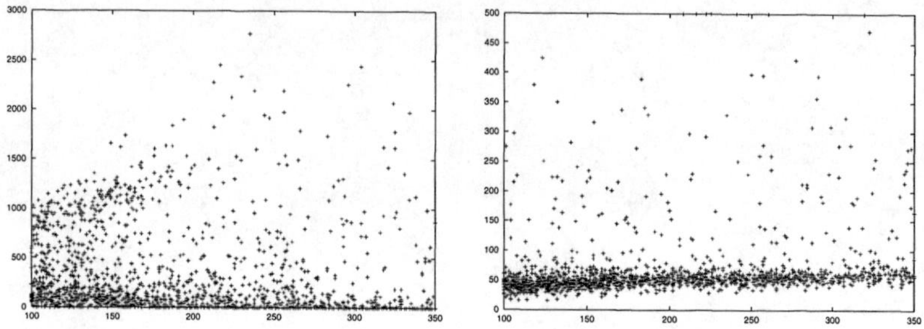

(*a*) (*b*)

Figure 8.15. Measures of fold complexity plotted against chain length (X-axis). (*a*) Using the value of S_{sum} (a high value indicates a simple fold) and (*b*) the number of smoothing cycles to reduce the chain to a straight line (a high value indicates a complex fold). A maximum of 500 cycles was used and those that retain that value are generally knotted.

by a Baker-like measure the jelly roll all-β fold has a large number of non-local contacts yet it can be described as a simple double helix (figure 10.5). Further investigation, and comparison with other measures (and even folding studies) will be required before any definitive choice can be made.

The value S_{sum} was used as a measure of complexity and calculated for the single domain proteins in the SCOP database (figure 8.15(*a*)). The simplest folds were highly repetitive structures (such as SCOP class 118) including leucine-rich repeat structures (see section 10.1). Overall, all-α proteins dominated which may reflect a bias in the fairly *ad hoc* values chosen for equation (8.6) or may be because the chain is 'compacted' within an α-helix, giving less effective length to construct complex folds. The $\beta\alpha$ protein class, as might be expected, were completely dominated by TIM-barrels at the top of the list of simple folds. The all-β proteins, however, were much more complex than either the all-α or the $\beta\alpha$ classes, again possibly because the extended β-strand gives more chain length to 'play' with and create complex folds. Among the simpler all-β folds were the repeated propellor folds and simple barrels in which the chain simply zig-zags up and down (e.g. PDB codes: `1swu` and `1d2b`). The overall complexity of the different structural classes can be judged from their mean ranking in the list to 1600 SCOP domains:

- all-α SCOP class 'a' = 209. (279 proteins).
- α+β SCOP class 'd' = 860. (387 proteins).
- β/α SCOP class 'c' = 916. (483 proteins).
- all-β SCOP class 'b' = 1020. (372 proteins).

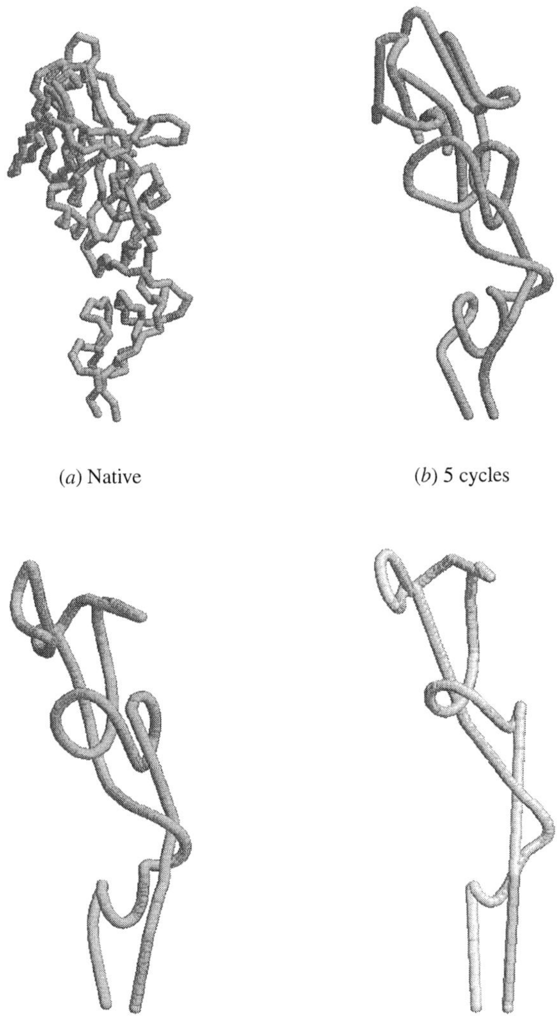

(*a*) Native (*b*) 5 cycles

(*c*) 15 cycles (*d*) 25 cycles

Figure 8.16. Smoothed protein structures. A small β/α protein (flavodoxin-like) is shown as a backbone (α-carbon) trace as: (*a*) the native structure; (*b*) with 5 cycles of backbone smoothing; (*c*) 15 smoothing cycles, and (*d*) 25 cycles. In each cycle the central atom in a triple of α-carbon atoms is replaced by their average position. The amino terminus (bottom-right) is held in a 'strangle hold' by the carboxy terminus. The figures were produced by the program `RASMOL`.

(Some minor classes are omitted). Some of these issues will be returned to in chapter 10.

It is clear from figure 8.15(*a*) that many proteins have very low scores (high complexity) and even when judged by the more generous measure S_{max}, 354 of the 1600 proteins have folds that are effectively inaccessible from any point along the chain ($S_{\text{max}} < 1$). Clearly, the distance parameters in equation (8.6) could be made more lenient but rather than 'fiddle' with parameters, an alternative approach for more complex folds will be considered in the next section.

8.3.3 Smoothing folds away

Section 2.1.1.1 described how protein α-carbon coordinate sets could be progressively smoothed (figure 2.2). This was extended in section 8.1.5.2 to remove redundant colinear points which means that when iterated to completion, all proteins end up as two points (unless they are knotted). It was stated previously that most unknotted chains reach this state in around 50 iterations and because the algorithm acts equally all along the chain, there is very little variation of this with the length of the chain (figure 8.15(*b*))

Although most folds reduce quickly to a straight line, there are still a reasonable number of unknotted folds that do not. Inspection of these indicated that the limiting factor in their reduction was having to resolve a twist in the structure. One of the worst 'offenders' of this type is shown in figure 8.16 in which the amino terminus is held in a 'strangle hold' by the carboxy terminus. Tangles like this take a long time to resolve as the smoothing algorithm can only make local moves and so cannot effect a concerted set of moves to untwist the structure. Although, such folds are not simple, they do not embody the quality of complexity discussed in the previous section and their effect may limit the use of the current algorithm as a suitable measure.

Chapter 9

Structure prediction and modelling

9.1 Random folds from distance geometry[1]

The method of distance geometry (DG) allows models to be constructed directly from a matrix of distances without any simulation of the kinetic folding pathway (section 2.2). To make a random model, in principle, all that is needed is a matrix of random numbers. However, some constraints can be placed on these numbers. Distances between sequential residues are exactly specified and those between residues close in sequence are bounded. Similarly, close contacts can be avoided by placing a lower bound. If a distinction is introduced between hydrophobic and hydrophilic residues, then some preference can also be given to pairs of like and mixed residue type and if secondary structure types are distinguished then many more constraints can be applied. These and other constraints will be described in the following sections and follow the implementation of this method in the modelling program DRAGON[2].

9.1.1 Outline of the projection strategy

When the interresidue distances are semi-random or have been predicted, there will be many that are inconsistent. For example, if one of the predictions is to place pairs of hydrophobic residues at, say, 5 Å, then it is clear that (for more than four residues) all pairs cannot attain their ideal separation. The resulting distance matrix of the model chain will therefore be highly non-metric and while some of these inconsistencies can be eliminated by processing the metric matrix (Aszódi and Taylor 1994a, Aszódi and Taylor 1994b, Aszódi *et al* 1995a), in general, it is likely that the metric matrix will give rise to negative eigenvalues.

In our strategy, negative eigenvalues were ignored so if $M < N - 1$ eigenvalues were positive, then the points were projected into an M-dimensional

[1] This section is reproduced in part from Aszódi and Taylor (1994a) with the kind permission of Oxford University Press.

[2] ***D***istance-geometry ***R***egularization ***A***lgorithm for ***G***eometric ***O***ptimizatio***N***.

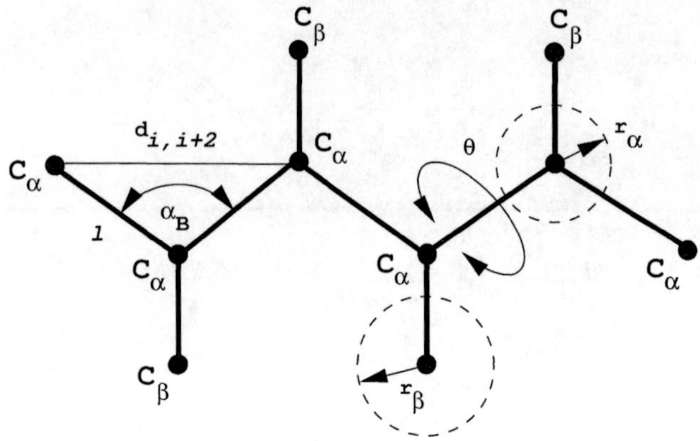

Figure 9.1. Model protein chain geometry. The molecular model used in DRAGON is based on α-carbon atoms with 'dummy' residue centroids (β-carbon) positioned 2 Å above the bisector of the α-carbon virtual bond angle (α_B). The main conformational flexibility is through the virtual torsion angle (θ). Both the α-carbon and β-carbon positions can have associated bump radii (r_α and r_β), the latter of which can vary with residue type.

subspace. If all eigenvalues were positive, then the M largest eigenvectors whose sum was a preset fraction of the sum of all eigenvalues were used (projection into less than 3D subspaces was not allowed). Further refinements were carried out in the embedding space, and, if the dimension was higher than three, a new distance matrix was constructed and the cycle repeated until a 3D embedding was achieved. (See also section 2.2 and specifically, section 2.2.4 for theory and computational details.)

9.1.2 Model specification

9.1.2.1 Polypeptide chain geometry

Polypeptide chains were modelled by a α-carbon backbone with dummy β-carbon positions representing side chains (figure 9.1). The coordinates of which were calculated from the local conformation of the backbone as a point lying 2 Å beyond the bisector of the α-carbon virtual angle. This construct has the advantage that the 'dummy' β-carbon coordinates or their distances could be calculated from the α-carbon positions or distances alone. In the simpler models, the monomers were geometrically identical, the only difference between them was their 'hydrophobicity', a binary property, which could either be present ('hydrophobic' monomers) or absent ('hydrophilic' monomers). (Aszódi and Taylor 1994a).

The possible set of conformations of the model chains were determined by the following interactions between the monomers.

(i) **Steric constraints:** the minimal and maximal distances between any two C_α atoms were determined by the van der Waals radii and the maximal length of the totally extended connecting chain, respectively.
(ii) **Hydrophobic interactions:** it was assumed that, in general, hydrophobic residues tend to cluster together, therefore their expected distances would be smaller than for hydrophilic residues.
(iii) **Hydrogen bonding:** the geometrical consequences of interpeptide hydrogen bonding was modelled by simple distance constraints.

These interactions can be applied to all residues in the model chain, regardless of their sequential positions.

The model chain also had global properties which could not be simply described as a sum of pairwise interactions. The global properties in the present model were as follows.

(i) **Packing density:** the number of α-carbon atoms per volume has a characteristic value for real polypeptides. This usually changed during projection steps and needed repeated readjustment.
(ii) **Shielding:** although solvent molecules were not included in the model, the effect of a uniform polar environment was simulated by maximizing the number of buried hydrophobic and exposed hydrophilic residues.
(iii) **Chirality:** since the model chains were built of symmetric monomers, the chirality of the chain conformation (especially for secondary structures) had to be adjusted separately.

9.1.2.2 Steric constraints

The maximal allowed separation between any two C_α atoms was the length of the chain connecting them in a fully extended conformation. To model repulsion and volume exclusion, the atoms were not allowed to approach each other any closer than the sum of their respective van der Waals radii. For any α-carbon—α-carbon pair, an adjustment factor was computed so that multiplying their distance by this factor would eliminate the steric clash, no matter whether it occurred between them or their respective β-carbon atoms. For the distance matrix values, this was sufficient but in Euclidean space, any bumping α-carbon atoms were moved apart symmetrically along the line connecting them.

9.1.2.3 Residue density

The average residue density of protein molecules is about $\rho = 6 \times 10^{-3}$ Å^{-3} (Gregoret and Cohen 1991). During the embedding calculation, the density of the model has to be readjusted to compensate for the shrinkages accompanying projection. It must be kept in mind that the residue density is defined for 3D space only and its value is meaningless in higher dimensions. Consequently, for adjustments performed in distance space and immediately after projection in higher dimensions, indirect scaling methods were applied based on the observed distribution of distances within 3D spheres of uniform density. In the 3D adjustment cycle, the residue densities were approximated by the inertial ellipsoid method (Taylor *et al* 1983). (See section 2.1.4 and section 2.1.5.)

9.1.2.4 Hydrophobic interactions

The binary nature of the 'hydrophobicity' of the monomers defined three possible kinds of pairwise interactions: hydrophobic/hydrophobic, hydrophobic/hydrophilic and hydrophilic/hydrophilic. To each interaction a desired distance ($d^{(\mathrm{des})}$) was assigned, and the entries in the distance matrix were updated in every cycle by these desired distances using the following linear combination:

$$d_{ij}^{(\mathrm{new})} = (1 - s_{ij})d_{ij}^{(\mathrm{old})} + s_{ij}d_{ij}^{(\mathrm{des})} \tag{9.1}$$

where the 'strictness' values $0 \leq s_{ij} \leq 1$ regulated the amount of desired modification. The hydrophobic/hydrophobic desired distances were set to a lower value at greater strictness than for the other two interactions to model the tendency of hydrophobic residues to cluster together inside the hydrophobic core. The actual choice of the distances was somewhat arbitrary as these served only as guidelines to the program rather than hard constraints to be strictly adhered to.

9.1.2.5 Solvent accessibility

The interaction with solvent was modelled by burying exposed hydrophobic residues and bringing buried hydrophilic residues to the molecular surface. The accessibility of a residue was determined by placing the apex of a cone on the α-carbon atom with the axis of the cone through the β-carbon (centroid). The exposure of the residue was then measured as the minimal solid angle necessary to enclose the rest of the molecule (or within a predefined cut-off distance) (figure 9.2(*a*)). Intuitively, an exposed residue would have a cone with a relatively narrow cone angle (less than 180°) whereas buried positions would give 'wide open' inverted cones that have large angles >180°, equivalent to narrow ordinary cones clear of any atoms (figure 9.2(*b*)). In distance space, pairs of exposed residues, which were far apart and at least one of the partners was hydrophobic, were moved closer by a factor depending on exposure and distance. In Euclidean space, exposed hydrophobics were moved towards the centroid, while buried hydrophilics were moved outward by a suitable factor.

9.1.3 Hydrogen bonds

9.1.3.1 Hydrogen bond geometry

The backbone of our model does not represent the peptide groups involved in main chain hydrogen bonding in real polypeptides. Instead, fake hydrogen

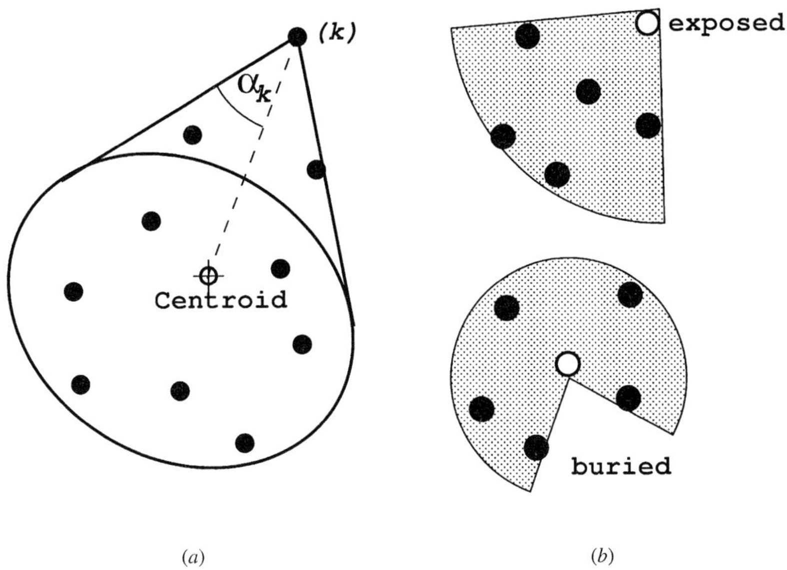

Figure 9.2. Conic accessibility measure. (*a*) A fast estimate of solvent exposure can be obtained from the solid angle (α_k) subtended at a residue (k) that contains all other atoms; (*b*) for an exposed residue (top) this angle forms a cone while for a buried residue (bottom) the cone is inverted.

bonds were allowed to form between the mid-points of the segments connecting two neighbouring α-carbon atoms and each mid-point could participate in two hydrogen bonds (figure 9.3(*a*)). A twist of the correct chirality was given to each pair corresponding to the twist found both between strands in a β-sheet and also between adjacent turns on an α-helix(figure 9.3(*b*)).

The definition of the desired geometry of these fake H-bonds involved the specification of the positions of two pairs of neighbouring strands (figure 9.3(*c*)). This unit was used to refine the expected approximate co-linearity of the bonds as shown in figure 9.3(*d*).

9.1.3.2 Prediction of bonding topology

H-bond formation exhibits positive cooperativity in two directions: first, if a H-bond is formed, giving rise to a unit-length ladder, the neighbouring bonds (extending the ladder) are more easily formed along the chain, producing longer parallel or antiparallel ladders. (Helices could be treated as special cases of parallel ladders where the two strands overlap in the sequence.) Second, an

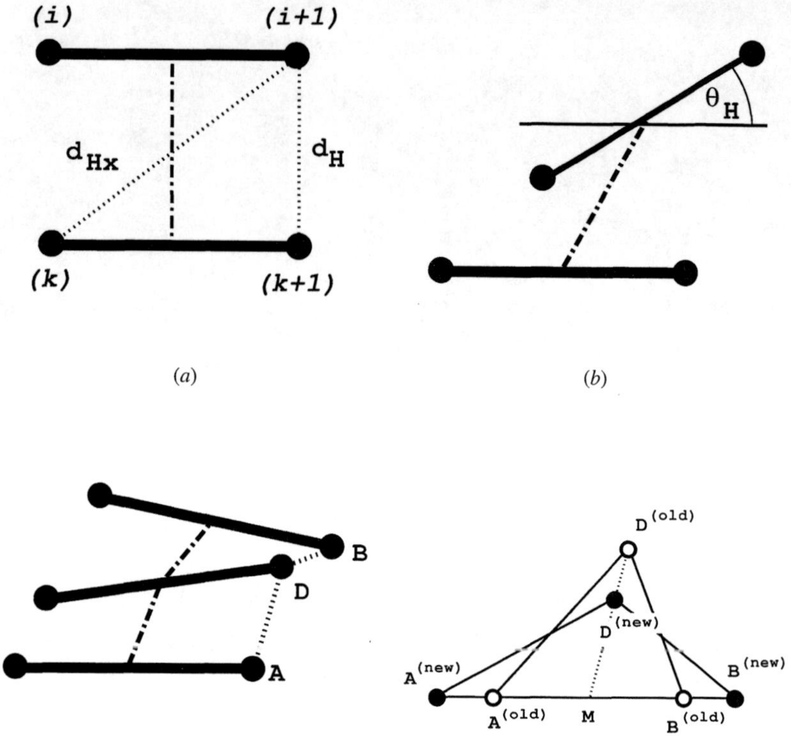

(a) (b) (c) (d)

Figure 9.3. Model hydrogen bond geometry. (*a*) A virtual hydrogen bond between the mid-points of two α-carbon virtual bonds. (*b*) Chirality imposed with an angle θ_H. (*c*) Extended H-bonding and (*d*) the shifts (old $\rightarrow$ new) implemented to refine colinearity along residues A–D–B.

existing non-helical ladder also aligns the peptide units so that it becomes easier for other ladders to join both sides, thus forming sheets across chain segments. For our model chains, first an initial set of candidate ladders were chosen. The 'rungs' of these ladders were all shorter than a preset threshold value, the candidate length. The construction of candidate ladders automatically included the simulation of positive cooperation of H-bond formation along the chain. The non-helical ladders were then joined by a single-linkage clustering algorithm, equivalent to that used in multiple sequence alignment (Taylor 1988), to select the best set of self-consistent ladder arrangement.

9.1.4 Chirality

The source of asymmetry in real proteins is the chiral α-carbon atoms, while in our model chains these centres are symmetric. The asymmetric influences that led to the preference of one mirror image over the other were incorporated in the handedness adjustment of hydrogen bonds. The steric arrangement of the four α-carbon atoms flanking a main chain hydrogen bond is chiral and the handedness of the configuration is consistent among right-handed helices and β-sheets. The asymmetry can be described by negative H-bond torsion angles $\theta_H < 0$ as defined in figure 9.3(*b*). In our model the configuration of these four α-carbon atoms around the H-bonds were set so that their handedness was consistent, i.e. all torsion angles should have the same sign.

Because of the problems with distance geometry in refining handedness (section 2.2.4.4), the inversion of handedness in helices required special treatment. A seven-residue long ideal right-handed α-helix was centred on each residue in helical strands which was locally better buried than its neighbours, then the model coordinates were moved towards the ideal coordinates (Taylor 1993a). This operation did not change the position of the locally buried helical residues, ensuring that the hydrophobic moment of the helix remained unchanged and the local geometry was preserved.

9.1.5 Scoring

The quality of the final 3D conformations could be assessed according to several criteria and it would be very hard to create a single score that reflected accurately all aspects of the structure. In the current implementation the following scoring schemes were used.

(i) Constraint score: the relative violations of the van der Waals distance constraints were averaged in this score (its ideal value was 0).
(ii) Hydrogen bond score: similar to the constraint score, the relative violations of distance constraints imposed upon the structure by the hydrogen bond geometry were averaged.

Both scores were simultaneously minimized during refinement. In addition to these, the fraction of hydrogen bonds with incorrect handedness was also monitored to assess the chirality of the model.

9.1.6 Simple models

Simulations of 90 model chains (all 100 residues long) with hydrogen bonds and 90 model chains without hydrogen bonds were carried out and various structural properties of the two sets were compared to each other and to a representative set of non-homologous (less than 25% homology), well-resolved (better than 2.6 Å) monomeric protein structures. Examples of the structures obtained are shown in figure 9.4.

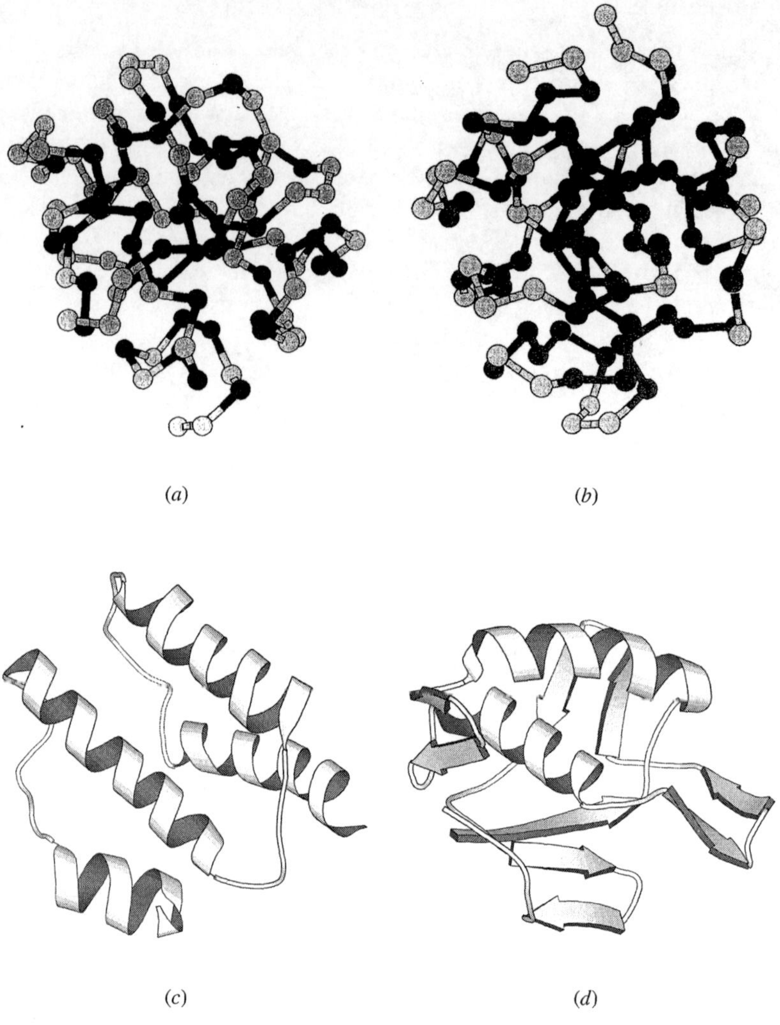

(a) (b) (c) (d)

Figure 9.4. Random protein models made by DRAGON; (*a*) and (*b*) without any H-bonding constraints, and (*c*) and (*d*) with H-bonding.

9.1.6.1 *Geometric properties*

The model chains folded into compact globules which had almost the same average residue densities as real proteins. The density distribution of the simulated chains (irrespective of H-bonding) was similar to that of the reduced reference set containing real proteins of chain length between 80 and 120 amino acids.

The molecular shape of the simulated chains was studied by fitting an ellipsoid to the molecule (Taylor *et al* 1983). The ellipsoids were then represented as points on the plot of the normalized semi-axes. The majority of the protein molecules in the representative set were found to be in the prolate region, which was well reproduced by the H-bonded model chains. On the other hand, the non-H-bonded molecules populated the spherical region as well, showing less preference for the prolate ellipsoidal shape (figure 2.7).

9.1.6.2 Hydrophobic core formation

The assessment of hydrophobic core formation in the artificial chains was carried out by calculating the *hydrophobic radius of gyration* (Robson *et al* 1987). To facilitate comparison with the binary hydrophobicity of the monomers in the model, in real proteins the amino acids A, C, F, G, I, L, M, P, V, W and Y were regarded as 'hydrophobic' and the rest 'hydrophilic'. The hydrophobic radius of gyration of real proteins was almost always slightly less than the overall radius of gyration. The H-bonded model chains followed this general tendency well. In the case of non-H-bonded chains, more marked hydrophobic cores were formed as judged by the comparatively lower hydrophobic radii of gyration.

9.1.6.3 Secondary structure formation

Although the non-H-bonded model chains folded into compact globules with distinct hydrophobic cores, no secondary structure was formed in these simulations. The H-bonded chains, on the other hand, always contained H-bond networks which were visually very similar to those found in real proteins. No special sequence patterns which would be prerequisites of secondary structure formation were found; all random sequences in our simulations folded up into conformations containing helices and/or sheets.

For real proteins, the expected number of main chain H-bonds (as defined by the `DSSP` program of Kabsch and Sander (1983)) varies approximately linearly with the chain size. The mean density of hydrogen bonds in our model chains was 41.2 bonds per 100 residues, less than the expected value of about 68 bonds. Helices were formed more easily than extended structures: 61% of all H-bonds were found in α-helices, π-helices contained an additional 14%, whereas antiparallel and parallel β-sheets accounted for 15% and 10% of H-bonds, respectively. The various secondary structure types occurred in almost all combinations, except for all-β proteins. The handedness of secondary structure elements were well reproduced in the simulations: most β-sheets exhibited a pronounced left-handed twist (viewed across the strands), and the overwhelming majority of helices were right handed.

9.2 Modelling with distance geometry

The distance geometry method described above combined with the various simple distance constraints are sufficient to make reasonable protein-like models. They did not, however, attempt to make any specific structure for a given sequence. In this section some further details of the method will be described that develop it into a practical modelling method. A more recent reincarnation of distance geometry modelling will also be described in which the basic DRAGON method has been 'embedded' within a genetic algorithm approach, resulting in a program called GADGET (for ***G***enetic ***A***gorithm and ***D***istance ***G***eometry for ***E***xploring ***T***opology).

9.2.1 Generic preferences

9.2.1.1 Hydrophobic packing

As discussed in the introduction (section 1.2), the propensity for hydrophobic residues to pack together is probably the most basic interaction found in globular proteins. It can be modelled in a very simple way as a binary effect (which is either present or not) and when present in a pairwise interaction implies that the pair of residues should have a shorter ideal separation which should be shortest for hydrophobic/hydrophobic pairs and greatest for hydrophilic/hydrophilic pairs (Aszódi and Taylor 1994a).

A slightly more sophisticated model might take into account the degree of hydrophobicity of each residue (using one of the many physicochemical or empirical scales) and if multiple aligned sequences are available then the hydrophobicity can also be averaged over the sequence family or, in addition, modified by the degree of conservation (Taylor 1993b).

Any of the above schemes, using different scales with different numbers of sequences of differing degrees of similarity, will give rise to a different range of values (or packing preferences). To convert these values to distances that are in the right range for the expected size of the protein requires a general scaling method. This can best be achieved by choosing a scaling method that transforms the packing preferences into a set of distances that has the same distribution as would be expected for a spherical compact protein of the same size as that being modelled. Most simply this can be done by refining the parameters of the transform function to match the moments of the calculated and theoretical distance functions. Previously, this was achieved matching only the first two moments (Taylor 1993b), however, a more general method has been described which can match the complete distribution (Aszódi and Taylor 1995).

9.2.1.2 Empirical potentials

If only short-range interactions are considered then it is not necessary to perform any scaling as these will be independent of the size of the protein in which they occur. At its most simple this might be a single value specifying a generic ideal packing distance for all hydrophobic residues. However, it can easily be elaborated to take account of, initially, the size of the residue and, subsequently, to consider the packing neighbourhood of the residue. Such preference (or empirical) potentials have been derived from the databank of known protein structures and used to assess the stability of folds—either native or modelled (Sippl 1990, Jones *et al* 1992).

9.2.2 Specific interactions

9.2.2.1 Conserved motifs

Recurring substructures in proteins (often referred to as *motifs*) constitute potential units of structure that might be recognized in a sequence and incorporated into a model. Motifs can range in uniqueness from an element of secondary structure to a whole protein domain. They might, respectively, have a theoretically known ideal form or be derived from a single occurrence. More typically, the situation will be intermediate, with no exact theoretical form but rather an average derived from several up to hundreds of examples. Embodiment of the average structure in a practical form is, in principle, not difficult and might be achieved by the superposition of the occurrences of the motif followed by the averaging of the coordinates of equivalent positions.

With remotely related motifs, however, sufficient differences might have accumulated to make superposition difficult. This problem has been overcome using locally defined frames of superposition—effectively defining a local structural environment for each residue. This environment takes the form of a set of interatomic vectors (in the local frame) to every other atom in the protein (Taylor and Orengo 1989b, Orengo and Taylor 1993) (see chapter 6). With the multiple structure superposition of equivalent frames each vector becomes a bundle of vectors (one for each protein) and the coherence of this bundle indicates whether the interaction is conserved or variable (Taylor *et al* 1994a). This can be made exact by measuring the sum of the squares of the deviations of the vectors from their average and the resulting value can then be used directly as a weight in the refinement of a segment of the model towards the motif (as described above).

9.2.2.2 Stick folds

The motifs described above have been derived from either complete domains or parts of real proteins. However, the method can equally well be used with outline sketches of protein folds—indeed, this was the application to which it was originally applied (Taylor 1991b, 1993b). The outline folds used in that application take the form of simple ‘stick’ figures where each stick represents the axis of a secondary structure (Cohen *et al* 1980, 1981). This simple representation has the advantage that it allows many folds to be combinatorially generated—

which might be viewed as a secondary structure-based lattice approach (see section 3.2 and section 8.2).

To convert a simple stick fold into a more realistic representation at the level of residue (α-carbon) resolution using distance geometry, a distance matrix was constructed from a linear combination of the matrix of hydrophobic preferences with a distance matrix derived from the ideal (stick) fold. The component from the stick model gives a bias to maintain the overall fold while local hydrophobic packing is encouraged by the hydrophobic distance matrix. In regions of secondary structure (or other motifs), the ideal substructure distances were also combined. As the refinement cycles progressed, the effect (weight) of both the generic hydrophobic packing propensity and the stick distances were reduced relative to the weight on motifs, resulting in a final model with ideal motifs and geometry (this method was used to refine the models constructed in section 8.2).

9.2.2.3 *Correlated residue changes*

It is sometimes thought that specific pairwise interactions can be predicted from the covariance of residues at different positions in a multiple sequence alignment. This belief has been based mainly on isolated examples—and while some of these involving functional residues in active or binding sites may be correct (Pazos *et al* 1997), it is difficult (if not impossible) to separate the effect from the general strong trend of conserved residues to lie close together anyway. A recent survey of a reasonably large number of proteins has found little support for any signal associated with correlated (compensating) amino acid substitutions when the effect of conservation is taken into account (Taylor and Hatrick 1994, Pollock and Taylor 1997).

9.2.3 Sources of real data

While the development of the methods described here have been motivated largely to incorporate predicted sources of data, they can equally be used to construct models to incorporate physical- or biochemical-based distance estimates.

9.2.3.1 *NOE distance estimates*

The nuclear overhauser effect (NOE) is exploited in NMR spectroscopy to obtain evidence of sequentially long-range interactions. However, these data do not specify a fully connected ($N \times N$) set of distances and generally the distance matrix is quite sparse. The problem has been overcome by a lengthy pre-estimation of the missing values, given the constrains of those that are known (Havel *et al* 1979). However, it is now more common for the problem to be solved by a standard refinement approach such as a simulated annealing method (deVlieg *et al* 1988).

The problem can also be solved using the gradual projection approach described above although this does introduce some added uncertainty as the

NMR data specifies distances between hydrogens which must be converted to estimates of interresidue distances to be used directly in our modelling program. Nevertheless, reasonable results can still be obtained (Aszódi *et al* 1995b).

9.2.3.2 Surface estimates

Estimates of surface exposure can be obtained from NMR spectroscopy by inclusion of a paramagnetic relaxation agent in the solution. These estimates can either be indirect (generic for residue type) or made specific through site-directed mutagenesis. However, under certain conditions, even the generic measure can yield specific exposure information (Scarselli *et al* 1999).

Both natural and artificial post-translational residue modification provide an additional source of information on surface exposure. Polysaccharide attachment sites and protease cleavage sites provide examples of these sources and in their incorporation in the method described above, such sites can be treated as very hydrophilic residues.

9.2.3.3 Disulfide bonds and other cross-links

Specific distance estimates can be obtained from cross-linking data. Some of these occur naturally through the post-translational modification of the protein: the most common being disulfide links between cysteine residues. Typically, there may only be a few of these but because of the short side chain length of the cys residue, the resulting distance constraints are quite powerful.

Other natural cross-links exist involving lysing or tyrosine but these are rare and the length of the side chains involved weaken the constraints. Artificial cross-links can also be introduced but again, the length of the linking molecule itself adds uncertainty to the constraint and these also typically attack lysine side chains.

9.3 Protein tertiary structure prediction

The question we pose in this section is to what degree methods can generate a specific protein fold given minimal constraints and ideally, given only the protein sequence. This problem is often referred to as the 'protein folding problem' since the first attempts to predict protein tertiary structure ran lengthy molecular simulations of the motions of the protein chain with the expectation that it would fold into its native form (Levitt 1976, Levitt 1983b, Skolnick *et al* 1989, Abagyan 1993). As these methods were often using only basic physics and chemistry, the approach also became known as the *ab initio* method. As neither of the methods that we have used previously attempt to simulate actual protein folding (and neither use basic atomic interactions), the more general problem is better referred to simply as the tertiary structure prediction problem.

9.3.1 *Ab initio* prediction

The folding of the protein sequence into its 3D form is neither random nor can it be directed by other factors since this would imply a source of information external to the system. Thus, by necessity, the protein must be self-assembling. This does not imply that proteins are forbidden to help each other fold (indeed there is growing evidence that this occurs), however, biochemical experiments have shown that many protein chains are capable of spontaneously folding to their functional form (Anfinsen 1973). An implication of this is that the form of the 3D fold is cryptically encoded in the protein sequence. To break this code and predict the 3D form of the protein from sequence is a challenge to molecular biologists and is a major unsolved problem in the field.

The enigmatic relationship between the protein sequence and its tertiary fold has been an endless topic of speculation and research ever since the first protein structures were determined and, allowing some limited successes, the basic problem is as great a puzzle today. The frustrating aspect of the problem is that, fundamentally, there is no mystery since the system is completely determined: we know the exact chemical structure of the protein, the nature of the solvent, and (in quantum mechanics) have a very successful theory that, for smaller systems, can predict chemical interactions with great accuracy.

So, in effect, the 'folding rules' of Monod are already known (see opening quote to chapter 1). Why then do they not work for proteins? Conventional wisdom answers that, in principle, they will but, as yet, we do not have the computer resources to run an accurate simulation for long enough. A rough calculation based on the internal rotational freedoms of the polymer imply an 'astronomic' time for the protein to search its conformational space to find the global minimum (Levinthal 1969). The implications of this are that proteins cannot fold (Levinthal's paradox) or that they do not search all of conformational space. Given that we observe folded proteins, then they must either be guided to the global minimum (by an evolved kinetic folding pathway) or exist in a metastable equilibrium.

Further difficulties arise from modelling the hydrophobic effect which is possibly the most important component in determining the overall tertiary fold. This is an entropic effect requiring sufficient bulk solvent around the protein chain to be simulated for perhaps tens of seconds (of the order of real folding times). Given that present molecular dynamics simulations (Brooks *et al* 1983) are measured in tens of picoseconds, there is a gap of 12 orders of magnitude, giving little hope for answers from this approach in the near future.

The above rationale presents a very common practical difficulty—time: but in essence, implies that the problem is solvable within a classical model. This view, however, denies the quantum nature of the molecules and before proceeding to pragmatic approaches it is worth considering that the problem may not be solvable in the conventional sense. In trying to reconcile the, often counterintuitive, relationship between the observer and observed in quantum mechanics, Penrose (1989) proposed that the indeterminate evolving quantum wave function is not collapsed by observation (as is commonly supposed) but by creating sufficient (gravitational) disturbance in the world—which he equated with the energy of one graviton. Thus, like a bubble blown too large, the function simply 'bursts', leaving a well-defined (observable) drop of soapy water.

The effect on the world created by a folding protein is roughly within this energy scale and one might imagine the protein folding (or partially folding) in a quantum phase where it is free to explore in parallel-time all possible pathways, so avoiding Levinthal's paradox. Unfortunately, this view obviates any direct computational approach to the problem—unless quantum computers (which exploit the same principle) become practical devices. Even with such advanced technology available, the most succinct formulation and solution of the calculation might be obtained using a single molecule—an 'analogue quantum computer' of the exact chemical nature as the protein!

Given the above critique of direct or (*ab initio*) methods of protein structure prediction, it might not be surprising that little practical advance has been made along these lines. The following sections outline a more empirical approach to the problem based on exploiting analogies with known structures, rather than deriving these structures directly. This approach lacks the full understanding of the system implicit in the *ab initio* methods but despite this unsatisfactory aspect, it currently represents the only approach capable of obtaining practical results.

9.3.2 Empirical methods

The approach that is pursued in this section is a development of the 'combinatoric approach' to structure prediction. This approach was based on a detailed analysis of secondary structure packing in proteins (Cohen *et al* 1979, Cohen *et al* 1981) and began what was later to be called the *empirical approach* to protein structure prediction, being based on the observation and analysis of known structures, rather than on fundamental physical principles. The essence of the approach was to generate possible structures and select the good ones using a series of filters. By dealing with large units of structure, this approach avoided many of the computational difficulties and was the first tertiary structure prediction method to achieve reasonable success on realistic (typical) proteins of biological interest. Although initially applied only to a subclass of proteins, the method was later extended to other classes (Cohen *et al* 1982, Taylor 1991b), including membrane proteins (Taylor *et al* 1994b).

The initial combinatorial approach was based on folds generated over a secondary structure lattice of the type described in section 3.2. In section 8.2 and section 9.2 of this chapter we have seen how both random-walk-based approaches and distance geometry (DG) can be used to elaborate and extend these models to generate a variety of protein-like folds. With the former method, the size of the proteins that can be constructed is almost unlimited whereas with DG, the behaviour of the method deteriorates with increasing size (due to the quadratic increase in unspecific pairwise interactions). As we have seen in the last section, however, DG has the capacity to incorporate a wide variety of constraints and synthesize these in a reasonably unbiased way.

The success of the combinatorial method depends on the volume and resolution with which fold space can be sampled and on the power of the filters to reject poor models. These two aspects are intimately linked: if the filters are able to recognize good protein models even when these are only roughly made, then fold space can be sampled on a course grid. On the other hand, if almost exact interactions must be present in the model before it can get a good score, then fold space would need to be sampled at a fine level. In this section, we will assume that the α-carbon models generated by the random walk and distance geometry methods are of sufficient resolution to provide a good coverage of fold space and will concentrate on whether it is possible to select the correct fold from these, given just sequence data.

9.3.2.1 Packing filters

Considering the polyhedral stick models of Murzin and Finkelstein (1988), the register of the sequence on the framework is set by the secondary structures with each structure being placed on alternate edges of the polyhedron as the winding progresses (section 3.2.5). Computationally, this can be achieved by the application of a recursive routine which chooses a path from each node until there is no further secondary structure units or a dead-end is encountered. On each of these conditions the procedure 'backtracks' to the preceding node and takes and alternative path. Exhaustive application of this procedure eventually enumerates every possible path.

On the icosahedral model of six helices, there are 1264 distinct paths (the smaller but less symmetric model for five helices can generate slightly more). This can be contrasted with the alternative approach (applied to the same size of problem) of simply adding one helix onto another and allowing the fold to grow through accumulated pairwise interactions. This generates many millions of possibilities (Cohen *et al* 1979) most of which infringe obvious steric constraints that are never encountered when the chain is constrained to an idealized framework.

9.3.2.2 Motif incorporation

The possible structures generated by an unconstrained combinatoric trace over all possible windings can be greatly reduced if a distance constraint can be placed on even a pair of structural elements. In a previous study on myoglobin (Cohen and Sternberg 1980b), the constraint implied by hæm binding was imposed after relatively detailed models had been built. However, it is more cost effective to apply any such constraints at an early stage. This might be done during the search over the tree of possibilities and if a forbidden pairing is encountered then all remaining combinations following that node on the tree can be neglected. This 'tree-pruning' strategy is most effective when the interactions being tested are sequentially local—such as the hand of $\beta\alpha\beta$ units of super secondary structure.

The single constraint of haem binding in the globins provides a relatively weak constraint on the possible folds, however, well-defined substructures—such as the EF-hand calcium binding motif— can provide powerful constraints when used as a filter. For example: the protein parvalbumin contains six helices, two pairs of which constitute EF-hands. Applying these as a constraint (independently of each other) reduced the possible structures from over 1200 to three—one of which corresponded to the native fold while the other two were trivial variants (Taylor 1991a).

In the $\beta\alpha$ protein class, possible folds are constrained by connection chirality and the avoidance of loop-cross-over. An example is shown in figure 9.5 for the chemotaxis Y protein (PDB code `3chy`), frequently used as an example (figure 3.16). The secondary structures of this protein can reasonably be predicted and from considering the hydrophobicity of the β-strands, two of the five can be identified as the probable edge strands of the sheet. The topologies of all six possible arrangements of the three core strands can then be easily enumerated (figure 9.5). Of these, one involves an unavoidable crossing of connecting loops while another forms a knot. Both of these might reasonably be excluded—although the latter with some reservation. (For a further discussion, see section 8.1.5.1.)

9.3.3 Model evaluation

Many different approaches have been taken to find a solution to the problem of model evaluation at different levels: ranging from detailed atomic-based energy calculations, through generic residue interactions to abstracted secondary structure packing interactions. These evaluation functions are also applied in different ways: either after the construction of a model (or many alternative models) as a selection filter or during the construction of the model by assessing different alignments (as in threading) of different packings (as in some *ab initio* prediction methods). To a large extent, the model evaluation is independent of the method of generating the models to be evaluated and, if a better evaluation function can be found, then all these approaches will benefit. This does not even require that the new evaluation function need be incorporated into any existing method: providing a modelling method can generate a variety of alternative structures then these can be ranked in a post-processing evaluation.

With the rise in popularity of threading methods, many new evaluation functions have been devised over the last ten years. To avoid the problem of constructing residue side chain positions, these have tended to concentrate on main chain interactions combined with generic side chain interactions, that is: all, say, Phe/Leu interactions will be evaluated from a single function representing an average (generic) interaction. Even with its disregard for residue side chain

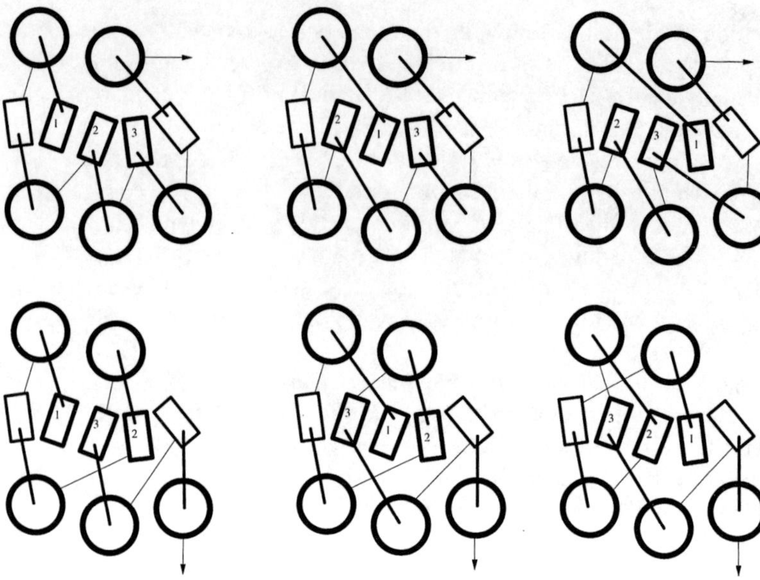

Figure 9.5. Possible folds for a small $\beta\alpha$ protein. The secondary structures for the protein 3chy can be predicted as: $\beta\alpha(\beta)\alpha\beta\alpha\beta\alpha(\beta)\alpha$, with those in parentheses being confined to the edge of the β-sheet. Labelling the remaining three core strands as 1,2,3, then their possible arrangements are: 123, 213, 231; 132, 312, 321. These are constructed in the manner of figure 1.4 so as to preserve a right-handed topology of connection between strands. The 123 variant is the native ('correct') fold; folds 213, 231 and 132 infringe no known constraint (213 is found in adenylate kinase); fold 321 has an unavoidable cross-over of connections and fold 312 forms a knot.

packing details, this approach can still lead to considerable complexity: for example, in the method of Sippl (1990) (20×20) residue pairs have an interaction specified for different spatial separations and different sequence separations, all of which are duplicated for different main chain interactions. Other approaches have simplified these terms by using abstracted features such as secondary structure state and residue exposure state (Ouzounis *et al* 1993, Taylor 1997b) and more recently, artificial neural nets have been used (Lin *et al* 2002).

9.3.3.1 Evaluating rough folds

A function is required, which, given all possible windings on an ideal framework, can recognize that which corresponds to the native fold. However, given the simplifications that are inherent in the idealized model such a function is unlikely to be reliable and attempts to specify it in terms of stick packing have yielded little, unless specific distance or motif constrains can be incorporated (Taylor 1991a). The length of chain connecting secondary structures might be used as a constraint, but this is also not very effective, given the relatively small dimensions of the packed globule and the uncertainty in secondary structure prediction. However, the fundamental problem is that the range of interactions between pairs of secondary structures is not great since one pair of packed hydrophobic surfaces looks much like any other.

A more realistic initial step has been to apply an evaluation function to models generated form known structures. A number of methods based on empirical energy potentials allow model protein structures to be evaluated without the need to fully specify side chain locations (e.g. Sippl (1990)). Such methods are effective at recognizing protein sequences matched—or *threaded*—onto correct homologues of known tertiary structure (Jones *et al* 1993). In principle, it is only necessary to apply the method to matching a sequence against a sufficiently realistic representation of a combinatorially generated structure to recognize the native fold. Two practical problems barring this simple solution are the accuracy and generality of the empirical potentials used in evaluating different threadings and the realism achieved by the models generated from 'stick' structures.

Results based on the globins indicate that while the methods are improving, the native fold cannot be recognized as a unique fit. The reason for this may simply be that the stick models are systematically different from 'real' proteins, so introducing an additional source of noise. Alternatively, there are many folds among the possible 'fake' proteins that, when viewed only at a detailed level, incorporate interactions that are more similar to the globin fold than anything encountered in the databank of 'real' proteins (for example; a 'mirror-image' globin fold). The elimination of these as candidate native folds may be impossible without full specification of (chiral) side chain interactions.

9.3.4 Genetic algorithm approach

When optimizing a complex (nonlinear) function of many variables, the DG approach is to allow all constraints to be fully satisfied but in more than three dimensions and concentrate on reducing the dimensionality of the model down to three (Crippen 1991, Aszódi *et al* 1995a). Alternatively, the dimensionality can be set at three and multiple attempts made to satisfy the constraints. Depending on the degree of randomness in the attempts, these methods range from Monte Carlo, through simulated annealing to simple (steepest-descent) minimizations (Levitt 1983b, Skolnick *et al* 1989, Abagyan 1993): and all can be repeated from different starting conditions.

An approach that is particularly attractive for protein structure modelling is to run multiple simulations in parallel and 'harvest' the best solution in the population. Given the hierarchic nature of protein structure, some individuals in the population may have attained a good substructure for one part of the sequence while others may be well formed in a different part. It is therefore desirable to let the individuals in a population 'trade' substructures, which by chance, should allow the emergence of individuals with a complete set of good substructures.

This approach has been developed for optimization where it is known as the genetic algorithm (GA) (Holland 1975). In this method, the system is encoded in a string which can be 'mutated' and sections swapped between individuals (like genetic cross-over). In the domain of protein structure, the parameter string can become a string of backbone torsion angles (which can be 'mutated' within an individual) but when a mutation or 'cross-over' (swap) is made between two individuals, the result is almost invariably disruptive: resulting in steric clashes and tangles, or a non-compact structure. With the speed of modern computers, the approach can still be applied in the face of these difficulties (Dandekar and Argos 1994), however, it is best to avoid them if possible.

PART 4

SYMMETRY

Perhaps the most remarkable features of the [myoglobin] molecule are its complexity and lack of symmetry. The arrangement seems to be almost totally lacking in the kinds of regularities which one instinctively anticipates, and it is more complicated than any theory of protein structure.

John Kendrew *et al* (1958)

Chapter 10

Structural symmetry

10.1 Symmetry

Despite the analysis of Kendrew and colleagues when describing the first protein structure (see opening quote), it has become apparent in the intervening years, (and, hopefully, also through reading the preceding pages) that proteins are not without internal order and often symmetry. Regularities in their structure span all levels of structural organization from the individual residue, through secondary structure (and super-secondary structure) to the overall fold of the protein. However, perhaps the greatest degree of symmetry is attained at an even higher level of the assembly of distinct protein chains (referred to as the *quaternary* structure in the hierarchy introduced in section 1.2). The symmetry seen in these assemblies (which may involve one or many distinct protein types) follow general 'rules', exhibiting a variety of symmetry operations (Blundell and Srinivasan 1996), but mostly simple two-, three-, and four-fold axes or extended helical arrangements—where the distinction from the large internally repeating (fibrous) protein structures becomes slight. The only symmetry operator not seen is, of course, any mirror plane. Although fascinating, and often strongly linked to function, quaternary structure will not be pursued in this work which will stay focused on the internal organization of proteins.

Within a protein chain, symmetry extends from the obvious to the obscure. The former comprises repeats that are closely similar both in structure and sequence and have undoubtedly been derived through a process of gene duplication, often with each repeat forming a distinct domain. The obscure type of symmetry is often so subtle that it is not clear whether it is a true property of the protein or a result of our innate ability to extract patterns from noise. Between these extremes is a range where the symmetry is clear but it is unclear whether it has arisen as a consequence of evolution or as a constraint of chain packing. This is where the interesting problems lie and we have already touched upon some of the issues when considering the complexity of a protein fold (section 8.3).

In the following sections, we will examine these types of symmetry, beginning with those that have a more obvious evolutionary origin.

10.1.1 Symmetry from domain duplication

There are symmetries within some proteins that have clearly arisen through the duplication and fusion of the protein chain (strictly, the underlying genetic code is duplicated). (For review, see Bajaj and Blundell (1984).) Duplication events are often manifest as two spatially and sequentially distinct domains. However, if the original protein existed in a dimeric form, then the fused dimer can still maintain its evolved interface in the new fusion protein. This probably happened in the aspartyl proteases (Tang *et al* 1978) and serine proteases (McLachlan 1979a) (see figure 5.8 and figure 5.6). The situation can be further complicated if two symmetric parts of the original dimer or the fusion protein have exchanged (or swapped) places (Bennet *et al* 1995) (see Heringa and Taylor (1997) for a review). This may have occurred in the knotted domain of `1yve` (section 8.1.5.3). With this degree of rearrangement, the form of the original protein becomes obscured and it is often difficult to decide if the symmetry has its origins in an evolutionary event or is a consequence of purely structural pressures.

In the following subsections we will review the known types of repeats from a structural viewpoint following a progression of all-α proteins through β/α to all-β. Some highly symmetric folds are seen in the β-trefoil[1] and β-propeller folds, some of which bring us to the limit of any certain evolutionary inferences concerning the origin of the symmetry. (For additional information on sequence relationships and function, see Andrade *et al* (2001)). Finally, we discuss the more obscure symmetry where there is no obvious sequence repeat. Such an ambiguous example can be seen in the Rossmann fold (discussed above and in figure 1.4(*b*)).

10.1.1.1 All-α super-helices

In our assessment of fold complexity in section 8.3, the simplest folds were repeated α-helices that packed together into a helix of helices with varying degrees of super-helical twist. Unlike the coiled-coils of α-helices in which the helices twist almost parallel to the super-helix axis, in these structures there is a repeating unit of two or three helices packing almost orthogonal to the super-helix axis. They can have varying degrees of twist from almost straight through various pitches (figure 10.1) to a completely closed barrel (figure 10.6, discussed below).

The all-α super-helices include proteins associated with nuclear transport proteins (exportins) and structural proteins that constitute part of the surface membrane (clatherin coated pits). The common feature of these is that they use their large surfaces for multifaceted protein–protein interactions (Andrade *et al* 2001). A general distinction can be made between two-helix and three-helix repeating units. The former include the widespread HEAT repeats (an

[1] In this section, the use of 'trefoil' is unrelated to the trefoil knot considered in section 8.1.5.1.

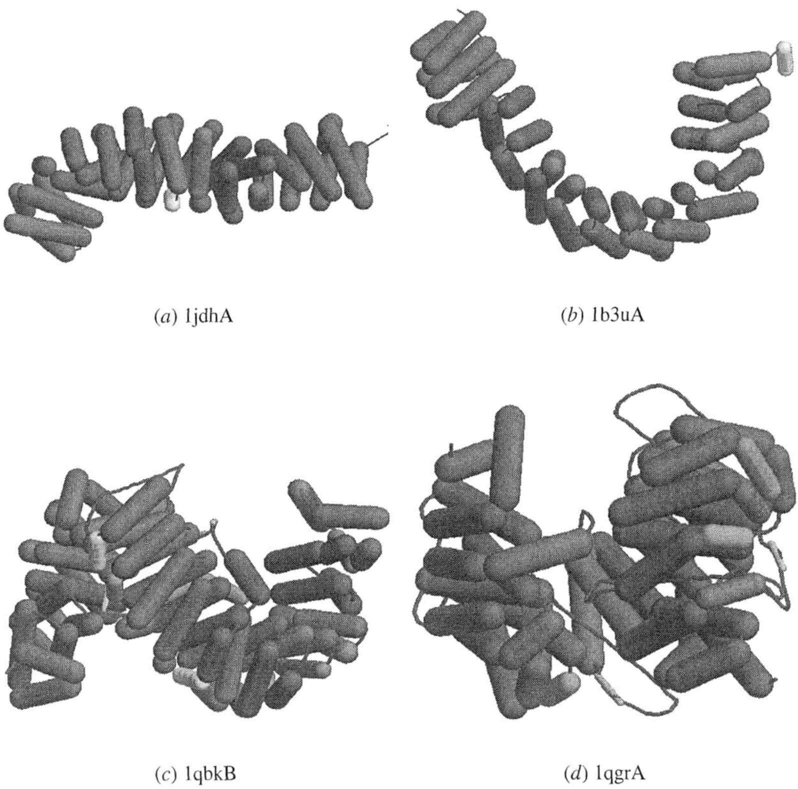

Figure 10.1. α-super helices. A gallery of helices of helices ranging from open helices of helices (*a*) to tight helices of helices (*d*). (PDB codes are given under each picture.) Individual α-helices are represented as 'sausages' (automatically defined as described in section 3.1) connected by a smoothed backbone.

acronym derived from the names of the proteins in which they were first found) while the latter include the Armadillo repeats (named from a fly protein, not the armoured beast). The more versatile interactions possible with a three-helix unit may explain why the Armadillo repeat proteins (figure 10.1(*c*) and (*d*)) may be able to form tighter super-helices than seen with the two-helix (HEAT) repeat unit (figure 10.1(*a*) and (*b*)).

10.1.1.2 Mixed helices in horseshoes

The structures described in the previous section are based just on α-helices but it is possible to find other secondary structure types in the repeating unit. The ankaryn repeat develops the two-helix helix by having a β-hairpin decorating

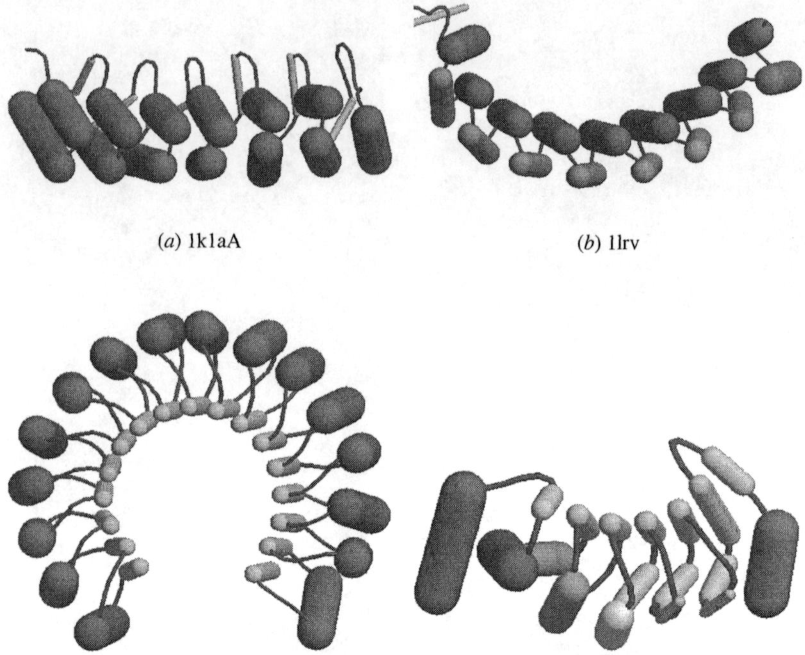

(*a*) 1k1aA (*b*) 1lrv

(*c*) 2bnh (*d*) 1a9nA

Figure 10.2. Mixed α helices and other SSEs. (*a*) An ankaryn $\alpha\alpha$-helix, (*b*) a LRR $\alpha 3_{10}$ helix, (*c*) a LRR $\alpha\beta$ helix and (*d*) a mixed $\alpha\beta 3_{10}$ helix. (PDB codes are given under each picture.) Individual α-helices are represented as 'sausages' (automatically defined as described in section 3.1) connected by a smoothed backbone.

one side (figure 10.2(*a*)). As the name suggests, this protein is also involved in protein–protein interactions. In the bacterial protein `1lrv` (figure 10.2(*b*)) a layer of α-helices packs with a layer of 3_{10} helices. The structure forms a horseshoe shape but, unexpectedly, with the inner (smaller) radius formed from the larger α-helix layer. This protein is a member of a large class of repeats referred to as leucine-rich repeats (LRR) from their preference to incorporate leucine in their hydrophobic packing. A more common type is based on a $\beta\alpha$ repeat unit (figure 10.2(*c*)) (Kobe and Deisenhofer 1993) and because of the strict distance imposed between the β-strands by hydrogen bonding, the structure forms a marked horseshoe shape with the β-sheet on the inner surface. (See Kobe and Deisenhofer (1995) for a review of the LRR motifs.) More complex mixtures of structure can also be found in layers (figure 10.2(*d*)) leading to a variety of structures formed from just β-structure. (Considered in the next section.)

10.1.1.3 β solenoids, prisms and trefoils

The type of simply wound (solenoid) fold continues from those seen in figure 10.2 without interruption into folds composed entirely of β-strands. Some, like that shown in figure 10.3(*a*) form a very regular triangular structure while others such as `1ulo` (not shown) are larger and less regular.

A different triangular fold is found with similar geometry to a solenoid but completely different connectivity in which each face of the triangular prism is formed by a more self-contained β-sheet. The exception is the amino-terminal segment which runs across the edge of all three sheets. (Figure 10.3(*b*)). An orthogonal switch in strand direction (as distinct from connection) generates yet another triangular architecture in which the sheets form the sides of a prism but with the strands aligned with the axis (figure 10.3(*d*)). The example shown is a domain from a larger protein and there is much less regularity in this aligned prism compared to the others.

β-trefoils consist of an unusual β-sheet formed by six β-hairpins arranged with three-fold symmetry in which the connections between strands fold into three very similar units adopting 'Y'-like structures (McLachlan 1979b, Murzin *et al* 1992). An assembly of three 'Y's form one half of the molecule which is then completed by an equivalent assembly for the other half (figure 10.3(*a*)). This combination of three-fold and two-fold symmetry allows the trefoil to be represented on an icosahedron (one of the Murzin and Finkelstein (1988) models described in section 3.2.5) in which a β-hairpin takes the place of an α-helix along an edge. (See Murzin *et al* (1992), figure 7). The trefoil fold also forms a closed barrel of six strands and can be analysed with the McLachlan (1979a) nomenclature ((Murzin *et al* 1992) and section 4.2). From a detailed analysis of their sequences, it seems likely that the β-trefoil hairpin units have arisen through gene duplication (Ponting and Russell 2000).

10.1.1.4 β-propellers

β-propeller structures are stacked arrays of β-sheets in which the edges of the sheets form a hub from which the sheets radiate. Because of the twist of the β-sheet, this gives the appearance of a ship's propeller (Murzin 1992). Over the years, proteins with different numbers of sheets have been found and there is now a structure from four up to eight sheets (the 4–7 sheet propellers are shown in figure 10.4). Murzin (1992) has analysed in detail the geometric constraints on the construction of propellers.

Each sheet in the propeller is self-contained and there is a clear sequence motif implying that the structure arose through tandem duplication of a single sheet. Depending on the family, the repeats are variously referred to as the WD40 motif (after conserved tryptophan and aspartate residues in the motif of 40 residues) or the Kelch repeat. As with some of the other β repeats described

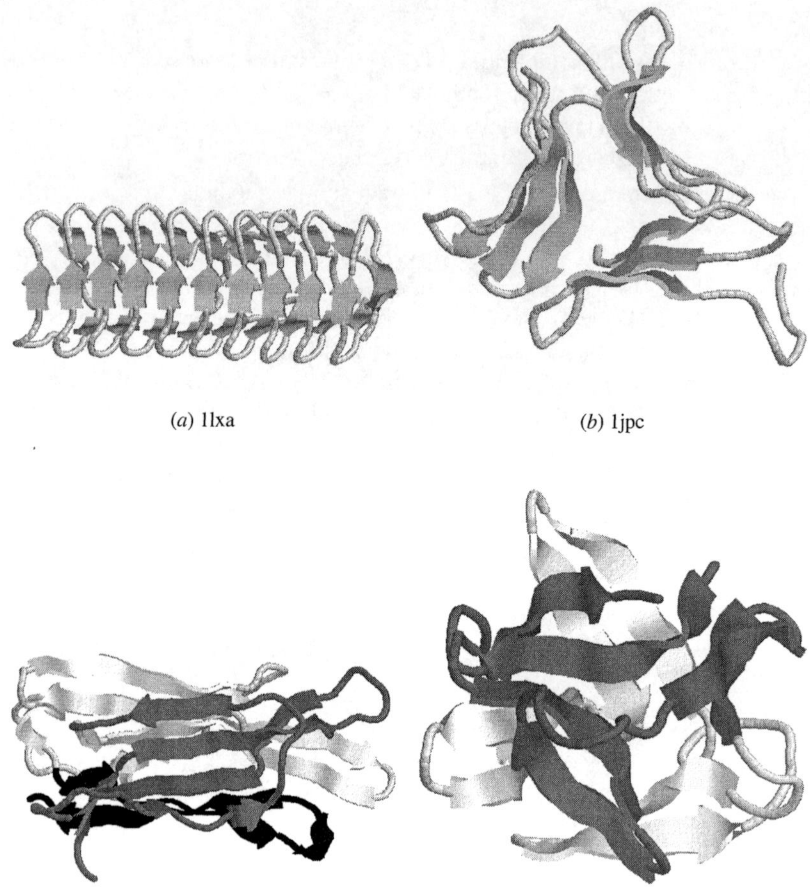

(*a*) 1lxa (*b*) 1jpc

(*c*) 1dlc (*d*) 1hcd

Figure 10.3. Triangular all-β proteins. (*a*) Triangular prism wound as a solenoid. (*b*) Triangular prism with almost distinct sheets. (The N-terminus at the back runs across all three sides.) (*c*) Aligned β-prism with strands running orthogonal to (*b*) (each face is shaded differently for clarity). (*d*) β-trefoil fold with the two symmetric halves shaded differently for clarity. Within each half, the 'Y' shape of three-linked hairpins can be seen. The figures were drawn with RASMOL and their PDB codes are given under each picture.

previously, the terminus crosses from the last to the first domain, perhaps acting to tie the structure closed.

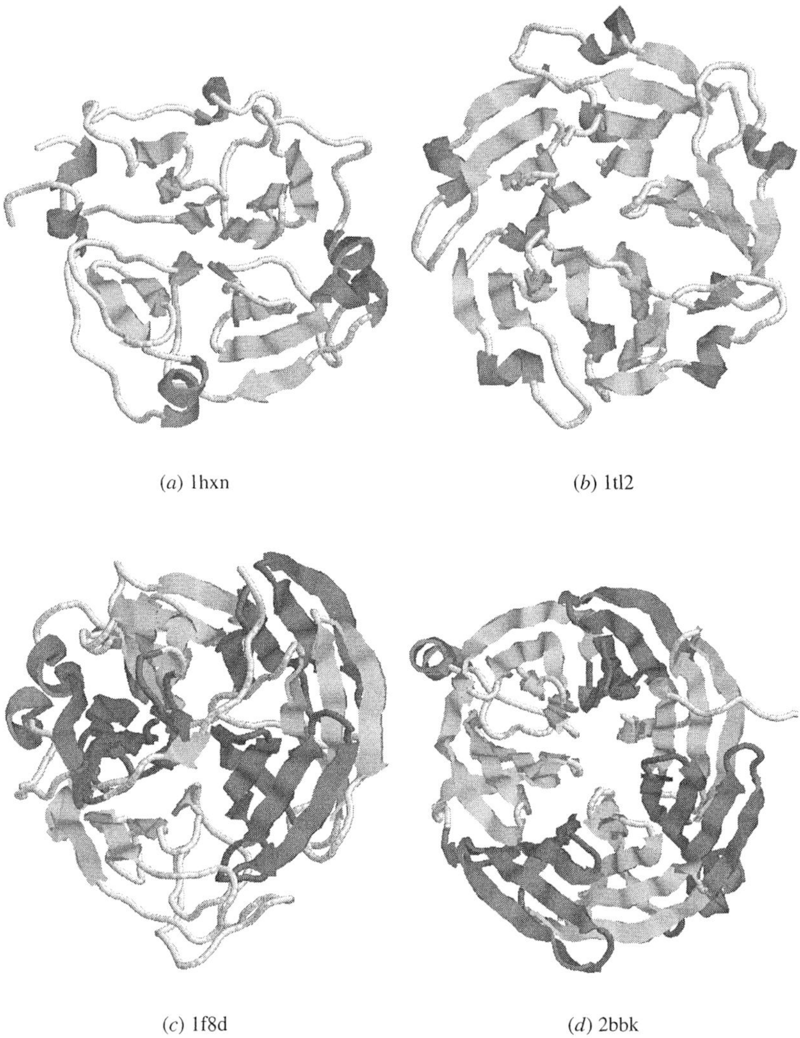

(*a*) 1hxn (*b*) 1tl2 (*c*) 1f8d (*d*) 2bbk

Figure 10.4. β-propeller structures for 4–7 sheets (*a–d*). In the larger structures, some sheets are shaded darker for clarity. The figures were drawn with RASMOL and their PDB codes are given under each picture.

10.1.2 Symmetries from secondary structure

With some exceptions, the structures considered above derive their symmetry from the helical (including linear and circular) arrangements of structural units that have been derived as a result of gene duplication. These units often remained distinct as domains or subdomains. In this section we consider the lower level of symmetry seen in protein folds within single domain proteins.

10.1.2.1 $\beta\alpha$-class

The clear chiral preference in connections between secondary structure units—the connection β-α-β is almost never left handed (see section 1.2 and figure 1.6)—can provide a strong source of symmetric structures. Imagine a protein consisting of consecutive units of α-helices and β-strands. Because local handedness is determined, all α-helices must lie on the same side of the β-sheet. However, as the α-helix is much wider than the β-strand, the structure must be curved to accommodate their differing bulk. This can result in a horseshoe (figure 10.2(*c*)) or a closed β-barrel surrounded by a ring of helices (Banner *et al* (1975) figure 1.5). Alternatively, if the end to which β-α units are added is reversed, the helices then fall on the opposite face of the β-sheet forming a structure with approximate two-fold symmetry (Rao and Rossmann 1973) (figure 1.4).

These structural constraints imply that the different $\beta\alpha$ folds we see cannot be random as there will always be the possibility of finding some symmetric arrangements of secondary structures. This is particularly likely when both the degree of symmetry (two-, three-, four-fold) is not specified beforehand and there is scope to neglect arbitrary 'disordered' parts of the structure. This problem is returned to in section 10.2 where a more quantitative approach to measuring structural symmetry will be considered.

10.1.2.2 $\beta\beta$-class

Equally intriguing symmetries can be found in the all-β class of structure. Typically, these are seen in structures consisting of a β-sheet (or sheets) with a closed connection forming barrel structure. If the barrels were opened-up (as in a Mercator projection of the world), the whole can be depicted in two dimensions. In this representation, some of the chiral symmetries resemble the decorative motif commonly used in classical Greece and was accordingly named the Greek key (Richardson 1977). The extension of this spiral has been called a 'jelly roll' (also by Richardson) and consists of eight strands in a closed barrel with two connections across the top and two below.

It has been suggested that the Greek key motif (and the jelly roll) might have arisen from the symmetric folding of an elongated hairpin β-structure in the form of a double helix (figure 10.5)). Similar ideas can be applied to the all-α class of structure also (Finkelstein and Ptitsyn 1987).

10.1.2.3 $\alpha\alpha$-class

Folding symmetries are also found in the α/α class but their relationship to the local chiral preferences of the substructures are less clear. Much of the apparent

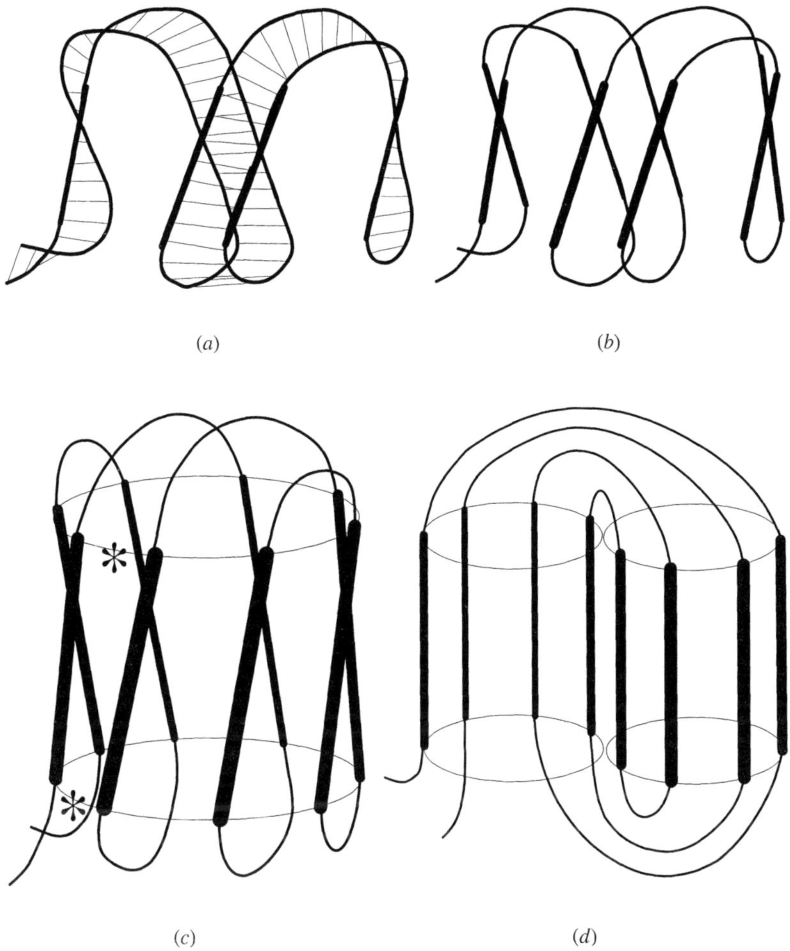

Figure 10.5. Various representations of an all-β protein. (*a*) Emphasizing the double helical nature of the chain which may have played a rôle in folding. (*b*) The final double-wound structure. (*c*) Hydrogen bonds between vertical strands creates a cylindrical β-sheet (arrows). (*d*) Opening the sheet (at the '$*$'s) produces a two-dimensional representation emphasizing the spiral that would be seen looking down the helix axis in (*a*) or (*b*) from the left. The centre describes a *Greek key* motif while the extended spiral is referred to as a *jelly roll*. Most β-sheets are less regular.

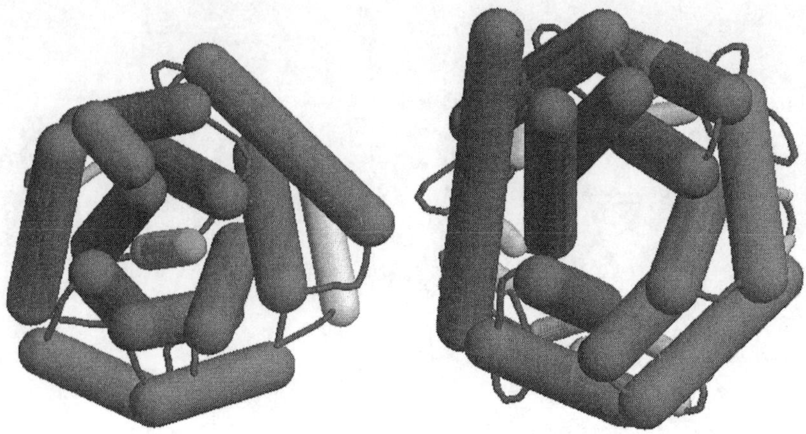

(*a*) 1c3d (*b*) 1fp3A

Figure 10.6. α-helical barrels. to tight helices of helices (*d*) (PDB codes are given under each picture). Individual α-helices are represented as 'sausages' (automatically defined as described in section 3.1) connected by a smoothed backbone.

symmetry within this class probably results simply from the more limited packing arrangements available with fewer secondary structures—a bundle of four or five helices will have some regularity almost no matter how they pack.

Symmetry becomes more apparent in the all-α helical barrels (figure 10.6). These have a simple solenoid fold like the super-helices discussed above but, in contrast to the super-helices, the barrels are all enzymes and to not have a repeating motif. As with most interactions involving α-helices, these barrels do not have an exact register that can be analysed in the same way as β-barrels (section 4.2).

10.2 A Fourier analysis of symmetry[2]

It should be apparent from the menagerie of structures surveyed in the previous section, that some structures contain repeated substructural elements often to a remarkable degree. Most analysis has taken the approach of dissecting out and variously classifying and counting the different substructures (Flores *et al* 1993, Salem *et al* 1999) or larger assemblies (Blundell and Srinivasan 1996). While this gives a good qualitative overview, it is difficult, using this approach, to get

[2] This section is reproduced in part from Taylor *et al* (2002) with the kind permission of Oxford University Press.

quantitative values that can be compared across structures with different motifs and across the range of size from secondary structure to domains.

The equivalent problem in the one-dimensional world of the sequence has been approached both by direct internal comparisons (Heringa and Argos 1993) and by Fourier transform (McLachlan and Stewart 1977), which is particularly suitable for the sequences of long fibrous proteins (McLachlan 1977, McLachlan 1983). The former approach can be directly transposed to 3D structures simply by repeated superposition of fragments (McLachlan 1979a, Matthews and Rossmann 1985). However, while the Fourier transform is a natural tool to adopt for the analysis of repetition in the one-dimensional sequence, its extension to structural data is not straightforward. In Cartesian coordinates (with each atomic point modelled by a Gaussian density) the 3D Fourier transform cannot be interpreted easily. Attempts have also been made using spherical coordinates (harmonics) (Duncan and Olson 1993) but addressing protein topography and not topology.

In this section we describe a Fourier analysis at an intermediate, two-dimensional, level based on the structural similarity matrices used in protein structure comparison. In this representation, symmetric[3] structures appear as off-diagonal 'ridges' of high score (as in the sequence 'dot-plot') and the periodicity of these ridges can be analysed by the Fourier transform. To allow for the quantitative interpretation of the resulting spectra, we also describe a normalization method based on 'random' structures.

10.2.1 Internal protein structure comparison

The `SAPit` program (Taylor 1999b) can be used to calculate the similarity between two proteins and if the two proteins are identical, then the program will perform a self-comparison (or an internal comparison). This method uses the double dynamic programming (DDP) algorithm (Taylor and Orengo 1989b). In the `SAPit` program, a pair of positions (one residue from each structure) is compared by matching their interatomic vectors to all other residues in their respective structures. This is achieved by constructing a matrix, defined by the sequence of one structure against the other, which is composed of a measure based on the relative geometry of the pair of pairs (see figure 6.3 and Taylor (1999b) for details). In structure comparison, this matrix (referred to as the *low-level matrix*) provides the base from which an alignment is extracted (using a standard dynamic programming algorithm). The alignments over all pairs of residues (between the two structures) are summed, forming another matrix (*high-level matrix*) from which the final alignment is extracted.

[3] In the following sections, the term '*repeat*' will be largely used to refer to sequence repeats, while the term '*symmetry*' will be used mainly for structural repeats. This distinction is made to emphasize that the measure of structural similarity used here is based on the comparison of 3D structures and captures more of the structural context than just the linear arrangement of secondary structures. For example, if the comparison matches two $\beta\alpha\beta\beta$ substructures, then these will have the same internal structure and will also hold a similar 3D relationship to the rest of the structure. This implies symmetry but does not specify what sort.

In the `SAPit` program, a small number of selections are made and these are then increased over successive iterations. For the current application, there was no iteration and the number of initial selections was correspondingly increased to equal the length of the protein (equivalent to the number in the final cycle of the iterated version). To prevent the selections from clustering on any strongly similar local substructures, a random perturbation factor was included in the choice of selected pairs. At the end of the comparison calculation, the high-level score matrix was normalized to attenuate elements with a value greater than three standard deviations (as calculated over the whole matrix).

10.2.2 Smoothing the score matrix

In a protein structure that contains repeated substructures, these will form diagonal ridges across the matrix parallel to the main diagonal ($i = j$) and it is the strength and periodicity of these undulations that can be analysed using a Fourier analysis. However, it was found that often these ridges are sharply defined and the resulting spikes or peaks generated a high 'background' level in the Fourier transform spectrum. (It should be remembered that the transform of a 'spike' gives rise to an equal signal across the full frequency range.)

To make the frequency spectrum easier to interpret, the high-level matrix was smoothed. Simple averaging was avoided as this tended to eliminate locally strong similarities (for example, the peak from two corresponding secondary structures would be averaged with the weak signals from their flanking coil regions). Instead the 'averaging' was made by taking the maximum over adjacent values—equivalent to a dynamic programming algorithm, as follows:

$$b_{i,j} = \tfrac{1}{2}a_{i,j} + \tfrac{1}{2}\max\begin{cases} a_{i-1,j-1} \\ a_{i-1,j}/2 \\ a_{i,j-1}/2. \end{cases} \tag{10.1}$$

The factor of a half applied to off-diagonal contributions is equivalent to a gap penalty and dampens smoothing in the antidiagonal direction while still allowing the propagation of some signal across small insertions.

This operation was applied to each cell in the matrix $\boldsymbol{A}$ (with increasing i and j) creating the matrix $\boldsymbol{B}$. As $\boldsymbol{B}$ will be asymmetric, the smoothing operation was then applied to $\boldsymbol{B}$ with decreasing i and j, recreating a new matrix $\boldsymbol{A}$. This flip-flop smoothing was repeated twenty times resulting in a matrix that still contained substantial detail. The effect of this repetition can be pictured by considering a single diagonal line ($i = j + n$) in the matrix, which for half its length has a value one and is otherwise zero. After 20 iterations, the sharp edge will be smoothed to a sigmoid curve which makes most of the transition from 1 to 0 over a range equal to the number of smoothing iterations. More generally, a single point will be smoothed into a Gaussian (bell-shaped) curve of the form exp($-x^2/n$), where n is the number of iterations. By analogy with optics, this means that two 'spikes' closer then 6.5 residues will not be resolved. In terms of protein structure, short-range features will be smoothed away, but typical secondary structures will remain distinct. It should be remembered, however, that this smoothing takes place after comparison by the `SAPit` program so all features, whatever their size, are fully considered in establishing internal similarities.

10.2.3 Fourier transform

The remaining problem with the data generated as described above, is the direction in which to calculate the Fourier transform. As the matrix is symmetric, this leaves a choice between the direction parallel to an axis or parallel to the minor diagonal ($i + j = N$, where N is the length of the protein: referred to below as the antidiagonal). The former has the problem that the diagonal itself constitutes an anomaly in the scores since the similarity of a residue to itself (plus environment) is very high. In the latter direction, however, there is only one line in this direction that covers the full length of the protein (the antidiagonal itself).

10.2.3.1 Fourier calculation

As proteins are short and computers are fast, the fast Fourier transform (FFT) algorithm was not used as this requires special treatment of the signal to render it in lengths of powers of two (Press *et al* 1986). Instead, the direct approach was used of multiplying sine and cosine waves over a range of frequencies and plotting the power of each component as a spectrum. Each term in the spectrum was normalized for the length of the protein (N) as: $100(s^2 + c^2)^{\frac{1}{2}}/N$, where s and c are sine and cosine (real and imaginary) components of the transform (see figure 10.7 for examples and a comparison to the FFT method).

The Fourier method detects a signal from the relative spacing of features, and not from their number, so it should be noted in these spectra, that a peak at five means ridges occur in the score plot with a spacing of $1/5$ the protein length—but there are not necessarily five of them. In addition, the size of the components giving rise to the repeat is not easily found in absolute terms as the same period will contain differing numbers of residues in proteins of different sizes while the substructures might be the same size but have longer ‘loops’ between.

For the calculation of the transform in the antidiagonal direction, the use of the antidiagonal alone may be unrepresentative and a band of twenty antidiagonals was taken. Such a construction, however, creates outer edges that are shorter than the antidiagonal and to correct this, the ‘missing’ corners were padded by repeated reflection of the diagonal at each level.

For the calculation in the direction of the rows, two variants were tried: first with the untouched score matrix and second, with the row shifted by one in each column and wrapped so as to shift the diagonal to the edge.

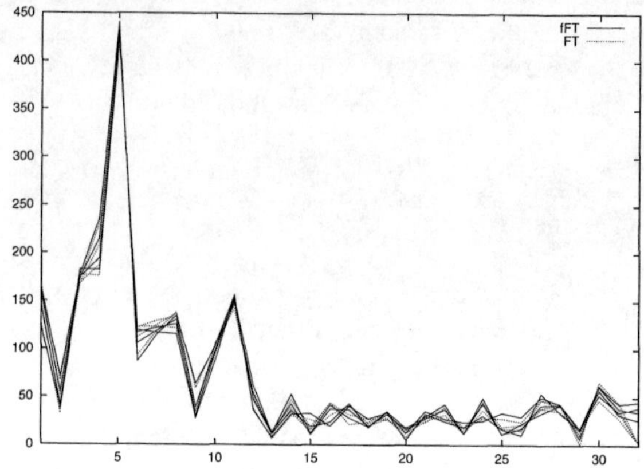

(*a*) Five-fold

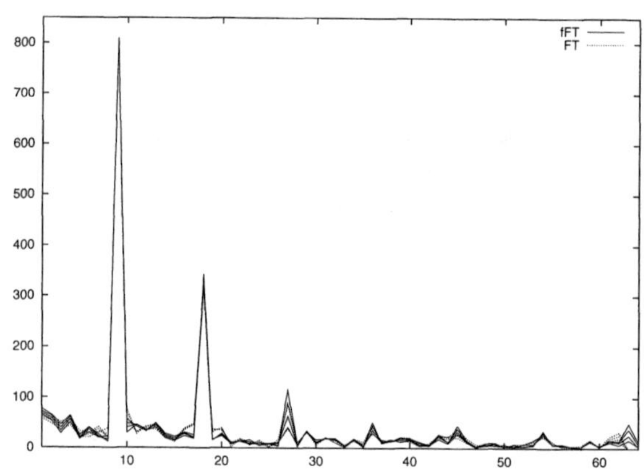

(*b*) Nine-fold

Figure 10.7. Example Fourier transform power spectra. The power spectra are plotted for the middle five rows from the score matrices of two proteins. (*a*) The small $\beta\alpha$ protein cheY (`3chy`, 128 residues) and (*b*) the last nine repeats of the ribonuclease inhibitor (`2bnh`, 256 residues) (see figure 10.2(*c*)). `3chy` has five $\beta\alpha$-units while the `2bnh` fragment has a clear nine-fold repeat. The latter repeats were strong enough to give rise to harmonics (peaks at multiples of nine). The spectra are plotted for the current implementation (full) and also for the fast Fourier transform algorithm (broken) using the routine `realft` from the numerical recipes collection (Press *etal*, 1986). The choice of proteins with lengths equal to powers of two allowed the unbiased comparison of the two algorithms.

10.2.4 Visualizing repeats

The power spectrum from a protein structure may indicate the period of the repeated structures but does not provide an exact definition of their location or boundaries[4]. These aspects must be extracted using more conventional structure comparison, but this can be guided by the Fourier analysis.

10.2.4.1 Biasing `SAPit` *to overlap repeats*

In the score matrix calculated by the `SAPit` program, each off-diagonal 'ridge' represents a region in which repeats can be aligned. The `SAPit` comparison was focused on each solution in turn by convoluting a Gaussian (bell-shaped) function over the off-diagonal ridge in the score matrix. The width of the bell curve was set as a function of the period of the repeat such that all neighbouring ridges were essentially eliminated, as follows:

$$s'_{ij} = s_{ij} \exp\left(-\left(i - j - \frac{rN}{p}\right)^2 \Big/ \left(\frac{N}{2p}\right)^2\right) \qquad (i \geq j) \qquad (10.2)$$

where N is the length of the protein, r is the rth ridge away from the diagonal ($r = 0$) and p is the frequency of the repeat. Equating the above bell curve with the normal distribution would give a standard deviation of $(N/2p)$. This means that the neighbouring ridges are two 'standard deviations' away, at which point the value of the curve is: 0.0183—sufficiently small to prevent the alignment path from jumping between ridges.

In a protein containing five repeated substructures, $A \ldots E$, the solution obtained with $p = 5$ and $r = 3$ is the alignment of substructures A, B, C with C, D, E.

10.2.4.2 Delineating repeat boundaries

The same masking approach can be used to identify repeat boundaries by focusing on the main diagonal ($r = 0$) and extracting the alignment path, then cyclicly shifting the original score matrix rows by one period (N/p) and repeating the alignment. After a full set of shifts, the points at which each of the p alignment paths cross the edge of the matrix gives the repeat boundaries.

[4] It might be thought that this information could be extracted using wavelet analysis, however, this technique can only be used where the feature is much smaller than the length of the signal (such as a short pattern of residues in a long protein sequence). In the current application the features constitute a large fraction of the length of the protein, and indeed, at this scale, wavelet and Fourier transform methods converge.

10.2.5 Analysis of total power

10.2.5.1 Model repeat proteins

The repeating $\beta\alpha$-horseshoe structure of the ribonuclease inhibitor 2bnh (figure 10.2(*c*)) was used to generate a double series of model structures. For one, the last strand was deleted, leaving exactly 16 $\beta\alpha$-units. These were then successively deleted in turn from the amino terminus (being the least regular) giving 16 coordinate sets each with a different number of repeats. For the second series, this was repeated but with the last strand deleted giving a series of 16 $\alpha\beta$-units.

The total power of the spectrum obtained when transforming each construct in these series of structures is plotted against the number of residues in the structure in figure 10.8(*a*). The power rises quickly followed by a slower, linear increase. The combined trend can be reasonably modelled with the combination of a linear component and an inverted Gaussian function:

$$p = ax + b(1 - \exp(-cx^2/10^4)) \tag{10.3}$$

giving the total power (p) from the number of residues x with the coefficients $a = 1$, $b = 2400$ and $c = 3$.

10.2.5.2 Ideal $\beta\alpha$ proteins

Using the same $\beta\alpha$-unit lengths as in 2bnh, model protein structures were generated with 4, 6, 8, 10 and 12 $\beta\alpha$-units (plus a terminal β-strand) and a sample of 15 different folds were taken from each. These are plotted on figure 10.8(*a*) where it can be seen that these structures have a very high power, with only the weaker members being comparable to the 2bnh series.

10.2.5.3 'Random' proteins

A series of random walk ('fake') proteins from 50 to 450 residues was generated in steps of 50 residues. A sample of 20 from each length was transformed and the resulting power of the spectrum plotted in figure 10.8(*a*).

This series shows a slight dependence on length, with the power level rising from around 1200 ($\pm$200) for structures with 100 residues. The upper limits of this distribution can again be modelled by the same function as described above (equation (10.3)) with values for the coefficients $a = 2$, $b = 1400$ and $c = 4$. This curve provides a good base-level against which the power of 'real' proteins can be assessed.

10.2.5.4 'Real' proteins

The program was run on a selection of proteins from the PDB (as used previously by Jonassen *et al* (2000)) and the power of their spectra plotted against length

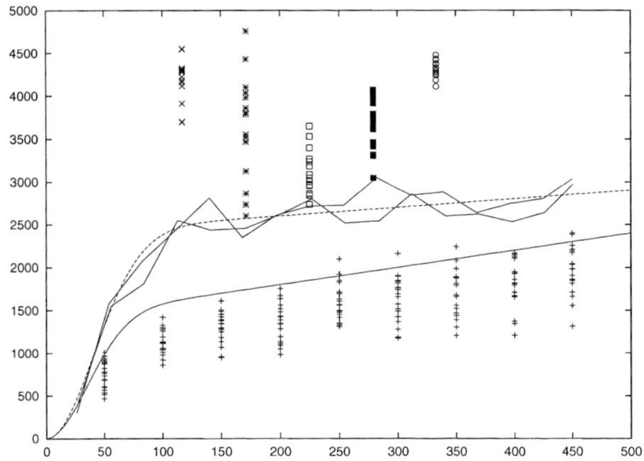

(*a*) Models

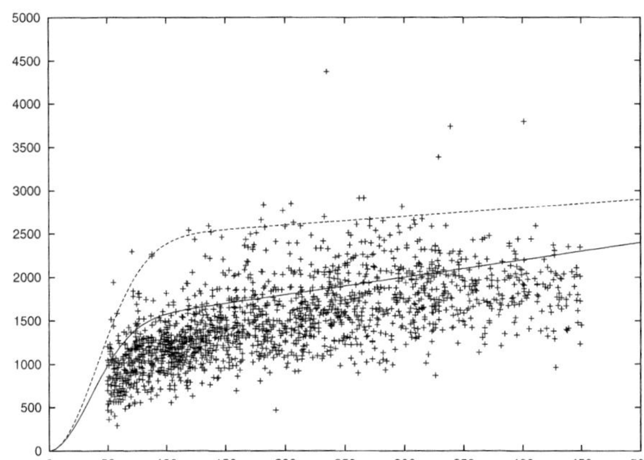

(*b*) Native

Figure 10.8. Total power against protein length. (*a*) Model proteins, including: the deletion series for the 2bnh structure based on its $\beta\alpha$-unit and $\alpha\beta$-unit ('zig-zag' lines) with a summary curve $N + 2400(-3N^2/10^4)$ (broken curve); random walk ('fake') proteins (+s) with their upper boundary summarized by $2N + 1400(-4N^2/10^4)$ (lower curve) based on a sample of 20 structures for each length (see figure 8.8). Ideal $\beta\alpha$ models are plotted as symbols, based on $(\beta\alpha)_n\beta$ with $n = 4$ ($\times$), $n = 6$ ($*$), $n = 8$ ($\square$), $n = 10$ ($\square$), $n = 12$ ($\circ$). (*b*) Native protein data with the summary curves from the model data replotted for reference.

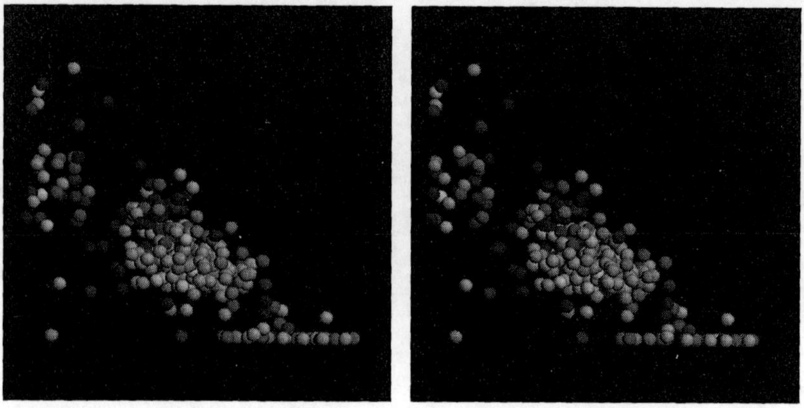

(*a*) (*b*)

Figure 10.9. Fourier results plotted with secondary structure composition. The percentage of β-structure (X-axis) and the percentage of α-structure (Y-axis) lie in the plane of the paper with the total power of the frequency spectrum plotted out of the page (Z-axis, shown as a stereo pair). Each protein that had a power above random ($1400 + 2N$, where N is the number of residues) is plotted as a sphere shaded by the frequency of the highest peak in the spectrum (light = high frequency, dark = low). The figures were produced by the program RASMOL.

(figure 10.8(*b*)). The majority of proteins lie below the upper limits attained by the 'fake' proteins but a considerable number (17%) still lie above this line and even above values attained by the series of 2bnh repeats (1%), with a few approaching the highest values seen for the ideal model proteins.

Inspection of the highest scoring structures revealed a collection of symmetric folds, dominated by the very regular β-propellor structures (Murzin 1992) but also containing many regular $\beta\alpha$ proteins, including TIM barrels. These results can be summarized by plotting the total power for each protein along with its secondary structure composition, while colouring the protein as a function of the most dominant period in its spectrum. (Figure 10.9).

In this plot, the 'bright' central region corresponds to the $\beta\alpha$ proteins (TIM barrels, Rossmann folds and leucine-rich repeats) while the bright edges are the more repetitive all-α and all-β folds.

10.2.6 Removing expected symmetries

Some of the proteins identified as symmetric, contain internally duplicated domains that are apparent from sequence comparison. To select the more unexpected symmetries, these were filtered using the REPRO program (Heringa and Argos 1993). This adjustment was made by taking the normalized power (s) divided by the REPRO score (r) as: $t = s\exp(-r^2/10^5)$. The new score t is typically $s/1000$ for the strongest sequence repeats ($r = 831$), falling to $s/10$ for clear repeats ($r = 480$) and $3s/4$ for weak repeats ($r = 170$).

The greatest change between the rankings is the disappearance of the highly repetitive folds: β-propellors, β-prisms and $\beta\alpha$-horseshoe proteins (see figures 10.2, 10.3 and 10.4). The single example of this type that holds its place in the top 50 is the four-fold β-propellor `1gen` and although its internal repeat was recognized by REPRO, the sequence identity over the repeats is less than 20%. Those remaining in the filtered list were overwhelmingly of the globular $\beta\alpha$ fold class and are dominated by the Rossmann fold types (which contain a pseudo-two-fold) and the $\beta\alpha$-barrel (TIM barrel like) fold which have eight-fold cyclic symmetry (but also incorporate many deviations). The only globular $\beta\alpha$ class protein to drop markedly in the rankings was the von Willebrand factor protein `1atzA` which as well as a very symmetric fold ('classic' Rossmann fold[5]) has sufficient sequence similarity in the two halves for REPRO to pick up a repeat.

A fold that makes a stronger appearance in the filtered list is the $\alpha\beta\beta\alpha$ layer protein `1ryp1` from the proteasome which contains an internal structural duplication. This symmetry runs through three of the four layers of secondary structure and, although it is not clear to the 'eye', it was identified as a two-fold repeat in the Fourier spectrum. The sequence identity over the repeats is less than 10% which would not be seen by any sequence-based method.

The more obvious structural repeat seen in the double domain γ-crystalins (`1bd7A` = 1.4 Å/84 res. 36% seq.ID; `1gcs` = 1.6 Å/84 res. 36% seq.ID) was easily eliminated by REPRO but the more distant internal (Greek key) repeat within each individual domain did not score enough to be downgraded and the structure `1dsl` holds its position after filtering.

As would be expected, there are no sequence repeats detected by REPRO that do not have a corresponding structural repeat. Although this simply reflects the principal that structure is better conserved than sequence, it is interesting to examine the proteins that approach the violation of this principle most closely. These are both artificial constructs, being linked dimers of a globin (`1abwA`) and the HIV protease (`1hvp`). Their exact internal sequence repeat gives a very high REPRO score while the single structural duplication, although strong, constitutes only one off-diagonal ridge in the score matrix. The closest native proteins to these are the annexins (`1aei` and `1axn`).

10.2.7 Assessment of the Fourier approach

The Fourier transform is a simple way of extracting the periodicities in a signal and in its application above is able to utilize all the information in the two-dimensional score matrix, also allowing statistics to be gathered on

[5] The original Rossmann fold was half a dinucleotide-binding domain $(\beta\alpha)_3$. It is used here to refer to the double fold: $2 \times (\beta\alpha)_3$, which constitutes an intact domain.

the significance of the peaks in the spectrum. Even without using the fast Fourier transform algorithm, the calculation takes very little time relative to the calculation of the structure comparison.

As mentioned in the introduction, a disadvantage of the approach is that it is not possible to tell the number or (within limits) the size of the substructures that give rise to any periodicity without returning to examine the original score matrix. Given that some information must be lost in the reduction of a complex structure to a few numbers, this is perhaps not unexpected and (as was outlined in the methods section 10.2.4), the transformed signal can still be used to help extract the repeating substructures.

A further ramification of this loss of information is when the substructures occur in a range of sizes—as is often seen with the $\beta\alpha$-units in the eight-fold $\beta\alpha$-barrel folds. In this situation, rather than the expected sharp peak at frequency $N/8$, the peak becomes spread (or sometimes split) over adjacent frequencies. Similarly, if the symmetric domain has a large insert or terminal addition, then the peak will again be displaced from the expected frequency. An example of this can be seen where the number of repeats for the seven-fold β-propellor structure `1gotB` is recorded as eight. Figure 10.10 shows the obvious explanation for this is the long N-terminal α-helix and loop.

When native proteins were ranked by the power of their spectrum (normalized for length) the β-propellor folds occupied the top positions but, otherwise, a wide variety of fold types were seen to score highly. However, the majority of proteins have no more symmetry in their structures than would be expected from a compact random walk. When this ranked list was corrected to downweight proteins that contained detectable internal sequence similarity, the top proteins became almost exclusively composed of globular $\beta\alpha$ class proteins. This result was unexpected as there is no obvious structural reason why more globular $\beta\alpha$ proteins should not be found with a clear internal sequence duplication or why more symmetric all-β or all-α protein should not exist without an accompanying signal in the sequence.

10.2.8 Origin of structural symmetry

A structural explanation of the dominance of the globular $\beta\alpha$ proteins might be based on the relative sizes and degrees of structural freedom that are available to the different super-secondary structure types. Those composed of all-β structure have a geometric regularity imposed by the plane of the β-sheet but are otherwise relatively topologically unconstrained, so giving rise to few symmetries by chance. The all-α protein structures lack the spatial register imposed by a hydrogen bonded sheet and so will naturally be less symmetric in their packing. However, as the α-helix is a relatively large structure, smaller proteins (with less than six helices) will stand a good chance of acquiring a symmetric arrangement. The $\beta\alpha$ unit combines both symmetry-inducing attributes of the previous types,

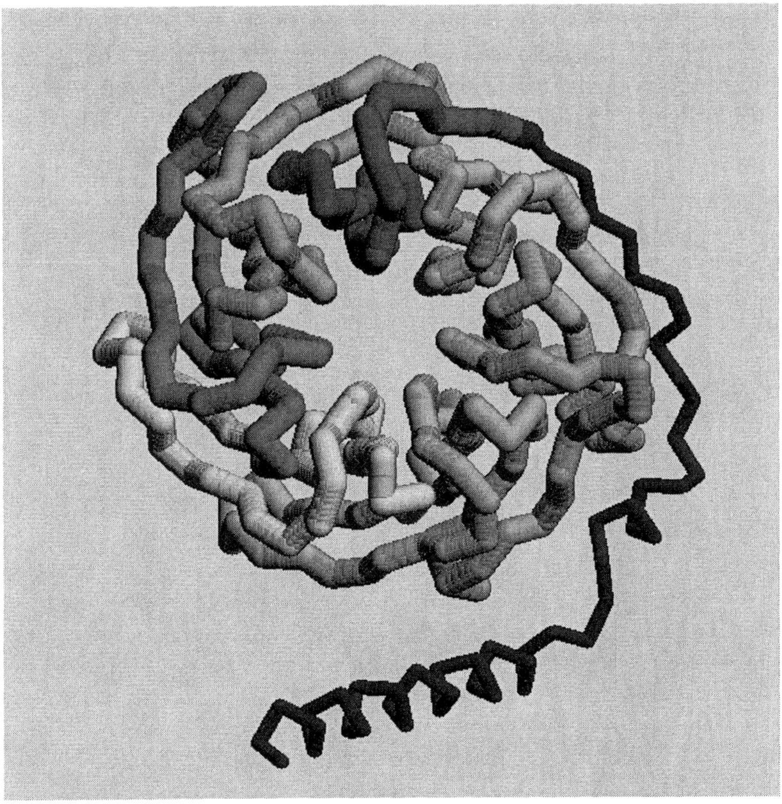

Figure 10.10. Repeats in a seven-fold β-propellor fold. The protein `1gotB` was identified as a very repetitive structure but with an eight-fold repeat rather than the expected seven-fold corresponding to its propellor structure. However, counting the N-terminal helix and loop (drawn thinner), each propellor 'blade' is roughly one eighth the length of the protein. This illustrates that a peak in the Fourier spectrum does not necessarily correspond to the number of repeated substructures. The repeats identified by the visualization method described in the text (methods section 10.2.4) are each shaded differently.

having the spatial register of the β-sheet while being relatively large, so that there will not be too many unsymmetric arrangements in a typical sized protein.

An alternative explanation for the symmetry of the $\beta\alpha$ class might lie in their history (Phillips *et al* 1978, Lupas *et al* 2001). The more obviously repeating structures are relatively recently evolved (within the last 500 million years) and so retain their sequence signal while those in the $\beta\alpha$ class tend to be ancient metabolic enzymes often common to all known life. This suggests that their structural symmetry may be a relic from duplications in the far distant past, far enough back in time that no trace of detectable sequence similarity remains. Such ideas are difficult, if not impossible, to prove (Phillips *et al* 1978). However speculative, they nonetheless provide one of the few glimpses into the distant origins of protein structrures. Some of these ideas will be pursued even further beyond the reach of verification in the next chapter.

Chapter 11

Evolution and origins

11.1 Evolution of structure and function[1]

The rich variety of protein sequence and structure observed today has resulted from a long process of evolution. Taxonomists, who consider the equivalent problem in the biological world, have a great advantage over their molecular counterparts as they have examples of intermediate forms in the fossil record, in addition to the 'living relics' (like the ceolecanth). In the molecular world, however, an evolutionary model must be developed through direct observation of genetic events (such as recombination, splicing, etc) as there is, effectively, no fossil record[2]. 'Living relics' can also be found in the protein world (Cammack *et al* 1971) but it must be remembered (as in the biological world) that such 'relics' have a lineage equally as long as any other living thing (Bains 1987). This is emphasized by the observation that the divergence of equivalent proteins among the bacteria is as great as between any bacterium and a 'higher' organism.

To understand protein sequence and structural similarities it is necessary to have a model for the processes that have given rise to the current forms. This can be only partly gained through a comparative study of the current sequence and structural databanks and to interpret fully these interrelationships, a consideration of the underlying mechanisms of evolution is needed. The distribution of expected protein folds (and hence the significance of similarity) is tightly bound with assumptions of the underlying mechanisms of molecular evolution. For example; how frequent are duplications? can transpositions be tolerated within a fold? Much of the evidence on which our model for these phenomena is based has come from the recognition of events in the relatively recent evolutionary past based on clear sequence similarity. In the same way that Lyell interpreted geological history based on current geographical phenomena, so we too must assume that

[1] This section is reproduced in part from Taylor (1997a) with the kind permission of Oxford University Press.

[2] Contrary to the impression created by Hollywood, the amount of ancient DNA known is trivially small.

the basic mechanisms have remained the same and can be used to account for more distant evolutionary events.

As we have seen in the previous chapters, the observed shapes of proteins are a result not only of their history, but also of the physicochemical constraints imposed by their constituent components (e.g. the strength of covalent and hydrogen bonds), their environment (aqueous or lipid, intra/extracellular) and the tasks that they are required to perform (catalysis or recognition). It is difficult to separate the forms imposed by these constraints from those that have been inherited, but this is, nevertheless, a problem worth tackling, because, if the physicochemical constraints could be quantified, then the evolutionary component would be better understood. For example, the wings of vertebrates must maintain a certain relationship to the size of the animal to allow flight. This is a purely physical constraint. By contrast, there is no physical law which dictates that feathers are the only material from which wings can be constructed; so finding that all birds use feathers leads us to believe that they had a common ancestor while finding that bats and birds have wings of an equivalent relative size does not inspire the same conclusion[3]. A parallel situation in molecular evolution might involve a general enzymatic reaction that requires a certain juxtaposition of chemical groups (supported by a sufficiently stable framework). If it could be shown that only one protein chain fold is able to achieve this, then no evolutionary inference can be made about equivalent enzymes using this catalytic mechanism (having the same fold), however, if the necessary groups can be supported by, say, 50 different folds, then a group of enzymes with the same fold appears much more likely to be related.

11.1.1 Gene duplication and fusion

11.1.1.1 Genetic mechanisms

As we have seen in section 10.1.1, a simple way to generate novel structure and function (without too many fatal mistakes) is to duplicate a gene into a double-length copy. The resulting double domain protein can then accumulate different mutations in each half, allowing the function and even structure of the two domains to diverge. A related strategy is to combine independent structures, either two the same or two different, into a single protein.

There are many ways to generate duplications and translocations: translocations can arise through the incorrect religation of broken double stranded DNA, while an easy route to generate duplication involves staggered (double) strand damage combined with 'fill-in' repair of the broken (single strand) ends before religation (see (Shapiro *et al* 1977), and the following papers). There

[3] Bats and birds, of course, had a common ancestor but this was sometime around the early Triassic period. One branch lead to archaeopteryx, perhaps through small raptors who lived by slitting the throats of saurapods and needed 'wings' to leap from trees to necks (or break their fall when flicked off), hence to birds. Bats came later, with their flight probably being just an extended form of jumping from tree to tree.

are many clear examples indicating that duplication (with fusion) has occurred extensively, both in the recent and remote past (for a review, see Bajaj and Blundell (1984)). These include single duplication (McLachlan 1979a, Schulz 1980), through triple- and double-duplication (Nojima 1987), to multiplication (McLachlan 1983) and explosion (Higgins *et al* 1994). (See chapter 10.)

In addition to fused genes, stop codons can also be copied, leading to multiple gene copies. These provide an important route to the evolution of new function from existing proteins (Piatigorsky and Wistow 1991), and can lead to a variety of unexpected genetic effects such as reversion to earlier gene copies (Dover 1982).

11.1.1.2 Dimeric precursors

Protein structures that function as dimers would be the most likely candidates for gene duplication into a fused protein as they have already evolved complementary interacting surfaces. A probable example of this process is seen in the aspartyl proteases. The form found in higher organisms has two (remotely related) domains with considerable differences in loop lengths and subdomain packing but the same chain fold. Each domain, however, contributes an aspartic acid to the active site which in the high-resolution structures can be seen to have an almost exact two-fold (180°) relationship (Blundell *et al* 1990). In addition the same two-fold axis closely corresponds to the symmetric relationship of the two domains suggesting a precursor molecule which functioned as a dimer (Tang *et al* 1978). An example of such a protein was identified in the retroviral proteases (Toh *et al* 1985), and the HIV protease was modelled as a dimeric enzyme (Pearl and Taylor 1987)—a result later confirmed by many crystallographic studies (Wlodawer *et al* 1989).

The dimers most susceptible to duplication and fusion would be those in which the two ends to be joined (the N-terminus of one subunit with the C-terminus of its symmetric half) lie close together. Without this, some unwinding of the chain at each terminus would be necessary, or an additional linking segment would be needed. Both would give rise to new interactions with the probability of these being unfavourable. A direct implication of this is that the remaining free ends (now the termini of the fused gene product) must, because of the two-fold symmetry, lie close together. Interestingly, this would explain the frequent proximity of termini in protein domains (Thornton and Sibanda 1983), a phenomenon that is largely unexplained by other effects.

11.1.2 Introns and exons

The general strategy of strand exchange as a mechanism for generating low-error diversity could be made more efficient if the recombination (or crossover) points, which are a source of added error, avoided the locally optimized sequence regions. In terms of protein structure this would entail introducing a bias for crossover to occur in the regions of sequence coding for surface loop regions. However, as the main mechanism of recombination (including sexual) involves random strand breakage (Wilson 1985) it is difficult to envisage any method of control.

The discovery of genes split by regions of non-coding sequence revealed a potential mechanism (Gilbert 1978b, Blake 1978). The intervening regions of 'junk' DNA (introns) are typically much longer than the coding regions (exons) making it more likely that a random recombination site will occur in an intron. By selection, the introns have come to lie in regions of the gene corresponding to the less critical features in protein structure, such as surface loops (Gō and Nosaka 1978). The only problem with this simple strategy is that the introns must be removed (spliced out) before the RNA message is translated. This vital task is carried out by a complex protein/RNA mechanism (Newman 1994, Steitz 1988).

Since their discovery, there has been much speculation on the origins of introns and their significance to evolution both at the level of protein structure (Gilbert 1978a, Artymiuk *et al* 1981) and at higher levels of organization (Doolittle 1987). A major puzzle was their absence in the prokaryotes. This was originally explained as removal due to the pressures of rapid replication, implying that the ancestral organism had introns (Doolittle (1987) and most of the papers above). The 'old intron' hypothesis never explained why the clearest examples of exons corresponded to domains found in proteins of relatively recent origin (Doolittle 1985, Patthy 1985, Blake *et al* 1978) and doubts about the fundamental nature of exons were expressed (Doolittle *et al* 1986), despite the growing dogma. With the discovery of introns in two prokaryotes (Belfort 1993, Logsdon and Palmer 1994) ideas have shifted again with less credibility now being given to the 'old intron' idea.

From evidence that introns can be inserted into coding regions (Dibb and Newman 1989, Belfort 1993), supported by a variety of arguments (Patthy 1991, Palmer and Logson 1991), it is now thought that the well-ordered splicing of introns seen today may be an evolutionarily recent mechanism (Patthy 1994). Indeed, Patthy argues that the exon/intron mechanism (if not 'invented') was exploited at the time of the metazoan radiation (or 'big bang') in the Cambrian (500MyBP). This recent importance is supported by both the phylogenetic distribution of exon containing proteins and their common occurrence in extracellular proteins associated with cell/cell communication. Although the latter is indirect evidence, the argument that a novel source of protein diversity was needed at this time is persuasive as the complex body forms that arose in the early Cambrian would have required many new cell surface receptors, both in embryogenesis and function, specific for each new tissue and their interactions. This would have been especially true in the vertebrates where the rapid evolution of complex nervous systems was required to coordinate the new body movement and process sensory data.

While the importance of recent introns can now no longer be doubted, it still does not account for their origin. The modern splicing apparatus (the spliceosome) is assembled from snRNPs which contain a catalytic RNA component (Newman 1994, Steitz 1988), implying an origin in the RNA world (3.5ByBP—long before the Cambrian). This raises the question of what introns were doing before the metazoan radiation. While still uncertain, the origin of introns is probably linked with retroviruses, transposable elements and principally, self-splicing introns[4], suggesting an earlier 'selfish' past (Newman 1994, Cavalier-Smith 1989).

11.1.3 Models of structure evolution

11.1.3.1 Wandering is sequence space

While it is clear that structure sets the boundaries within which evolutionary exploration is confined, the ultimate selection pressure (at the molecular level) is on protein function. Some of the 'options' available to proteins in different situations in an arbitrary sequence space are illustrated in figure 11.1. Any move in this space represents a change in sequence, with small moves corresponding to point mutations and large jumps to splicing (or frame shifts). Each area labelled 'fold' encloses sequences with a distinct protein fold (a superfamily)—such as the $(\beta/\alpha)_8$ fold (TIM barrel).

Within each fold region, sequence subsets are enclosed representing proteins with different functions. There can be different functions with the same fold (e.g. A and C in fold 1), or different folds with the same function (A in folds 1 and 2). If, in the latter situation, the two folds perform their function using the same mechanism, they would probably be held up as an example of convergent evolution. There has been much sterile discussion in the literature on whether particular similarities are divergent or convergent evolution. This can be avoided by adopting the simple maxim that: if the folds are the same, the sequences have diverged from a common ancestor but, if the folds are different and the mechanism is the same then they have converged. As there is generally no independent means of verification, all further debate is unscientific (Bajaj and Blundell 1984).

It is also possible to have a function that does not require a folded chain, or a sequence that can adopt more than one fold, but for simplicity, it will be assumed that functional proteins are uniquely folded. A more relevant possibility is that functions might overlap, with their intersection subset representing proteins carrying out both functions. This situation might arise either through an overlap in catalytic specificity at a single active site or through dual (or multiple) sites on the same protein. Each function in the latter situation would probably be associated

[4] The introns discussed here are those found in eukaryotic nuclei which are removed from mRNA by the spliceosome and referred to as spliceosomal introns. The other, non-spliceosomal introns are self-splicing (auto-catalytic RNA) and found both in eukaryotic and prokaryotic organisms, not only in mRNA but also (indeed, primarily) in tRNA and rRNA. They are divided into three groups (I, II and III) depending on splicing mechanism and cofactors and their reliance on a catalytic RNA function strongly suggests an ancient origin in the RNA world. The three groups should not be confused with the three codon reading frames in which spliceosomal introns can be inserted.

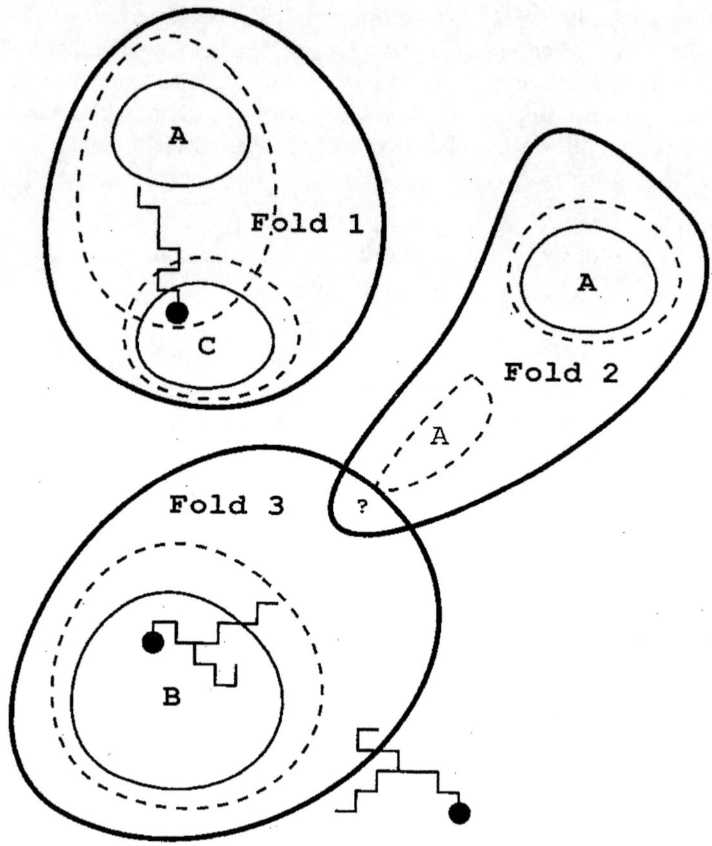

Figure 11.1. Evolutionary options for proteins. An imaginary sequence space is depicted in which sequences evolve by short steps. Some sequences (dots) correspond to folds (fold 1, . . .), which can have various functions (A, B, . . .). Partial functions are enclosed by broken curves and where these overlap a path is created for the evolution of one function into the other (see the main text for a further discussion). The fold-1 C-ase enzyme is also a weak A-ase, allowing divergence of a copy towards an optimal A-ase. The fold-3 B-ase evolves and duplicates with one copy retaining and the other losing function. An unfolded 'random' sequence similarly wanders with one copy attaining a fold (3). The interesting possibility of sequences which can form two different folds is also indicated as '?'.

with distinct domains and as these effectively correspond to smaller proteins, the possibility will be ignored. A more realistic situation for two functions at one site is that one function is dominant over the other. In general, both fold and function would have different degrees of stability and efficiency (respectively), giving a continuously varying (contoured) field over the sequence space.

On to this model, selection pressure must be applied, which most simply, can be assumed to 'quickly' kill-off any sequences without a function. A vital protein will, therefore, be tightly 'confined' within the subset of sequence space where it retains its function. To explore outside this space it is necessary to make a new copy on which to experiment—just as any good programmer would not experiment with their only copy of the computer code of a working program. This can be achieved by gene duplication, allowing one copy to avoid selection pressure—either completely as an unexpressed pseudo-gene or partially, as a redundant (expressed) member in a multiple-copy gene. Still assuming a relatively quick death for functionless proteins means that mutants can only survive for a limited time away from an 'island' of function. Even if mutations in the protein are now non-lethal, its expression carries some additional load, putting the organism at a selective disadvantage. This implies that the most probable evolutionary paths between functions are across the closest gaps where the functions overlap.

11.1.3.2 Leaping in sequence space

The sequence space depicted in figure 11.1 is clearly highly complex. Although each fold is depicted as a single smoothly bounded area, it is more likely to be a discontinuous (probably fractal) multidimensional hyper-volume. In addition, the boundaries are not fixed but will move under different physiological conditions (e.g. pH, temperature, etc). In optimization problems where the function is too complex to be minimized by conventional methods, the (computer) method called the *genetic algorithm* can be used, and as discussed previously (section 9.3.4), it is particularly suited to sequence-based applications (Dandekar and Argos 1992, Jones 1994, Petersen and Taylor 2003). The method incorporates a random mutation model to generate diversity in its population of sequences, however, this alone is insufficient to produce good convergence on an optimal solution. The key feature that makes the genetic algorithm effective is its ability to exchange pieces of sequence between members of the population. These exchanges, called *crossovers* take place between one or more exchange points and mimic the strategy exploited at the level of the genome in sexual reproduction.

Evolution is also a process of optimization, generally under continually changing conditions and although the genetic algorithm method derives from evolutionary ideas, it provides a useful model of how proteins might evolve. Swapping genetic material in this way allows locally optimal solutions, developed independently in different sequences, to be recombined to produce a super-sequence. The process is wasteful, since for every good/good combination there will be many good/poor and poor/poor combinations but the best combinations survive and multiply, enriching the population for the next cycle. If proteins had to make only stepwise (residue-by-residue) changes, then many wasteful or lethal 'dead-ends' would be investigated that would become increasingly likely to undo the good solutions as they approach the optimum.

The effectiveness of the crossover strategy relies on the existence of locally optimal solutions and if the stability of the minimum solution was equally dependent on every element of the sequence then the crossover strategy could not work. Proteins lie somewhere between these extremes and while the stability of the fold is, to some extent, dependent on the interaction of every amino acid, there is also a clear hierarchy of structure from secondary, through super-secondary to domains, which is ideally suited to GA optimization, both in computer simulations (Dandekar and Argos 1994, Petersen and Taylor 2003) and in nature.

11.1.4 Evolution of function

There are many examples of different enzymes with the same fold but there are not many where the sequences can be shown to be related (to the degree of implying a common ancestor). In these few examples, the proteins retain their key catalytic residues and so support the idea that enzymes can evolve different specificities while retaining the same catalytic mechanism (Ollis *et al* 1992, Murzin 1993). A good example is the similarity of mandelate racemase to muconate lactonising enzyme. These enzymes have similar structures (TIM barrels) and similar active sites but use different divalent metals. Interestingly, the lactonising enzyme can function with either metal, which together with other active site differences, suggests how an ancestral enzyme might have had low level racemase activity which could have provided an evolutionary pathway across to the new function (Petsko *et al* 1993).

The evolution of a new enzymatic function is useless in isolation as any product would simply accumulate, probably to a lethal extent. This implies the 'coordinated' evolution of whole pathways—an unlikely event if the changes had to be made simultaneously. However, the existence of low-level alternative substrates across a number of enzymes allows for many possibilities, as all combinations of steps involving a common product/substrate are potential metabolic pathways. If one combination produces something useful, this will provide sufficient evolutionary pressure for the evolution of a new set of enzymes. Evidence that this can occur comes from further study of the mandelate pathway where it has been shown, either by structure or sequence comparison, that each step in the pathway has evolved from an enzyme in another pathway (Petsko *et al* 1993). Such studies open the possibility of reconstructing the evolution of metabolism by 'peeling-off' successive layers of pathways. Alternatively, the comparison of the metabolism of different organisms can give some indication of what was operational in the ancient 'proto-organisms', suggesting a core from which later pathways evolved (Benner and Ellington 1990, Benner *et al* 1993).

The previous examples have illustrated the evolution of one enzyme into another (or one specific function to another specific function). Examples can also be found where an enzyme has moved into an unspecific functional rôle. This has been observed in the structural eye-lens proteins which include a diverse collection of enzymes (Wistow and Piatigorsky 1987, Piatigorsky and Wistow 1991), some of which retain some functional activity. While the possible importance of residual function cannot be ruled-out, the prime function of these proteins appears to be to maintain the optical properties of the lens. This structural function does not greatly conflict with enzymatic function and an example has been seen where the same gene is 'shared' without duplication (Piatigorsky and Wistow 1991). As in the evolution of enzymes, this may represent an early stage in divergence before the conflicting pressures of the combined functions become sufficiently severe to favour duplication and functional uncoupling.

11.1.5 Selection on random folds

If the prediction of tertiary structure from just sequence data were 100% accurate, then many of the problems raised above concerning the number and symmetry of protein folds could be solved simply by generating all possible sequences, predicting their structure and comparing the results. This would provide not only the number of folds but also their underlying frequencies. Comparison of these distributions to the observed would then distinguish physicochemical constraints from evolved biases. Unfortunately, the required prediction methods are not remotely accurate enough or, if they were, it is unlikely that computers would be fast enough to apply them to an even useful number of sequences.

Despite this pessimistic critique, some limited speculations can be made in this direction by generating random sequences, folding them (in a computer simulation) and then assessing the degree to which the sequences are able to adopt a stable fold. At one extreme, if all sequences can fold, then new proteins would be likely to arise 'spontaneously', say, from frame shifts or translated introns (Taylor 1997a). Away from this extreme, the dependence on evolution from existing structures would be more important, increasing the persistence of evolutionary relics.

For computational tractability, such calculations are best carried out using a simplified protein model such as those described in section 8.2. The sequence is typically a binary string of hydrophobic and hydrophilic residues while the chain is confined to lie on a lattice. The lattice model has variously been two- (Unger and Moult 1993) or (more commonly) three-dimensional (Baker *et al* 1993, Godzik *et al* 1993), with each residue reduced to a point. A further level of simplification can be attained by adopting fixed secondary structures which can then be reduced to line segments (Taylor 1991b) (section 3.1). The computational technique of distance geometry (section 2.2) provides an alternative to the artificial constraints of a lattice, while still avoiding the long calculation times of conventional molecular dynamics. This approach has been used to create realistic 'random' proteins with secondary structure (Aszódi and Taylor 1994b). In addition to these simulation approaches, the problem has been treated more theoretically, applying the principles of statistical mechanics (Ptitsyn 1983, Finkelstein 1991), 1994.

Despite their widely different basis, most of these studies conclude that many 'random' sequences appear capable of forming stable folds, a conclusion now supported by experimental results (Davidson and Sauer 1994). This implies that totally new proteins might well appear spontaneously, or that large random insertions might be tolerated, as they would have greater probability of forming an independent folding unit which would not disrupt the existing structure. This latter possibility might easily occur through the mutation of an intron splice site, leading to the translation of the 'random' intron sequence. Such domains of random origin would not require any immediate function to survive, but could 'hitch-hike' on the strength of the function of their attached protein. If they later acquired a function they might even escape their parent (by random processes of intron insertion) and become autonomous proteins (or functional domains). By whatever mechanism they arose, these novel proteins would carry no evolutionary imprint of internal pseudosymmetry and may account for some of the less regular or complex folds.

11.1.5.1 Structure in Shakespeares sonnets

As a limited test of the hypothesis described in the previous section, a semi-random sequence was taken from the literature:

```
SHALLICQMPARETHEETQASUMMERSDAY
THQUARTMQRELQVELYANDMQRETEMPERATE
RQUGHWINDSDQSHAKETHEDARLINGBUDSQFMAY
ANDSUMMERSLEASEHATHALLTQQSHQRTADATE
SQMETIMETQQHQTTHEEYEQFHEAVENSHINES
ANDQFTENISHISGQLDCQMPLEXIQNDIMMD
ANDEVERYFAIRFRQMFAIRSQMETIMEDECLINES
BYCHANCEQRNATURESCHANGINGCQURSEUNTRIMMD
*
BUTTHYETERNALSUMMERSHALLNQTFADE
NQRLQSEPQSSESSIQNQFTHATFAIRTHQUQWEST
NQRSHALLDEATHBRAGTHQUWANDERSTINHISSHADE
WHENINETERNALLINESTQTIMETHQUGRQWEST
SQLQNGASMENCANBREATHEQREYESCANSEE
SQLQNGLIVESTHISANDTHISGIVESLIFETQTHEE
```

Its secondary structures were predicted using the program `PSIPRED` (Jones 2000) and 'tidied up' using the rules of Taylor and Thornton (1984). Examination of the predicted structures combined with indications from the literature, suggested that the sequence could be broken into two domains (indicated by the asterisk above), forming a larger N-terminal $\beta\alpha$ domain and a smaller almost all-α C-terminal domain.

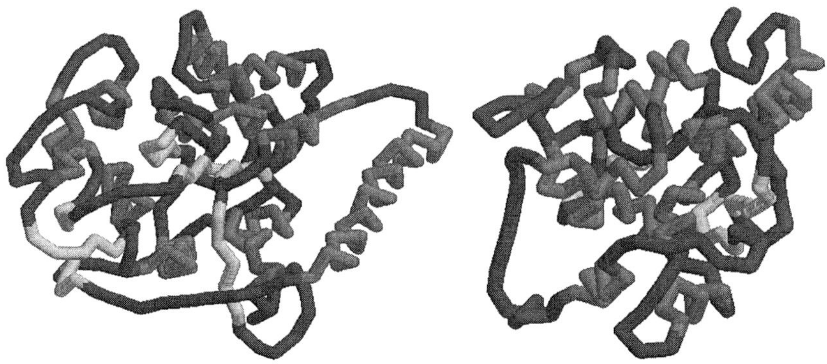

(*a*) N-domain (*b*) C-domain

Figure 11.2. Structure of a Shakespeare sonnet. Predicted globular domains for Shakespeares 18th sonnet. (See text for details.)

These sequences were presented individually to the `RAMBLE` program (Taylor *et al* (2003a) and section 8.2), and while not unique, plausible globular structures were obtained for both domains (figure 11.2).

11.2 The origins of proteins

11.2.1 The emergence of proteins in an RNA world

It is now generally accepted that, before the first living cells (just under four gigayears ago), 'life'—or rather the assemblies of self reproducing macromolecules—were ribonucleic acids (RNA). Circumstantial evidence for this can be found in 'relic' pieces of RNA that still hold a few of the most central functions in the processes of life. The most convincing observation is that in the synthesis of proteins on the ribosome, the key chemical event of peptide bond formation is catalysed by RNA (Nissen *et al* 2000).

In this 'RNA world', a single type of molecule performed both the functions of active (catalytic) agent and repository of its own description—the 'blueprint' from which further copies could be taken. The former function is a property of the folded molecule while the later is a property of the linear polymer sequence, and the two functions need not necessarily be compatible. One can imagine a situation in which, say, for more efficient catalysis, an extra chemical activity was needed at a particular point: however, a modification of the RNA structure to achieve this (such as the chemical modification of a part of the molecule) might leave it incapable of duplication or folding. It seems that RNA circumvented this problem by recruiting cofactors that could augment its chemical reprotire without compromising its ability to make copies of itself. Some of these cofactors were probably peptides and a development can be imagined in which the peptide cofactors became more complex as the functional role of the RNA diminished.

In this simple world, however, RNA would rely on the chance synthesis of suitable peptides which would limit both the size of the peptides and the number of these that could be involved with the RNA. This fundamental problem was overcome through the establishment of a synergistic loop in which the RNA was able to act as a template to guide the synthesis of the peptides that it needed. With the limitation on the chance synthesis of the right peptides now removed (or limited only by the fidelity in the translation of RNA into peptide sequences), the system was free to become much more complex[5]. The details of how this key event in life became established are very vague but some plausible hypothes are described in the opening chapters of '*The RNA World*' (Gesteland and Atkins 1993). This transition marked the escape from the error-prone world of self-replicating macromolecules to a system with unlimited scope to control its own metabolism and replication. It also began the divergence of function: with peptides/proteins taking over the active (catalytic) activity while RNA became more inert with its main function now being to encode proteins, which would then periodically help replicate the RNA itself.

From this state, the introduction of the third major molecular component of life—DNA—is almost incidental. With RNA free from most of its structural constraints and under strong evolutionary pressure to maintain the reproductive fidelity of the increasingly complex protein/RNA machine: in computer terms, a back-up facility was required. This was found in DNA, which is only a slightly modified form of RNA but has much greater stability—especially when 'locked' away in its famous double helical structure. This subsidiary role for DNA is maintained in all present-day life: and although proteins can interact directly with DNA, there is no direct link from DNA to protein except via RNA intermediates.

11.2.2 Functions for protoproteins

The protoproteins described above were not only short, the fidelity of their production must have been very poor and they may also have been limited to a few types of amino acids (Crick 1968): possibly only three[6] (Arg, Leu and Ser) (Gibson and Lamond 1990). What use can be made of short 20–30 residue length peptides of limited composition? One suggestion is that, being quite basic, they simply served to stabilize the structure of the early ribozymes, including a protoribosome (Nissen *et al* 2000), or that they were able to form pores in membranes around the first cells using a Ser/Leu amphipathicity (Gibson and Lamond 1990). With a preponderance of Leucine, it is probable that many of

[5] The short time span between the impact that created the Moon and the first cell have led some to suspect that there was not enough time for this complexity to develop on Earth.

[6] It is not possible to say whether just a limited variety of amino acids were used or a wide variety were all sharing a limited number of codons.

the peptides would adopt an α-helical conformation. Since specific sequences cannot have been reliably produced, a function that is specific to the α-helix itself may have been used. One such property that is independent of sequence is the propensity of the α-helix dipole to attract and bind phosphate (Hol *et al* 1978). In a world of sugar-phosphate nucleotides, this function could have found many interaction partners, either binding RNA itself or catching free nucleotides for polymerization on a ribozyme.

It does not then take a large leap to imagine that these α-helix helpers would function better first as multimers or later as linked chains. The latter state has the advantage that the charges on the multiple termini would be eliminated and to keep the dipoles aligned (for polynucleotide binding), the termini would need to make an antiparallel connection between the helices, creating the familiar repeating $\beta\alpha$-unit. As we have seen many times in the previous chapters, the potential folds for repeating $\beta\alpha$-units are limited, with the simplest being the simply wound solenoid fold producing a TIM barrel (or part-barrel) and those incorporating a chain jump giving rise to the large family of Rossmann fold-like proteins (figure 1.4). In addition, a class of small $\beta\alpha$-proteins, the KH-domains interact with RNA. Although protein function has not been mentioned much in the previous chapters, it is significant that most of these proteins bind sugar-phosphates or nucleotides. In addition, the $\beta\alpha$ class covers almost every protein that can be traced back to ancient origins, including most of those involved in basic metabolism and nucleic acid chemistry.

If these larger protoproteins, some 50–100 residues long, are to be formed from the chance fusion of their individual ($\beta\alpha$-unit genes), then we are beginning to assume that there is quite a sophisticated protein synthesis (and RNA replication) apparatus. However, these longer proteins do not necessarilly imply that the equivalent nucleic acid message exists. It is possible that chain initiation and termination were also not well controlled in these early times so multiple short genes could be read randomly from a pool, giving rise to a wide variety of proteins that are larger than might be expected for the fidelity of the machinery. The functionality of these longer proteins would then provide the selection pressure towards longer messages (and greater fidelity) just as may have happened at an earlier stage in the symbiosis of RNA and peptides.

With a reasonably stable RNA-directed protein synthesis, many subsequent developments can be postulated (Lupas *et al* 2001) but as discussed in chapter 10 and above, it is very difficult to decide how much of these are the result of ancient relics and how much are imposed by the unavoidable symmetries induced by packing and chirality constraints. The application of the approach described in section 7.4 may shed some light on this difficult problem.

11.3 The secret of life

> *'What is the secret of life?' I asked. [...] 'Protein', the bartender declared. 'They found out something about protein.'*
>
> Kurt Vonnegut, from *Cat's Cradle*

The storyteller in Vonnegut's '*Cat's Cradle*' (Vonnegut 1963), having posed the ultimate biological question, unfortunately, did not pursue the bartender's insightful lead but instead continued to seek for truth in the '*Books of Bokonon*' (Johnson 1925–1963), which despite being a source of great wisdom, have very little to say about proteins and (in all their fourteen volumes) nothing at all about protein structure. This brief, but deeply meaningful, encounter was supposed to have taken place in the early sixties and the event to which the bartender was obviously referring was the elucidation of the first globular protein structure by x-ray crystallography.

Is an understanding of protein structure then the key to the 'secret of life?' As the motivating agents of the cellular machinery, there are few alternative candidates among the other biological macromolecules. The only possible candidate are the nucleic acids, but DNA is an inert material used and manipulated by proteins. Considering the origins of life as an RNA/protein 'symbiosis', perhaps equal weight should be given to RNA. Despite this, proteins hold a special position: not simply because of what they do but in how they do it. As described in section 1.1.2, although only single molecules, proteins act as individual agents (like little machines), bridging the gap between the statistical behaviour described by chemistry and the mechanical world in which we operate. It is their position at this interface between chemistry and biology that makes them unique.

From the opening quote to chapter 1, however, Monod would have us believe that knowing the structure and function of proteins is not enough. We must also know the rules by which they fold. Unfortunately, over the 30 years since that statement was written, there has been little progress on the *ab initio* prediction of protein structure. While there may be hope with bigger, faster computers, it might also be that life's ultimate secret is locked away in a code so obscure that our conventional approaches do not even recognize it. Take, for example, one of those clever 3D wooden puzzles in which all the bits fit neatly together pairwise but to make the complete shape, all the pairwise interactions must lock simultaneously. Proteins may be like this and to predict a structure from its sequence, all the interactions within the structure must be considered simultaneously (like a Fourier transform of the sequence).

In this situation, the most practical way forward is to cheat: that is to copy, without understanding, from the sequences and structures that we can observe. Fortunately, over the past 30 years, we have become quite good at cheating; helped considerably by the increasing number of examples of proteins of known structure that we can crib from. Comparative analysis of these data give us indirect access to the evolutionary history of proteins and untangling these lines of descent, both within and between species, lets us see how proteins, and life, have evolved back through the most distant past, perhaps even close to its origin. Given sufficient sampling over this evolutionary space, we may begin to gain some idea of the structural envelope within which any given protein structure is able to be maintained. Although this activity has more affinity to Rutherford's 'stamp collecting' than to physics, it is to be hoped that increasing structural insight will be gained along the way.

References

Abagyan R A 1993 Towards protein folding by global energy optimization *FEBS Lett.* **325** 17–22

Adams C C 1994 *The Knot Book: An Elementary Introduction to the Mathematical Theory of Knots* (New York: Freeman)

Alexandrov N N and Fischer D 1996 Analysis of topological and nontopological structural similarities in the PDB: new examples with old structures *Proteins* **25** 354–65

Alexandrov N N and Go N 1994 Biological meaning, statistical significance and classification of local spatial similarities in nonhomologous proteins *Prot. Sci.* **3** 866–75

Alm E and Baker D 1999 Matching theory and experiment in protein folding *Curr. Opin. Struct. Biol.* **9** 189–96

Andrade M A, Perez-Iratxeta C and Ponting C P 2001 Protein repeats: structures, functions and evoltion *J. Struct. Biol.* **134** 117–31

Anfinsen C B 1973 Principles that govern the folding of protein chains *Science* **181** 223–30

Artymiuk P J, Blake C C F and Sippel A E 1981 Genes pieced together: exons delineate homologous structures of diverged lysozymes *Nature* **290** 287–88

Artymiuk P J, Rice D W, Mitchell E M and Willett P 1990 Structural resemblance between the families of bacterial signal-transduction proteins and of G proteins revealed by graph theoretical techniques *Prot. Eng.* **4** 39–43

Artymiuk P J, Bath P A, Grindley H M, Pepperrell C A, Poirrette A R, Rice D W, Thorner D A, Wild D J, Willett P, Allen F H and Taylor R 1992a Similarity searching in databases of three-dimensional molecules and macromolecules *J. Chem. Inf. Comput. Sci.* **32** 617–30

Artymiuk P J, Grindley H M, Park J, Rice D W and Willett P 1992b Three-dimensional structural resemblance between leucine aminopeptidase and carboxypeptidase A revealed by graph-theoretical techniques *FEBS Lett.* **303** 48–52

Aszódi A and Taylor W R 1993 Connection topology of proteins *Comput. Appl. Biol. Sci.* **9** 523–9

——1994a Folding polypeptide α-carbon backbones by distance geometry methods *Biopolymers* **34** 489–506

——1994b Secondary structure formation in model polypeptide chains *Prot. Eng.* **7** 633–44

——1995 Estimating polypeptide α-carbon distances from multiple sequence alignments *J. Math. Chem.* **17** 167–84

Aszódi A, Gradwell M J and Taylor W R 1995a Global fold determination from a small number of distance restraints *J. Mol. Biol.* **251** 308–26

——1995b Protein fold determination using a small number of distance restraints *Protein Folds: A Distance Based Approach* ed H Bohr and S Brunak (Boca Raton, Fl: Chemical Rubber Company) pp 85–97

Bachar O, Fischer D, Nussinov R and Wolfson H 1993 A computer vision based technique for 3-D sequence-independent structural comparison of proteins *Prot. Eng.* **6** 279–88

Bains W 1987 Evolutionary paradoxes and natural non-selection *TIBS* **12** 90–1

Bajaj M and Blundell T 1984 Evolution and the tertiary structure of proteins *Ann. Rev. Biophys. Bioeng.* **13** 453–92

Baker D, Chan H S and Dill K A 1993 Coordinate-space formulation of polymer lattice cluster theory *J. Chem. Phys.* **98** 9951–62

Bangham J A 1988 Data-sieving hydrophobicity plots *Anal. Biochem.* **174** 142–5

Banner D W, Bloomer A C, Petsko G A, Phillips D C, Pogson C I and Wilson I A 1975 Structure of chicken muscle triose phosphate isomerase determined crystallographically at 2.5 Å resolution *Nature* **255** 609–614

Barton G J and Sternberg M J E 1988 LOPAL and SCAMP: techniques for the comparison and display of protein structure *J. Mol. Graph.* **6** 190–6

Belfort M 1993 An expanding universe of introns *Science* **262** 1009–10

Bement T R and Waterman M S 1977 Locating maximum variance segments in sequential data *Math. Geol.* **9** 55–61

Benham C J and Jafri M S 1993 Disulphide bonding patterns and protein topologies *Prot. Sci.* **2** 41–54

Benner S A and Ellington A D 1990 Evolution and structural theory: the frontier between chemistry and biology *Bioorganic Chem. Front.* **1** 1–70

Benner S A, Cohen M A, Gonnet G H, Berkowitz D B and Johnsson K P 1993 Reading the palimpsest: contemporary biochemical data and the RNA world *The RNA World: The Nature of Modern RNA Suggests a Prebiotic RNA World* ed R F Gesteland and J A Atkins (Cold Spring Harbor, NY: Cold Spring Harbor Laboratory Press) pp 27–70

Bennet M J, Schlunegger M P and Eisenberg D 1995 3D domain swapping: a mechanism for oligomer assembly *Prot. Sci.* **4** 2455–68

Biou V, Dumas R, Cohen-Addad C, Douce R, Job D and Pebay-Peyroula E 1997 The crystal structure of plant acetohydroxy acid isomeroreductase complexed with NADPH two magnesium ions and a herbicidal transtition state analog at 1.65 Å resolution *EMBO J.* **16** 3405–15

Blake C C F, Harlos K and Holland S K 1978 Exon and domain evolution in the proenzymes of blood coagulation and fibrinolysis *Cold Spring Harbor Symp. Quant. Biol.* **LII** 925–9

Blake C C F 1978 Do genes-in-pieces imply proteins-in-pieces? *Nature* **273** 267

Blundell T L and Srinivasan N 1996 Symmetry, stability and dynamics of multidomain and multicomponent protein systems *Proc. Natl Acad. Sci., USA* **93** 14 243–8

Blundell T L, Jenkins J A, Sewell B T, Pearl L H, Cooper J B, Tickle I J, Veerapandian B and Wood S P 1990 X-ray analyses of aspartic proteinases: the 3-dimensional structure at 2.1 Å resolution of endothiapepsin *J. Mol. Biol.* **211** 919–41

Bowie J U 2000 Helix-bundle membrane protein fold templates *Prot. Sci.* **8** 2711–19

Brändén C-I and Tooze J 1991 *Introduction to Protein Structure* (New York: Garland)

Brenner S E, Chothia C and Hubbard T J P 1998 Assessing sequence comparison methods with reliable structurally identified distant evolutionary relationships *Proc. Natl Acad. Sci., USA* **95** 6073–8

Brint A T, Davies H M, Mitchell E M and Willett P 1989 Rapid geometric searching in protein structures *J. Mol. Graph.* **7** 48–53

Bron C and Kerbosch J 1973 Algorithm 457: finding all cliques of an undirected graph *Comm. ACM* **16** 575–7

Brooks B R, Bruccoleri R E, Olafson B D, States D J, Swaminathan S and Karplus M 1983 CHARMM a program for macromolecular energy, minimisation and dynamics calculations *J. Comput. Chem.* **4** 187–217

Brown N P, Orengo C A and Taylor W R 1996 A protein structure comparison methodology *Comput. Chem.* **20** 359–80

Bruce A and Wallace D 1992 Critical point phenomena: universal physics at large length scales *The New Physics* ed P Davies (Cambridge: Cambridge University Press) pp 236–67

Brünger A T and Nilges M 1993 Computational challenges for macromolecular structure determination by X-ray crystallography and solution NMR-spectroscopy *Quarterly Rev. Biophys.* **26** 49–125

Burkhard P, Stetefeld J and Strelkov S V 2001 Coiled-coils: a highly versatile protein folding motif *Trends Cell Biol.* **11** 82–8

Cahn J W and Gratias D 1987 Quasi-periodic crystals: a revolution in crystallography *Advancing Materials Research* ed P A Psaras and H D Langford (Washington, DC: National Acadamy Press) pp 151–60

Cammack R, Hall D and Rao K 1971 Ferredoxins: are they living fossils? *New Scientist Sci. J.* **23** 696–8

Carson M 1991 Ribbons 2.0 *J. Appl. Cryst.* **24** 958–61

Cavalier-Smith T 1989 Intron phylogeny: a new hypothesis *TIG* **7** 145–8

Chan H S and Dill K A 1990 Origins of structure in globular-proteins *Proc. Natl Acad. Sci., USA* **87** 6388–92

Chothia C and Finkelstein A V 1990 The classification and origins of protein folding patterns *Ann. Rev. Biochem.* **59** 1007–39

Chothia C and Janin J 1981 Relative orientation of close-packed β-pleated sheets in proteins *Proc. Natl Acad. Sci., USA* **78** 4146–50

——1982 Orthogonal packing of β-pleated sheets in proteins *Biochemistry* **21** 3955–65

Chothia C and Murzin A G 1993 New folds for all-β proteins *Structure* **1** 217–22

Chothia C, Levitt M and Richardson D 1981 Helix to helix packing in proteins *Proc. Natl Acad. Sci., USA* **78** 4146–50

Chothia C 1984 Principles that determine the structure of proteins *Ann. Rev. Biochem.* **53** 537–72

——1988 Protein structure: the 14th barrel rolls out *Nature* **333** 598–9

——1992 Proteins—1000 families for the molecular biologist *Nature* **357** 543–4

Chou K-C, Carlacci L and Maggiora G G 1990 Conformational and geometrical properties of idealised β-barrels in proteins *J. Mol. Biol.* **213** 315–26

Claessens M, van Cutsem E, Lasters I and Wodak S 1989 Modelling the polypeptide backbone with 'spare-parts' from known protein structures *Prot. Eng.* **2** 335–45

Clark D A, Shirazi J and Rawlings C J 1991 Protein topology prediction through constraint-based search and the evaluation of topological folding rules *Prot. Eng.* **4** 751–60

Cohen C and Parry D A D 1994 α-helical coiled coils: more facts and better predictions *Science* **263** 488–9

Cohen F E and Sternberg M J E 1980a On the prediction of protein structure: the significance of the root-mean-square deviation *J. Mol. Biol.* **138** 321–33

Cohen F and Sternberg M 1980b On the use of chemically derived distance constraints in the prediction of protein structure with myoglobin as an example *J. Mol. Biol.* **137** 9–22

Cohen F E, Richmond T J and Richards F M 1979 Protein folding: Evaluation of some simple rules for the assembly of helices into tertiary structures with myoglobin as as example *J. Mol. Biol.* **132**

Cohen F E, Sternberg M J E and Taylor W R 1980 Analysis and prediction of protein β-sheet structures by a combinatorial approach *Nature* **285** 378–82

——1981 Analysis of the tertiary structure of protein β-sheet sandwiches *J. Mol. Biol.* **148** 253–72

Cohen F E, Sternberg M J E and Taylor W R 1982 Analysis and prediction of the packing of α-helices against a β-sheet in the tertiary structure of globular proteins *J. Mol. Biol.* **156** 821–62

Colloc'h N, Etchebest C, Thoreau E, Henrissat B and Mornon J-P 1993 Comparison of three algorithms for the assignment of secondary structure in proteins: the advantages of a consensus assignment *Prot. Eng.* **6** 377–82

Connolly M L, Kuntz I D and Crippen G M 1980 Linked and threaded loops in proteins *Biopolymers* **19** 1167–82

Cozzarelli N R and Wang J C (ed) 1990 *DNA Topology and its Biological Effects* (Cold Spring Harbor, NY: Cold Spring Harbor Laboratory Press)

Crick F H C 1953 The packing of α-helices: simple coiled coils *Acta. Crystallogr.* **6** 689–97

——1968 The origin of the genetic code *J. Mol. Biol.* **38** 367–79

Crippen G M and T F Havel T F 1988 Distance geometry and molecular conformation *Chemometrics Research Studies Press* (New York: Wiley)

Crippen G M 1974 Topology of globular proteins *J. Theor. Biol.* **45** 327–38

——1975 Topology of globular proteins II *J. Theor. Biol.* **51** 495–500

——1978 The tree structural organisation of proteins *J. Mol. Biol.* **126** 315–32

——1991 Prediction of protein folding from amino-acid-sequence over discrete conformation spaces *Biochemistry* **30** 4232–7

Dandekar T and Argos P 1992 Potential of genetic algorithms in protein folding and protein engineering simulations *Prot. Eng.* **5** 637–45

——1994 Folding the main-chain of small proteins with the genetic algorithm *J. Mol. Biol.* **5** 637–45

Davidson A R and Sauer R T 1994 Folded proteins occur frequently in libraries of random amino acid sequences *Proc. Natl Acad. Sci., USA* **91** 2146–50

Denton M and Marshall C 2001 Laws of form revisited *Nature* **410** 417

deVlieg J, Scheek R M, van Gunsteren W F, Berendsen H J C, Kaptein R and Thomason J 1988 Combined procedure of distance geometry and restrained molecular dynamics techniques for protein structure determination from nuclear magnetic resonance data: Application to the DNA binding domain of Lac repressor from *Escherichia coli Prot. Struct. Funct. Genet.* **3** 209–18

Dibb N J and Newman A J 1989 Evidence that introns arose at proto-splice sites *EMBO J.* **8** 2015–21

Dickerson R E and Geis I 1969 *The structure and action of proteins* (New York: Harper and Row)

Dong Z and Corbet J D 1994 $Na_2K_{21}Tl_{19}$, a novel thallium compound containing isolated Tl_5^{7-} and Tl_9^{9-} groups: a new hypoelectric cluster *J. Am. Chem. Soc.* **116** 3429–35

Doolittle R F, Feng D-F, Johnson M S and McClure M A 1986 Relationships of human protein sequences to those of other organisms *Cold Spring Harbour Symp. Quant. Biol.* **LI** 447–55

Doolittle R F 1985 The genealogy of some recently evolved vertabrate proteins *TIBS* **2** 233–7

Doolittle W F 1987 What introns have to tell us: hierarchy in genome evolution *Cold Spring Harbor Symp. Quant. Biol.* **LII** 907–13

Dover G 1982 Molecular drive: a cohesive mode of species evolution *Nature* **299** 111–17

Duncan B S and Olson A J 1993 Approximation and characterization of molecular surfaces *Biopolymers* **33** 123–456

Efimov A V 1987 Pseudo-homology of protein standard structures formed by two consecutive β-strands *FEBS Lett.* **224** 372–6

——1991a Structure of $\alpha - \alpha$-hairpins with short connections *Prot. Eng.* **4** 245–50

——1991b Structure of coiled $\beta - \beta$-hairpins and $\beta - \beta$-corners *FEBS Lett.* **284** 288–92

——1993 Standard structures in proteins *Prog. Biophys. Mol. Biol.* **60** 201–39

——1997 Structural trees for protein superfamilies *Proteins* **28** 241–60

——1999 Complementary packing of α-helices in proteins *FEBS Lett.* **463** 3–6

Eidhammer I, Jonassen I and Taylor W R 2000 Structure comparison and structure patterns *J. Comput. Biol.* **7** 658–716

Eisenberg D, Weiss R M and Terwilliger T C 1982 The helical hydrophobic moment: a measure of the amphiphilicity of a helix *Nature* **299** 371–4

——1984 The hydrophobic moment detects periodicity in protein hydrophobicity *Proc. Natl Acad. Sci., USA* **81** 140–4

Eisenberg D, Wesson M and Wilcox W 1989 Hydrophobic moments as tools for analyzing protein sequences and structures *Prediction of Protein Structure and the Principles of Protein Conformation* ed G D Fasman (New York: Plenum) ch 16, pp 635–46

Feldman R J 1976 Atlas of protein structure on microfiche *Technical Report* Tracor Jitco Inc., Rockville, MD, USA ISBN 0-917934-01-6

Finer-Moore J and Stroud R M 1984 Amphipathic analysis and possible formation of the ion channel in an acetylcholine receptor *Proc. Natl Acad. Sci., USA* **81** 155–9

Finkelstein A V and Ptitsyn O B 1987 Why do globular proteins fit the limited set of folding patterns? *Prog. Biophys. Mol. Biol.* **50** 171–90

Finkelstein A V and Reva B A 1991 A search for the most stable folds of protein chains *Nature* **351** 497–9

Finkelstein A V 1991 Rate of beta-structure formation in polypeptides *Prot. Struct. Funct. Genet.* **9** 23–7

——1994 Implications of the random characteristics of protein sequences for their three-dimensional structure *Curr. Opinion Struct. Biol.* **4** 422–8

Finney J L 1970 Random packings and the structure of simple liquids *Proc. R. Soc.* A **319** 479–93

Fischer D, Bachar O, Nussinov R and Wolfson H 1992 An efficient automated computer vision based technique for detection of three dimensional structural motifs in proteins *J. Biomol. Struct. Dynam.* **9** 769–89

Fischer D, Wolfson H and Nussinov R 1993 Spatial, sequence-order-independent structural comparison of alpha/beta proteins—evolutionary implications *J. Biomol. Struct. Dynam.* **11** 367

Fischer D, Wolfson H, Lin S L and Nussinov R 1994 3-dimensional, sequence order-independent structural comparison of a serine-protease against the crystallographic database reveals active-site similarities—potential implications to evolution and to protein-folding *Prot. Sci.* **3** 769–78

Flores T P, Orengo C A, Moss D S and Thornton J M 1993 Comparison of conformational characteristics in structurally similar protein pairs *Prot. Sci.* **2** 1811–26

Flores T P, Moss D S and Thornton J M 1994 An algorithm for automatically generating protein topology cartoons *Prot. Eng.* **7** 31–7

Flory P J 1969 *Statistical Mechanics of Chain Molecules* (New York: Wiley)

Flower D R 1998 A topological nomenclature for protein structure *Prot. Eng.* **11** 723–7

Gesteland R F and Atkins J A editors 1993 *The RNA World: The Nature of Modern RNA Suggests a Prebiotic RNA World* (Cold Spring Harbor, NY: Cold Spring Harbor Laboratory Press)

Gibrat J-F, Madej T and Bryant S H 1996 Surprising similarities in structure comparison *Curr. Opinion Struct. Biol.* **6** 377–85

Gibson T J and Lamond A I 1990 Metabolic complexity in the RNA world and implications for the origin of protein synthesis *J. Mol. Evol.* **30** 7–15

Gilbert D, Westhead D, Nagano N and Thornton J M 1999 Motif-based searching in TOPS protein topology databases *Bioinformatics* **15** 317–26

——Motif-based searching in TOPS protein topology databases *Bioinformatics* in press

Gilbert W 1978a The exon theory of genes *Cold Spring Harbor Symp. Quant. Biol.* **LII** 901–5

——1978b Why genes in pieces? *Nature* **271** 501

Glunt W, Hayden T L, Hong S and Wells J 1990 An alternating projection algorithm for computing the nearest Euclidean distance matrix *SIAM J. Matrix Anal. Appl.* **11** 589–600

Glunt W, Hayden T L and Raydan M 1992 Molecular conformations from distance matrices *J. Comput. Chem.* **14** 114–20

Gō M and Nosaka M 1978 Protein architecture and the origin of introns *Cold Spring Harbor Symp. Quant. Biol.* **LII** 915–24

Godzik A, Kolinski A and Skolnick J 1993 Lattice representations of globular-proteins—how good are they *J. Comput. Chem.* **14** 1194–202

Godzik A 1996 The structural alignment between two proteins: is there a unique answer? *Prot. Sci.* **5** 1325–38

Govindarajan S, Recabarren R and Goldstein R K 1999 Estimating the total number of protein folds *Proteins* **35** 408–14

Green P, Lipman D, Hillier L, Waterston R, States D and Claverie J-M 1993 Ancient conserved regions in new gene-sequences and the protein databases *Science* **259** 1711–16

Green P 1994 Ancient conserved regions in gene sequences *Curr. Opinion Struct. Biol.* **4** 404–12

Gregoret L M and Cohen F E 1991 Protein folding: effect of packing density on chain conformation *J. Mol. Biol.* **219** 109–22

Grindley H M, Artymiuk P J, Rice D W and Willett P 1993 Identificaton of tertiary structure resemblance in proteins using a maximal common subgraph isomorphism algorithm *J. Mol. Biol.* **229** 707–21

Hadley C and Jones D T 1995 A systematic comparison of protein structure classifications SCOP, CATH and FSSP *Structure* **7** 1099–112

Hargittai I and Hargittai M 2000 *In Our Own Iamge: Personal Symmetry in Discovery* Kluwer Academic/Plenum Publishers

Harrison A, Pearl F, Mott R, Thornton J and Orengo C 2002 Quantifying the similarities within fold space *J. Mol. Biol.* **323** 909–26

Havel T F, Crippen G M and Kuntz I D 1979 Effects of distance constraints on macromolecular conformation. II Simulation of experimental results and theoretical predictions *Biopolymers* **18** 73–81

Havel T F 1991 An evaluation of computational strategies for use in the determination of protein structure from distance constraints obtained by nuclear magnetic resonance *Prog. Biophys. Mol. Biol.* **56** 43–78

Havlin S and Ben-Avraham D 1982 Fractal dimensionality of polymer chains *J. Phys. A: Math. Gen.* **15** L311–28

Hawkins D M and Merriam D F 1973 Optimal zonation of digitized sequential data *Math. Geol.* **5** 389–95

Heringa J and Argos P 1993 A method to recognize distant repeats in protein sequences *Prot. Struct. Funct. Genet.* **17** 391–411

Heringa J and Taylor W R 1997 Three-dimensional domain duplication swapping and stealing *Curr. Opinion Struct. Biol.* **7** 416–21

Higgins D G, Labeit S, Gautel M and Gibson T J 1994 The evolution of titin and related giant muscle proteins *J. Mol. Evol.* **38** 395–404

Ho K and Curmi P M G 2002 Twist and shear in β-sheets and β-ribbons *J. Mol. Biol.* **317** 291–308

Hol W G J, van Duijnen P T and Berendsen H J C 1978 The α-helix dipole and the properties of proteins *Nature* **273** 443–6

Holland J 1975 *Adaptation in Natural and Artificial Systems* (Anne Arbor, MI: University of Michigan Press)

Holm L and Sander C 1991 Database algorithn for generating protein backbone and side-chain co-ordinates from a C^{α} trace: Applications to model building and detection of co-ordinate errors *J. Mol. Biol.* **218** 183–94

——1993a Globin fold in a bacterial toxin *Nature* **361** 309

——1993b Protein-structure comparison by alignment of distance matrices *J. Mol. Biol.* **233** 123–38

——1994a Parser for protein-folding units *Prot. Struct. Funct. Genet.* **19** 256–68

——1994b Searching protein structure databases has come of age *Proteins* **19** 165–73

——1997 Dali/FSSP classification of three-dimensional protein folds *Nucleic Acids Res.* **25** 231–4

——1998 Touring protein fold space with dali/FSSP *Nucleic Acids Res.* **26** 316–19

Holm L, Ouzounis C, Sander C, Tuparev G and Vriend G 1992 A database of protein structure families with common folding motifs *Prot. Sci.* **1** 1691–8

Hou J, Sims G E and Kim S-H 2002 A global representation of protein fold space *Proc. Natl Acad. Sci., USA* **100** 2386–90

Hu H, Elstner M and Hermans J 2003 Comparison of a QM/MM force field and molecular mechanics force fields in the simulations of alanine and glycine 'dipeptides' (Ace-

Ala-Nme and Ace-Gly-Nme) in water in relation to the problem of modelling the unfolded peptide backbone in solution *Prot. Struct. Funct. Genet.* **50** 451–63

Hubbard T J P, Murzin A G, Brenner S E and Chothia C 1997 SCOP: a structural classification of proteins database *Nucleic Acids Res.h* **25** 236–9

Hutchinson E G and Thornton J M 1993 The greek key motif—extraction, classification and analysis *Prot. Eng.* **6** 233–45

Islam S A, Luo J and Sternberg M J E 1995 Identification and analysis of domains in proteins *Prot. Eng.* **8** 513–25

Jacobs S, Harp J, Devarakonda S, Kim Y, Rastinejad F and Khorasanizadeh S 2002 The active site of the SET domain is constructed on a knot *Nat. Str. Biol.* **9** 828–32

Janin J and Chothia C 1985 Domains in proteins: definitions location and structural principles *Meth. Enzymol.* **115** 420–40

Janin J and Wodak S J 1983 Structural domains in proteins an their role in the dynamics of protein function *Prog. Biophys. Mol. Biol.* **42** 21–78

Johannissen L O and Taylor W R 2004 Protein fold comparison by the alignment of topological strings *Prot. Eng.* **16** 949–55

Johnson L B 1925–1963 *The Books of Bokonon* (San Lorenzo, CA: McCabe)

Jonassen I, Eidhammer I and Taylor W R 1999 Discovery of local packing motifs in protein structures *Prot. Struc. Funct. Gene.* **34** 206–19

Jonassen I, Eidhammer I, Grindhaug S H and Taylor W R 2000 Searching the protein structure databank with weak sequence patterns and structural constraints *J. Mol. Biol.* **304** 599–619

Jones D T, Taylor W R and Thornton J M 1992 A new approach to protein fold recognition *Nature* **358** 86–9

——1994 A model recognition approach to the prediction of all-helical membrane protein structure and topology *Biochemistry* **33** 3038–49

Jones D T, Orengo C A, Taylor W R and Thornton J M 1993 Progress towards recognising protein folds from amino acid sequence *Prot. Eng.* **6** 124

Jones S, Stewart M, Michie A, Swindells M B, Orengo C and Thornton J M 1998 Domain assignment for protein structures using a consensus approach: characterization and analysis *Prot. Sci.* **7** 233–42

Jones D T 1994 *De novo* protein design using pairwise potentials and a genetic algorithm *Prot. Sci.* **3** 567–74

——2000 The PSIPRED protein structure prediction server *Bioinformatics* **16** 404–5

Kabsch W and Sander C 1983 Dictionary of protein secondary structure: Pattern recognition of hydrogen-bonded and geometrical features *Biopolymers* **22** 2577–637

Kajva A V 1992 Left-handed topology of super-secondary structure formed by aligned α-helix and β-hairpin *FEBS Lett.* **302** 8–10

Kauzmann W 1959 Relative probabilities of isomers in cystine-containing randomly coiled polypeptides *Sulfur in Proteins* ed R E A Benesch (New York: Academic) pp 93–108

Kendrew J C, Klyne W, Lifson S, Miyazawa T, Nemethy G, Phillips D C, Ramachandran G N and Scheraga H A 1970 *Biochemistry* **9** 3471–9

Kikuchi T, Némethy G and Scheraga H 1986 Spatial geometric arrangements of sisulphide-crosslinked loops in proteins *J. Comput. Chem.* **7** 67–88

Klapper M H and Klapper I Z 1980 The 'knotting' problem in proteins: loop penetration *Biochim. Biophys. Acta* **626** 97–105

Klopman G and Henderson R V 1991 A graph theory-based 'expert system' methodology for structure-activity studies *J. Math. Chem.* **7** 187–216

Kobe B and Deisenhofer J 1993 The structure of porcine ribonuclease inhibitor, a protein with leucine-rich repeats *Nature* **366** 751–6

——1995 Proteins with leucine-rich repeats *Curr. Opinion Struct. Biol.* **5** 409–16

Koch I, Kaden F and Selbig J 1992 Analysis of protein sheet topologies by graph theoretical methods *Proteins* **12** 314–23

Koch I, Lengauer T and Wanke E 1996 An algorithm for finding maximal common subtopologies in a set of protein structures *J. Comput. Biol.* **3** 289–306

Kraulis P J 1991 MOLSCRIPT: A program to produce both detailed and schematic plots of protein structures *J. Appl. Crystallogr.* **24** 946–50

Krzanowski W J 1988 *Principles of Multivariate Analysis: A User's Perspective* (Oxford: Clarendon)

Kuntz I D, Crippen G M, Kollman P A and Kimelman D 1976 Calculation of protein tertiary structure *J. Mol. Biol.* **106** 983–94

Kuntz I D, Thomason J F and Oshiro C M 1989 Distance geometry *Meth. Enzymol.* **177** 159–204

Kyte J 1995 *Structure in Protein Chemistry* (New York: Garland Publishing)

Lagrange J L 1870 *Oeuvres* vol 5, Paris

Lasters I, Wodak S, Alard P and van Cutsem E 1988 Structural principles of parallel β-barrels in proteins *Proc. Natl Acad. Sci., USA* **85** 3338–42

Lasters I, Wodak S and Pio F 1990 The design of idealised α/β-barrels: analysis of β-sheet closure requirements *Prot. Struct. Funct. Genet.* **7** 249–56

Lasters I 1990 Estimating the twist of β-strands embedded within a regular parallel β-barrel structure *Prot. Eng.* **4** 133–5

Le Nguyen D, Heitz A, Chiche L, Castro B, Boigegrain R and Coletti-Previero M 1990 Molecular recognition between serine proteases and new bioactive microproteins with a knotted structure *Biochimie* **72** 431–5

Lee B K and Richards F M 1971 The interpretation of protein structures: estimation of static accessibility *J. Mol. Biol.* **55** 379–400

Lesk A M and Chothia C 1980 How different amino acid sequences determine similar protein structures: The structure and evolutionary dynamics of the globins *J. Mol. Biol.* **136** 225–70

Lesk A M, Branden C I and Chothia C 1989 Structural principles of α/β-barrel proteins: the packing of the interior of the sheet *Prot. Struct. Funct. Genet.* **5** 139–48

Lesk A M 1979 Detection of three-dimensional patterns of atoms in chemical structures *Comm. ACM* **22** 219–24

Levinthal C 1969 *Mossbauer Spectroscopy in Biological Systems* ed P Debrumer, J C M Tsibris and E Munck (Urbana, IL: University of Illnois Press) pp 22–4

Levitt M and Chothia C 1976 Structural patterns in globular proteins *Nature* **261** 552–8

Levitt M and Gerstein M 1998 A unified statistical framework for sequence comparison and structure comparison *Proc. Natl Acad. Sci., USA* **95** 5913–20

Levitt M and Greer J 1977 Automatic identification of secondary structure in globular proteins *J. Mol. Biol.* **114** 181–293

Levitt M 1976 A simplified representation of protein conformations for rapid simulation of protein folding *J. Mol. Biol.* **104** 59–107

——1983a Molecular dynamics of native proteins: II. Analysis and nature of motion *J. Mol. Biol.* **168** 621–57

——1983b Protein folding by restrained energy minimization and molecular dynamics *J. Mol. Biol.* **170** 723–64

Li H Q, Chen S H and Zhao H M 1990 Fractal structure and conformational entropy of protein chain. *Int. J. Biol. Macromol.* **12** 374–8

Liang C and Mislow K 1994a Knots in proteins *J. Am. Chem. Soc.* **116** 11 189–90

——1994b Topological chirality of proteins *J. Am. Chem. Soc.* **116** 3588–92

——1995 Topological features of protein structures: knots and links *J. Am. Chem. Soc.* **117** 4201–13

Lin K, May A C and Taylor W R 2002 Threading using neural networks (TUNE): the measure of protein sequence-structure compatibility *Bioinformatics* **18** 1350–7

Logsdon J M J and Palmer J D 1994 Origin of introns—early or late? *Nature* **369** 526–7

Louie A H and Somorjai R L 1982 Differential geometry of proteins: a structural and dynamical representation of proteins *J. Theor. Biol.* **98** 189–209

——1983 Differential geometry of proteins: helical approximations *J. Mol. Biol.* **168** 143–62

Luo X, Taylor K and Mezey P G 1993 Vertex mobility of polyhedra *Bull. Math. Biol.* **55** 131–40

Lupas A N, Ponting C P and Russell R B 2001 On the evolution of protein folds: are similar motifs in different protein folds the result of convergence, insertion or relics of an ancient peptide world? *J. Struct. Biol.* **134** 191–203

MacKay A L 1986 Towards a grammar of inorganic structure *Comput. Math. Appl.* B **12** 803–24

Maiorov V N and Crippen G M 1994 Significance of root-mean-square deviation in comparing three-dimensional structures of globular proteins *J. Mol. Biol.* **235** 625–34

Mandelbrot B B 1982 (San Francisco, CA: Freeman)

Mansfield M L 1994 Are there knots in proteins *Nature Struct. Biol.* **1** 213–14

——1997 Fit to be tied *Nature Struct. Biol.* **4** 116–17

Mao B 1989 Molecular topology of multiple-disulphide polypeptide chains *J. Am. Chem. Soc.* **111** 6132–6

——1993 Topological chirality of proteins *Prot. Sci.* **2** 1057–9

Matthews B W and Rossmann M G 1985 Comparisons of protein structures *Meth. Enzymol.* **115** 397–420

May A C W and Johnson M S 1994 Protein structure comparisons using a combination of a genetic algorithm, dynamic programming and least-squares minimisation *Prot. Eng.* **7** 475–85

——1995 Improved genetic algorithm-based protein structure comparisons: pairwise and multiple superpositions *Prot. Eng.* **8** 873–82

May A C W 1996 Pairwise iterative superposition of distantly related proteins and assessment of the significance of 3-D structural similarity *Prot. Eng.* **9** 1093–101

McLachlan A D and Stewart M 1977 the 14-fold periodicity in α-tropomyosin and the interaction with actin *J. Mol. Biol.* **103** 271–98

McLachlan A D 1977 Analysis of periodic patterns in amino acid sequences: collagen *Biopolymers* **16** 1271–97

——1979a Gene duplication in the structural evolution of chymotrypsin *J. Mol. Biol.* **128** 49–79

——1979b Three-fold structural pattern in the soybean trypsin inhibitor (kunitz) *J. Mol. Biol.* **133** 557–63

——1983 Analysis of gene duplication repeats in the myosin rod *J. Mol. Biol.* **169** 15–30

——1984 How alike are the shapes of two random chains? *Biopolymers* **23** 1325–31

Mercier M, Mekenyan O, Dubois J E and Bonchev D 1991 DARC/PELCO and OASIS methods: I. Methodological comparison; modeling purine pk_a and antitumor activity *Eur. J. Med. Chem.* **26** 575–92

Michel G, Sauve V, Larocque R, Li Y, Matte A and Cygler M 2002 The structure of the RlmB 23S rRNA methyltransferase reveals a new methyltransferase fold with a unique knot *Structure* **10** 1303–15

Min J R, Zhang X, Cheng X D, Grewal S I S and Xu R-M 2002 Structure of the SET domain lysine methyltransferase Clr4 *Nature Struct. Biol.* **9** 833–8

Mitchell T J, Tute M S and Webb G A 1989 A molecular modeling study of the interaction of noradrenaline with the beta-2-adrenegeric receptor *J. Comput. Aided Mol. Des.* **3** 211–23

Mizuguchi K, Deane C M, Blundell T L and Overington J P 1998 HOMSTRAD: a database of protein structure alignments for homologous families *Prot. Sci.* **7** 2469–71

Murthy M R N 1984 A fast method of comparing protein structures *FEBS Lett.* **16** 97–102

Murzin A G and Finkelstein A V 1988 General architecture of the α-helical globule *J. Mol. Biol.* **204** 749–69

Murzin A G, Lesk A M and Chothia C 1992 β-trefoil fold: patterns of structure and sequence in the kunitz inhibitors interleukins-1β and 1α and fibroblast growth factors *J. Mol. Biol.* **223** 531–43

——1994a Principles determining the structure of β-sheet barrels in proteins: I a theoretical analysis *J. Mol. Biol.* **236** 1396–81

——1994b Principles determining the structure of β-sheet barrels in proteins: II the observed structures *J. Mol. Biol.* **236** 1382–400

Murzin A G, Brenner S E, Hubbard T and Chothia C 1995 SCOP: a structural classification of proteins database for the investigation of sequences and structures *J. Mol. Biol.* **247** 536–40

Murzin A G 1992 Structural principles for the propeller assembly of β-sheets: the preference for seven-fold symmetry *Prot. Struct. Funct. Genet.* **14** 191–201

——1993 Can homologous proteins evolve different enzymatic activities? *TIBS* **18** 403–5

Nagano K 1977 Logical analysis of the mechanism of protein folding: IV super-secondary structures *J. Mol. Biol.* **109** 235–50

Newman A 1994 Small nuclear RNAs and pre-mRNA splicing *Curr. Opinion Cell Biol.* **6** 360–7

Nishikawa K and Ooi T 1974 *J. Theor. Biol.* **48** 443–53

Nissen P, Hansen J, Ban N, Moore P B and Steitz T A 2000 The structural basis of ribosome activity in peptide bond synthesis *Science* **289** 920–30

Nojima H 1987 Molecular evolution of the calmodulin gene *FEBS Lett.* **217** 187–90

Novotny J, Bruccoleri R E and Newell J 1988 Twisted hyperboloid (*strophoid*) as a model of β-barrels in proteins *J. Mol. Biol.* **177** 567–73

Nureki O *et al* 2002 An enzyme with a deep trefoil knot for the active-site architecture *Acta Crystallogr.* D **58** 1129–39

Nussinov R and Wolfson H J 1991 Efficient detection of 3-dimensional structural motifs in biological macromolecules by computer vision techniques *Proc. Natl Acad. Sci., USA* **88** 10 495–9

Ollis D L *et al* 1992 The α/β hydrolase fold *Prot. Eng.* **5** 197–211

Orengo C A and Taylor W R 1990 A rapid method for protein structure alignment *J. Theor. Biol.* **147** 517–51

——1993 A local alignment method for protein structure motifs *J. Mol. Biol.* **233** 488–97

——1996 SSAP: sequential structure alignment program for protein structure comparison *Computer methods for macromolecular sequence analysis (Meth. Enzymol. 266)* ed R F Doolittle (Orlando, FL: Academic) pp 617–35

Orengo C A and Thornton J M 1993 Alpha plus beta folds revisited: some favoured motifs *Structure* **1** 105–20

Orengo C A, Brown N P and Taylor W R 1992 Fast protein structure comparison for databank searching *Prot. Struct. Funct. Genet.* **14** 139–67

Orengo C A, Flores T P, Jones D T, Taylor W R and Thornton J M 1993a Recurring structural motifs in proteins with different functions *Curr. Biol.* **3** 131–9

Orengo C A, Flores T P, Taylor W R and Thornton J M 1993b Identification and classification of protein fold families *Prot. Eng.* **6** 485–500

Orengo C A, Jones D T and Thornton J M 1994 Protein superfaimiles and domain superfolds *Nature* **372** 631–4

Orengo C A, Michie A D, Jones S, Jones D T, Swindells M B and Thornton J M 1997 CATH—a hierarchic classification of protein domain structures *Structure* **5** 1093–108

Orengo C A 1994 Classification of protein folds *Curr. Opinion Struct. Biol.* **4** 429–40

Ouzounis C, Sander C, Scharf M and Schneider R 1993 Prediction of protein-structure by evaluation of sequence-structure fitness—aligning sequences to contact profiles derived from 3-dimensional structures *J. Mol. Biol.* **232** 805–25

Palmer J D and Logson J M 1991 The recent origins of introns *Curr. Opinion Genet. Dev.* **1** 470–7

Park B and Levitt M 1996 Energy functions that discriminate X-ray and near native folds from well-constructed decoys *J. Mol. Biol.* **258** 367–92

Pastore A and Lesk A M 1990 Comparison of the structures of globins and phycocyanins—evidence for evolutionary relationship *Prot. Struct. Funct. Genet.* **8** 133–55

Patthy L 1985 Evolution of the proteases of blood coagulation and fibrinolysis by assembly from modules *Cell* **41** 657–63

——1991 Exons—original building blocks of proteins? *BioEssays* **13** 187–91

——1994 Introns and exons *Curr. Opinion Struct. Biol.* **4** 383–92

Pazos F, Helmer-Citterich M, Ausiello G and Valencia A 1997 Correlated mutations contain information about protein-protein interaction *J. Mol. Biol.* **271** 511–23

Pearl L H and Taylor W R 1987 A structural model for the retroviral proteases *Nature* **329** 351–4

Penrose R 1989 *The Emperor's New Mind: Concerning Computers, Minds and the Laws of Physics* (New York: Oxford University Press)

Petersen K and Taylor W R 2003 Modelling zinc-binding proteins with GADGET: Genetic algorithm and distance geometry for exploring topology *J. Mol. Biol.* **325** 1039–59

Petsko G A, Kenyon G L, Gerlt J A, Ringe D and Kozarich J W 1993 On the origin of enzymatic species *TIBS* **18** 372–6

Phillips D C, Sternberg M J E, Thornton J M and Wilson I A 1978 An analysis of the structure of triose phosphate isomerase and its comparison with lactate dehydrogenase *J. Mol. Biol.* **119** 329–51

Phillips D C 1966 The three-dimensional structure of an enzyme *Sci. Am.* **215** 78–90

——1970 The development of crystallographic enzymology *British Biochemistry, Past and Present (Biochem. Soc. Symp.)* (London: Academic) pp 11–28

Piatigorsky J G and Wistow G 1991 The recruitment of crystallins: New functions precede gene duplication *Science* **252** 1078–9

Plaxco K W, Simons K T and Baker D 1998 Contact order, transitionstate placement and the refolding rates of single domain proteins *J. Mol. Biol.* **277** 985–94

Pollock D D and Taylor W R 1997 Effectiveness of correlation analysis in identifying protein residues undergoing correlated evolution *Prot. Eng.* **10** 647–57

Ponting C P and Russell R B 2000 Identification of distant homologues of fibroblast growth factors suggests a common ancestor for all -trefoil proteins *J. Mol. Biol.* **302** 1041–7

Presnell S R and Cohen F E 1989 Topological distribution of four-α-helical bundles *Proc. Natl Acad. Sci., USA* **86** 6592–6

Press W H, Flannery B P, Teukolsky S A and Vetterling W T 1986 *Numerical Recipes: The Art of Scientific Computing* (Cambridge: Cambridge University Press)

——1992 *Numerical Recipes: The Art of Scientific Computing* 2nd edn (Cambridge: Cambridge University Press)

Ptitsyn O B and Finkelstein A V 1980 Similarities of protein topologies: Evolutionary divergence, functional convergence or principles of folding? *Quart. Rev. Biophys.* **13** 339–86

Ptitsyn O B 1983 Protein as an 'edited' statistical copolymer *Conformation in Biology* ed R Srinivasan and R M Sarma (New York: Academic) pp 49–58

——1995 Structures of folding intermediates *Curr. Opinion Struct. Biol.* **5** 74–8

Purisima E O and Scheraga H A 1984 Conversion from a virtual-bond chain to a complete polypeptide backbone chain *Biopolymers* **23** 1207–24

Randić M 1992 In search of structural invariants *J. Math. Chem.* **9** 97–146

Rao S T and Rossmann M G 1973 Comparison of super-secondary structures in proteins *J. Mol. Biol.* **76** 241–56

Rashin A 1985 Location of domains in globular proteins *Meth. Enzymol.* **115** 420–40

Rawlings C J, Taylor W R, Nyakairu J, Fox J and Sternberg M J E 1985 Reasoning about protein topology using the logic programming language PROLOG *J. Mol. Graph.* **3** 151–7

——1986 Using PROLOG to represent and reason about protein structure *Lecture Notes in Computer Science*

Reardon D and Farber G K 1995 Protein motifs. 4: the structure and evolution of α/β barrel proteins *Faseb. J.* **9** 497–503

Richards F M and Kundrot C E 1988 Identification of structural motifs from protein coordinate data: Secondary structure and first level supersecondary structure *Prot. Struct. Funct. Genet.* **3** 71–84

Richardson J S, Getzoff E D and Richardson D C 1978 The β-bulge: a common small unit of nonrepetitive protein structure *Proc. Natl Acad. Sci., USA* **75** 2574–8

Richardson J S 1977 β-Sheet topology and the relatedness of proteins *Nature* **268** 495–500

——1981 The anatomy and taxonomy of protein structure *Adv. Prot. Chem.* **34** 167–339

——1985 Describing patterns of protein tertiary structure *Meth. Enzymol.* **115** 341–80

Richmond T J and Richards F M 1978 Packing of α-helices: Geometrical constraints and contact areas *J. Mol. Biol.* **119** 537–55

Rippmann F and Taylor W R 1991 Visualization of structural similarity in proteins *J. Mol. Graph.* **9** 3–16

Robson B, Platt E, Fishleigh R V, Marsden A and Millard P 1987 Expert system for protein engineering: Its application in the study of chloramphenicol acetyltransferase and avian pancreatic polypeptide *J. Mol. Graph.* **5** 8–17

Rose G D 1979 Hierarchic organisation of domains in globular proteins *J. Mol. Biol.* **234** 447–70

Rossmann M G and Argos P 1976 Exploring structural homology of proteins *J. Mol. Biol.* **105** 75–96

Rost B and Sander C 1993 Prediction of protein secondary structure at better than 70-percent accuracy *J. Mol. Biol.* **232** 584–99

Rothbard J B, Townsend A, Edwards M and Taylor W R 1986 Pattern recognition among T-cell epitopes *Modern Trends in Human Leukemia* vol VII (Amsterdam: Elsevier)

Russell R B and Barton G J 1992 Multiple protein-sequence alignment from tertiary structure comparison: assignment of global and residue confidence levels *Prot. Struct. Funct. Genet.* **14** 309–23

Russell R B 1998 Detection of protein three-dimensional side-chain patterns: New examples of convergent evolution *J. Mol. Biol.* **279** 1211–27

Salem G M, Hutchinson E G and Orengo C A 1999 Correlation of observed fold frequency with the occurrence of local structural motifs *J. Mol. Biol.* **287** 969–81

Salemme F R and Weatherford D W 1981a Conformational and geometrical properties of β-sheets in proteins: I parallel β-sheets *J. Mol. Biol.* **146** 101–17

——1981b Conformational and geometrical properties of β-sheets in proteins: Ii. antiparallel and mixed β-sheets *J. Mol. Biol.* **146** 119–41

Salemme F R 1981 Conformational and geometrical properties of β-sheets in proteins: Iii. isotropically stressed configurations *J. Mol. Biol.* **146** 143–56

Šali A and Blundell T L 1990 Definition of general topological equivalence in protein structures: a procedure involving comparison of properties and relationship through simulated annealing and dynamic programming *J. Mol. Biol.* **212** 403–28

Sayle R and Milner-White E J 1995 RasMol: Biomolecular graphics for all *TIBS* **20** 374–5

Scarselli M, Bernini A, Segoni C, Molinari H, Esposito G, Lesk A M, Laschi F, Temussi P and Niccolai N 1999 Tendamistat surface accessibility to the TEMPOL paramagnetic probe *J. Biomol. NMR* **15** 125–33

Scheerlinck J-P Y, Lasters I, Claessens M, De Maeyer M, Pio F, Delhaise P and Wodak S J 1992 Recurrent $\alpha\beta$ loop structure in tim barrel motifs show a distinct pattern of conserved structural features *Proteins* **12** 299–313

Schulz G E 1980 Gene duplication in glutathione reductase *J. Mol. Biol.* **138** 335–47

Sedgewick R 1990 *Algorithms in C* (Reading, MA: Addison-Wesley)

Sela M and Lifson S 1959 On the reformation of disulphide bridges in proteins *Biochim. Biophys. Acta* **36** 471–8

Shapiro J A, Adhya S L and Bukhari A I 1977 Introduction: New pathways in the evolution of chromosome structure *DNA Insertion Elements, Plasmids and Episomes* ed A I Bukhari, J A Shapiro and S L Adhya (Cold Spring Harbor, NY: Cold Spring Harbor Laboratory Press) pp 3–11

Siddiqui A S and Barton G J 1995 continuous and discontinuous domains—an algorithm for the automatic generation of reliable protein domain definitions *Prot. Sci.* **4** 872–84

Sippl M J 1990 Calculation of conformational ensembles from potentials of mean force. an approach to the knowledge-based prediction of local structures in globular proteins *J. Mol. Biol.* **213** 859–83

Sklenar H, Etchebest C and Lavery R 1989 describing protein structure: a general algorithm yielding complete helicoidal parameters and a unique overall axis *Prot. Struct. Funct. Genet.* **6** 46–60

Skolnick J, Kolinski A and Yaris R 1989 Dynamic Monte Carlo study of the folding of a six-stranded Greek key globular protein *Proc. Natl Acad. Sci., USA* **86** 1229–33

Smith T F and Waterman M S 1981 Identification of common molecular subsequences *J. Mol. Biol.* **147** 195–7

Sowdhamini R and Blundell T 1995 An automatic method involving cluster analysis of secondary structures for the identification of domains in proteins *Prot. Sci.* **4** 506–20

Sowdhamini R, Rufino S D and Blundell T L 1996 A database of globular protein structural domains: clustering of representative family members into similar folds *Fold. Design* **1** 209–20

Sowdhamini R, Burke D F, Huang J-F, Mizuguchi K, Nagarajaram H A, Srinivasan N, Steward R E and Blundell T L 1998 CAMPASS: A database of structurally aligned protein superfamilies *Structure* **6** 1087–94

Steitz J A 1988 Snurps *Sci. Am.* **258** 36–41

Sternberg M J E and Thornton J M 1977a On the conformation of proteins: An analysis of β-pleated sheets *J. Mol. Biol.* **110** 285–96

——1977b On the conformation of proteins: The handedness of the connection between parallel β-strands *J. Mol. Biol.* **110** 269–83

Sternberg M J E, Taylor W R, Nyakairu J, Fox J and Rawlings C J 1985 Reasoning about protein topology using the logic programming language PROLOG *J. Mol. Graph.* **3** 108–9 (abstract)

Subbarao N and Haneef I 1991 Defining topological equivalences in macromolecules *Prot. Eng.* **4** 887–4

Subbiah S, Laurents D V and Levitt M 1993 Structural similarity of DNA-binding domains of bacteriophage repressors and the globin core *Curr. Biol.* **3** 141–8

Swindells M B 1995a A procedure for detecting structural domains in proteins *Prot. Sci.* **4** 103–12

——1995b A procedure for the automatic determination of hydrophobic cores in protein structures *Prot. Sci.* **4** 93–102

Tang J, James M N G, Hsu I N, Jenkins J A and Blundell T L 1978 Structural evidence for gene duplication in the evolution of the acid proteases *Nature* **271** 619–21

Taylor W R and Hatrick K 1994 Compensating changes in protein multiple sequence alignments *Prot. Eng.* **7** 341–8

Taylor W R and Jones D T 1993 Deriving an amino acid distance matrix *J. Theor. Biol.* **164** 65–83

Taylor W R and Lin K 2003 A tangled problem *Nature* **421** 25 (concept)

Taylor W R and Orengo C A 1989a A holistic approach to protein structure comparison *Prot. Eng.* **2** 505–19

——1989b Protein structure alignment *J. Mol. Biol.* **208** 1–22

Taylor W R and Thornton J M 1984 Recognition of super-secondary structure in proteins *J. Mol. Biol.* **173** 487–514

Taylor W R, Thornton J M and Turnell W G 1983 A elipsoidal approximation of protein shape *J. Mol. Graph.* **1** 30–8

Taylor W R, Flores T P and Orengo C A 1994a Multiple protein structure alignment *Prot. Sci.* **3** 1858–70

Taylor W R, Jones D T and Green N M 1994b A method for α-helical integral membrane protein fold prediction *Prot. Struct. Funct. Genet.* **18** 281–94

Taylor W R, May A C W, Brown N P and Aszódi A 2001 Protein structure: Geometry, topology and classification *Rep. Prog. Phys.* **64** 517–90

Taylor W R, Heringa J, Baud F and Flores T P 2002 A Fourier analysis of symmetry in protein structure *Prot. Eng.* **15** 79–89

Taylor W R, Munro R E J, Petersen K and Bywater R P 2003a *Ab initio* modelling of the N-terminal domain of the secretin receptors *Comput. Biol. Chem.* **27** 103–14

Taylor W R, Xiao B, Gamblin S J and Lin K 2003b A knot or not a knot? SETting the record 'straight' on proteins *Comput. Biol. Chem.* **27** 11–15

Taylor W R 1986a The classification of amino acid conservation *J. Theor. Biol.* **119** 205–18

——1986b Identification of protein sequence homology by consensus template alignment *J. Mol. Biol.* **188** 233–58

——1987 Multiple sequence alignment by a pairwise algorithm *Comput. Appl. Biol. Sci.* **3** 81–7

——1988 A flexible method to align large numbers of biological sequences *J. Mol. Evol.* **28** 161–9

——1991a Sequence analysis: spinning in hyperspace *Nature* **353** 388–9 (News and Views)

——1991b Towards protein tertiary fold prediction using distance and motif constraints *Prot. Eng.* **4** 853–70

——1993a Modelling protein structure from remote sequence similarity: an approach to tertiary structure prediction *Computational Methods in Genome Research* ed S Suhai (New York: Plenum) pp 317–28

——1993b Protein fold refinement: building models from idealised folds using motif constraints and multiple sequence data *Prot. Eng.* **6** 593–604

——1993c Protein structure prediction from sequence *Comput. Chem.* **17** 117–22

——1997a Evolution and relationships of protein families *Nucleic Acid and Protein Sequence Analysis: A Practical Approach* 2nd edn, ed M J Bishop and C J Rawlings (Oxford: IRL) pp 313–40

——1997b Multiple sequence threading: an analysis of alignment quality and stability *J. Mol. Biol.* **269** 902–43

——1997c Random models for double dynamic score normalisation *J. Mol. Evol.* **44** S174–80 (Special issue in memory of Kimura)

——1998 Dynamic databank searching with templates and multiple alignment *J. Mol. Biol.* **280** 375–406

——1999a The properties of amino acids in sequences *Nucleic Acid and Protein Databases: A Practical Approach (Second Edition)* 2nd edn, ed M J Bishop (New York: Academic) ch 5, pp 81–103

——1999b Protein structure alignment using iterated double dynamic programming *Prot. Sci* **8** 654–65

——1999c Protein structure domain identification *Prot. Eng.* **12** 203–16

——2000a A deeply knotted protein and how it might fold *Nature* **406** 916–19

——2000b Protein structure *Handbook of Statistical Genetics* (New York: Wiley) pp 209–35

——2000c Searching for the ideal forms of proteins *Biochem. Soc. Trans.* **28** 264–9

——2001 Defining linear segments in protein structure *J. Mol. Biol.* **310** 1135–50

——2002a A periodic table for protein structure *Nature* **416** 657–60

——2002b Protein structure comparison using bipartite graph matching *Mol. Cell. Proteomics* **1** 334–9

Thomas D J 1994 The graduation of secondary structure elements in proteins *J. Mol. Graph.* **12** 146–52

Thomazeau K, Dumas R, Halgand F, Forest E, Douce R and Bi V 2000 Structure of spinach acetohydroxyacid isomeroreductase complexed with its product of reaction dihydroxy-methylvalerate, manganese and ADP-ribose *Acta Crystallogr.* D **56** 389–99

Thornton J and Sibanda B 1983 Amino and carboxy-terminal regions in globular proteins *J. Mol. Biol.* **167** 443–60

Thornton J M, Edwards M S, Taylor W R and Barlow D J 1986 Location of 'continuous' antigenic determinants in the protuding regions of proteins *EMBO J.* **5** 409–13

Thornton J M 1981 Disulphide bridges in globular proteins *J. Mol. Biol.* **151** 261–87

Thouless D 1992 Condensed matter in less than three dimensions *The New Physics* ed P Davies (Cambridge: Cambridge University Press) pp 209–35

Toh H, Ono M, Saigo K and Miyata T 1985 Retroviral protease-like sequence in the yeast transposon ty1 *Nature* **315** 691

Torgerson W S 1958 *Theory and Methods of Scaling* (London: Wiley)

Trievel R C, Beach B M, Dirk L M A, Houtz R L and Hurley J H 2002 Structure and catalytic mechanism of a SET domain protein methyltransferase *Cell* **111** 91–103

Ullmann J R 1976 An algorithm for subgraph isomorphism *J. ACM* **23** 31–42

Unger R and Moult J 1993 Genetic algorithms for protein folding simulations *J. Mol. Biol.* **231** 75–81

Vonnegut K 1963 *Cat's Cradle* (New York: Dell)

Walther D, Eisenhaber F and Argos P 1996 Principles of helix–helix packing in proteins: the helix lattice superposition model *J. Mol. Biol.* **255** 536–53

Weber P C and Salemme F R 1980 Structural and functional diversity in 4-α-helical proteins *Nature* **287** 82–4

Wilson J R, Jing C, Walker P A, Martin S R, Howell S A, Blackburn G M, Gamblin S J and Xiao B 2002 Crystal structure and functional analysis of the histone methyltransferase SET7/9 *Cell* **111** 105–15

Wilson J H 1985 *Genetic Recombination* (Menlo Park, CA: Benjamin-Cummings)

Wistow G and Piatigorsky J G 1987 Recruitment of enzymes as lens structural proteins *Science* **236** 1554–6

Wlodawer A, Miller M, Jaskolski M, Sathyanarayana B K, Baldwin E, Weber I T, Selk L M, Clawson L, Schneider J and Kent S B H 1989 Conserved folding in retroviral proteases: crystal structure of a synthetic HIV-1 protease *Science* **245** 616–621

Young G and Householder A S 1938 Discussion of a set of points in terms of their mutual distances *Psychometrika* **3** 19–22

Zehfus M H 1987 Continuous compact protein domains *Proteins* **2** 90–110

Zhang X, Tamaru H, Khan S I, Horton J R, Keefe L J, Selker E U and Cheng X 2002 Structure of the neurospora SET domain protein Dim-5, a histone H3 lysine methyltransferase *Cell* **111** 117–27

Zimm B H and Bragg J R 1959 Theory of the phase transition between helix and random coil in polypeptide chains *J. Chem. Phys.* **31** 526–35

Znamenskiy D, Le Tuan K, Poupon J C and Mornon J-P 2000 β-sheet modelling by helical surfaces *Prot. Eng.* **13** 407–12

Index

$\alpha\beta\alpha$ architecture, 87
$\alpha\beta\alpha$ layer model, 70, 231
$\alpha\beta\beta\alpha$ layer model, 71
$\alpha\beta\beta\alpha$ layer protein, 289
α-helix, 9
α-helix dipole, 305
β structure, 9
$\beta\alpha$-barrel, 71, 85, 120, 289
$\beta\alpha$-horseshoe, 286, 289
$\beta\alpha$ protein, 11, 246, 265, 288, 290
$\beta\beta$ layer model, 70
β-barrel, 85, 118, 138
β-bulge, 95, 97
β-helices, 11, 194
β-prism, 194, 289
β-propellor, 288–290
β-sandwich, 85, 113
β-sheet, 93, 133
β-sheet geometry, 93
β-sheet bend, 111
β-sheet classification, 140
β-sheet curl, 111
β-sheet definition, 56, 134
β-sheet dimensions, 136
β-sheet model, 94
β-sheet residue packing, 106
β-sheet stacking model, 113, 116
β-strand, 93, 133
β-strand stagger, 120
β-structure, 93
β-trefoil, 194
γ-crystalin, 289
ϕ rotation, 23
ϕ, ψ values, 94
ψ rotation, 23
$\alpha\beta\alpha$ layer model, 85, 200
$\alpha\beta\beta\alpha$ layer model, 85
β-sheet diagonals, 103
β-sheet twist, 93
2D grid, 69
2D representation, 63, 68, 85
3D embedding, 250

ab initio prediction, 47, 90, 261, 265
accessible surface area, 252
accretion-of-spheres, 79
acetohydroxy acid isomeroreductase, 221
adenovirus tail-spike, 128
adenylate kinase, 237
Alexander polynomial, 217
all-α protein models, 71
all-α protein, 10, 230, 246, 290
all-β protein, 10, 246, 290
alternate alignments, 168
ancient conserved regions, 183
angle/coordinate conversion, 25
annexin, 289
antigenic epitopes, 34
aperiodic minimal surfaces, 127
architecture, 13, 50, 63, 182
artificial neural net, 266
aspartyl protease, 295
atomic coordinates, 19
atomic twisted sheet model, 109
axial vector, 164, 169

backbone fragment, 161

bacterial toxin, 179
bacteriophage repressor, 179
bacterium, 293
ball of string, 237
ball of string fold, 238, 243
barrel architecture, 121
barrel axis, 120
barrel structure, 11
basic principles, 5
Bernal 'lattice', 78, 234
Bernal, John Desmond, 78, 79
Big Mac, 75
binary hydrophobicity, 257
bipartite graph, 171, 191
bishop's crozier, 127
bit-string, 162
Blu-Tac, 79, 227
Books of Bokonon, 306
branching polymer, 214
burger bun, 75

calcium binding motif, 156, 265
calmodulin, 156
Cambrian big-bang, 296
CAMPASS database, 181
carbonic anhydrase, 221
cartoon representation, 19
catalytic mechanism, 300
catalytic RNA, 297
catenanes, 212
catenoid, 112, 118
CATH database, 181, 182, 191
ceolecanth, 293
chain reversal model, 179
chain tangles, 248
chaperone, 237
Chaucy's ellipsoid, 32
Chebychev polynomials, 20
chemical topology, 212
chemotaxis Y protein, 85, 88, 172, 195, 233, 244, 265
chirality, 47, 48
Christmas tree, 227
chymotrypsin, 153
circular knot, 219
circular polymers, 214
circular proteins, 217
classification databases, 180
classification of ideal forms, 187
classification of proteins, 180
classifying topology, 236
clique, 170, 171
close-packed hexagonal lattice, 77
coiled-coil packing, 91
colicin-A, 179
colipase, 215
combinatoric approach, 263
combined rise, 66
common core, 196
common subgraph, 170
COMPARER program, 168
comparison methods, 159
comparison score distributions, 173
complexity of a fold, 238, 243
conic sections, 97
conic solvent accessibility, 252
conic surface, 97
connectedness index, 239–241
connection chirality, 14, 265
connection topology, 144
connectivity graph, 212, 214
constructing cylindrical barrels, 121
constructing hyperbolic barrels, 125
contact map, 161
contact normal, 63, 85
contact order, 243
continuous secondary structure, 55, 61
convergent evolution, 297
convex hulls, 75
correlated residue changes, 260
coupling function, 145
covalent cross-linking, 216
covalent knot, 226
covalent topology, 215
covariance of residues, 260
Crippen, Gordon, 214, 216
Cro repressor, 73

cross-link, 216
cubic lattice, 214
cyclic strings, 89
cylinder of equal penetration, 91
cylindrical β-barrel, 121, 130
cylindrical barrel construction, 121
cylindrical projection, 91
cylindrical sheet, 137, 189
cysteine, 214

DALI program, 161, 176, 182
David Phillips, 27
deletion, 161
deltahedra, 73, 79
deltahedra models, 75
density of packing, 251
dihedral angle, 23, 170
dimeric precursors, 295
directed graph, 213
distance geometry, 34, 66, 86, 175, 185, 238, 249, 260, 264, 301
distance geometry modelling, 258
distance matrix, 26, 85, 86, 161, 164, 249, 250, 260
distance plot, 27, 223
distance transform function, 258
distance weighting, 46, 48
disulfide bond, 215, 216, 223, 261
disulfide bridge, 214
disulfide knot, 216
domain definition, 143, 188
domain granularity, 149, 153
domain movements, 34
domain phase-space, 153
domain structure, 15
double dynamic programming, 165, 167, 168, 191, 201, 281
Dowker knot notation, 220
down-curling diagonal, 103
down-down radius, 105
DRAGON program, 249, 258
DSSP program, 51, 57, 62, 134, 257
dynamic programming, 52, 164, 167, 171, 201, 281

E-value, 174
eccentricity, 33
eccentricity plot, 33
edit distance, 200
edit-operation, 217
EF-hand motif, 156, 265
effective chain length, 239, 240
Efimov, Alexander, 200
eigenvalues, 36, 40, 185, 250
electron orbital, 186, 189
ellipsoid, 97, 143
elliptic paraboloid, 97, 99
elliptical cylinder, 118
embedding ellipsoids, 42
emergence of proteins, 303
empirical energy potential, 258, 267
empirical prediction method, 263
equivalent ellipsoid, 31, 32
evaluation function, 265
evolution of function, 293, 300
evolution of structure, 293
evolutionary model, 161, 297
evolutionary steps, 198
exons and introns, 295
exponential switch, 244
extended hyperbolic barrels, 126
external barrel curvature, 120
external coordinate strings, 88
extreme value distribution, 174
eye-lens protein, 301

fake β/α models, 231, 286
fake all-α models, 230
fake fold space, 232
fake hydrogen bond geometry, 252
fake protein, 229, 230, 240, 286
fake protein topology, 236
families of folds, 183
fibrous proteins, 6, 91, 281
figure-of-eight knot, 221
figure-of-eight knots, 221
Finkelstein, Alexey, 69, 73
flat sheet, 189
Flemish knots, 221

flying-saucers, 32
fold combinatorics, 232
fold complexity, 238, 243, 302
fold families, 162, 183
fold island, 207
fold space, 180, 185, 187, 198, 207, 227, 237, 264
fold superfamilies, 183
fold tree, 200
folding rate, 243
folding rules, 262
forms, 63
Fourier analysis, 280
Fourier transform, 92, 280, 283, 306
fractal, 242
frequency spectrum, 282
FSSP database, 181
full barrels, 139
function evolution, 293, 300

GADGET program, 258
gene duplication, 208, 294, 295, 299
generalized crystallography, 79
genetic algorithm, 47, 258, 267, 299
genetic mechanisms, 294
GeoMag, 79
geometric hashing method, 162
geometric representations, 19
geometric sheet model, 94
globin, 179, 241, 265, 267, 289
globin fold, 177, 179, 267
globin similarities, 177
globular proteins, 7, 73
gradual projection, 45, 260
granularity level, 149
graph matching, 169, 170, 191
graph topology, 212
Greek key, 13, 289
growth model for a β-sheet, 96

helical β-sheet model, 94
helical net, 91
helical rise, 56
helicoid, 112
helix packing classes, 91
helix-coil transition, 144
helix/helix contacts, 82
helix/helix packing, 91
hexagonal lattice, 77
hexagonal side chain packing, 110
hexapeptide, 161
hierarchical classification, 182
high-level matrix, 165, 167, 281
histone lysyl methyltransferase, 223
HIV protease, 289, 295
HOMFLY polynomial, 217
HOMSTRAD database, 181
hydrogen bond, 9, 56, 94, 133, 223, 226, 234, 251, 252
hydrogen bond ladder, 253
hydrogen bond networks, 133, 253
hydrophilic residues, 8
hydrophobic core, 7, 257
hydrophobic interaction, 251, 252
hydrophobic packing, 258
hydrophobic radius of gyration, 257
hydrophobic residue, 8, 249, 258
hydrophobic surface, 110
hyperbolic β-barrel, 125, 130
hyperbolic barrel construction, 125
hyperbolic model, 120, 121
hyperbolic paraboloid, 97, 99, 111
hyperbolic sheet, 138
hyperbolic surface, 118
hyperboloid, 97
hyperboloid of one sheet, 112, 118

ICO index, 190
icosahedron, 75, 82, 264
ideal form, 171, 187, 191, 200, 207, 236
incomplete barrel, 85
inertial axes, 27, 52, 175
inertial ellipsoid, 52, 252
inertial embedding, 41
inertial frame, 29

inertial tensor, 30
insertion, 161
interatomic distances, 26
internal comparison, 281
internal coordinate strings, 89
internal coordinates, 26
interpoint distances, 26
intrinsic topology, 212
intron splice site, 302
introns and exons, 295
inverse overlap area, 64
Ising model, 144, 156

jelly roll, 13, 246
Jones polynomial, 217

Kendrew, John, 208
KH-domain, 305
kinetic folding pathway, 249, 262
kinetic traps, 47
knobs into holes, 91
knot, 211, 216, 217, 238, 265
knot formation, 226
knot projection, 220
knot theory, 211, 217
knot-like structure, 225
knots in proteins, 217
knots in random chains, 227
knotted topology, 214

ladder of hydrogen bonds, 253
Lagrange theorem, 37
largest common fold, 200
largest common fragment, 187
largest common substructure, 203
lattice model, 175, 301
layer architecture, 68
layer structure, 188
layer-based model, 70
layer-based nomenclature, 84
layers, 187
layers of secondary structure, 87, 200
length distributions, 57
lengths of α-helices, 57
lengths of β-strands, 59
leucine-rich repeat, 246, 288
Levinthal's paradox, 262
line segment, 49, 50, 52, 63, 64, 85, 170, 232, 301
line segment overlap, 63, 86, 171
linear polymer, 213
linked loops, 216
living relics, 293
local packing, 243
locomotive nomenclature, 84, 172, 193, 236
longitudinal packing, 82
loop penetration, 215, 216, 223, 238
loop protrusions, 34
loop topology, 241
LOPAL program, 164
low-level matrix, 165, 167, 281

Maiorov–Crippen line, 233, 235
mandelate racemase, 300
Markov process, 174
matching ideal forms, 191, 200
matrix smoothing, 282
mattress model, 100
maximal common subgraph, 170
Maxwell's demon, 2
McLachlan barrel model, 120, 121
McLachlan shear number, 120
McLachlan, Andrew, 120, 153
membrane proteins, 7, 71, 191, 263
metal atom clusters, 79
metazoan radiation, 296
metric matrix, 249
model chain, 249
model chain geometry, 250
model evaluation function, 265
model hydrogen bond geometry, 252
model of evolution, 297
molecular dynamics, 262, 301
molecular embedding, 35
molecular modelling, 258

Molscript, 19
molten globule state, 82
momental ellipsoid, 31
moments of inertia, 28, 97
Monod, Jacques, 3, 262
Monte Carlo minimization, 47
motif, 259, 264
muconate lactonising enzyme, 300
multidimensional scaling, 35, 185
multiple domain definition, 155
multiple sequence alignment, 260
multiple structure alignment, 165
multiple structure superposition, 259
Murzin, Alexey, 73
Murzin–Finkelstein model, 73, 77, 89
Murzin–Finkelstein nomenclature, 82
myoglobin, 240, 264

N-terminal folding, 237
negative curvature, 137
negative eigenvalues, 40, 249
NMR spectroscopy, 260
NOE distance, 260
nomenclatures for packing, 82
non-local packing, 243
non-metric data, 48
non-planar graph, 215
normal form, 97, 118
nuclear overhauser effect (NOE), 260
number of folds, 183, 301
numeric knot descriptor, 220

oblate ellipsoids, 33
octahedron, 75, 77, 79, 82
off-lattice fold, 232
open chain topology, 218
optimized barrel, 129
origin of proteins, 303
origin symmetry, 290
orthogonal barrel, 127, 140

P-value, 174
packed layers, 9
packing classes, 9
packing density, 251
packing nomenclature, 82
packing stacked β-sheets, 114
paraboloid, 97
part barrels, 139
partial barrel, 85
parvalbumin, 265
Penrose patterns, 79
pepsin, 152, 154
peptide bond, 133
periodic minimal surface, 127
periodic table, 189, 197
periodic table of proteins, 187
pet hamsters, 71
phi-psi angles, 22
Phillips, David, 27
phosphate, 305
phycocyanin, 177
picture square, 80
ping-pong balls, 79
platonic solid, 75
polyhedra, 75, 80
polyhedral stick models, 73, 230, 264
polyhedron, 77, 230
polymer topology, 212
positive curvature, 137
POSSUM program, 169, 170
post-translational modification, 216
power spectrum, 285, 286
predicting helix packing, 89
probabilistic domain definition, 156
products of inertia, 29
projection, 40, 185
projection approach, 46, 249
projection into 2D, 85
prokaryotes, 296
prolate ellipsoids, 33
propellor fold, 246
proteasome, 289
protein architecture, 87, 200

protein classification, 180
protein data bank (PDB), 180
protein density, 33, 251
protein domain, 143, 188
protein domain definition, 143
protein fold, 180, 200, 238, 293
protein fold complexity, 238
protein fold space, 185
protein folding, 262
protein folding problem, 261
protein knots, 225
protein origins, 303
protein periodic table, 187
protein structure classification, 180
protein structure comparison, 159, 281
protein symmetry, 280
protein tertiary structure prediction, 261
protein topography, 281
protein topology, 13, 195
PROTEP program, 170
protoproteins, 304
proximity of termini, 295
pseudo-fivefold symmetry, 79
pseudo-knot, 223, 226, 238
pseudo-links, 215
pseudo-random walk, 243
pseudo-topology of proteins, 216
PSIPRED program, 302
Ptitsyn, Oleg, 69, 198

QL algorithm, 41
quadratic functions, 97
quadratic surface, 97, 118
quasi-crystals, 79
quasi-regular solids, 79

radius of gyration, 32, 257
ragged ribbon, 97, 126
Ramachandran plot, 23, 94
RAMBLE program, 232, 234, 303
rambling walk, 232
random chain, 227
random chain knots, 227
random folds, 249, 301
random model, 175
random proteins, 229, 286
random structural model, 174
random walk, 227, 232, 237, 264, 286, 290
random walk algorithm, 229
randomized alignment, 175
RASMOL, 19
recent intron, 296
recombination, 295
recursive β-sheet model , 96
recursive growth algorithm, 111
recursive growth model, 96
reducible pattern, 215
refinement approach, 46
reflected chain, 175
Reidemeister moves, 214
repeat boundary, 285
REPRO program, 288
residue density, 251
residue packing β-sheets, 106
residue packing in barrels, 123
retrovirus, 297
reversed chain model, 175, 179
ribbon representation, 19
ribonuclease inhibitor, 286
ribonucleic acid, 303
ribosome, 237, 303
ribozyme, 304, 305
ridges into grooves, 91
RMS deviation, 171, 173, 221, 233, 236
RMS superposition, 75
RMSD, 171, 173, 175, 181
RNA methyltransferase, 223
RNA structure, 226
RNA world, 297, 303
root-mean-square deviation, 171, 173, 181, 191, 223
Rossmann fold, 13, 243–245, 288, 289, 305
rugby-balls, 32

S-adenoyslmethionine synthetase, 221, 223
saddle shape, 97, 99
`SAPit` program, 167, 233, 281, 283, 285
SCOP database, 181, 183, 246
scorpion neurotoxin, 215
secondary structure, 8, 49, 50
secondary structure definition, 55, 134
secondary structure graph, 169
secondary structure in fake proteins, 230, 257
secondary structure lattice, 70, 232
secondary structure layer, 70, 172, 187, 200
secondary structure packing, 66
secret of life, 306
segment packing, 65
segmentation, 56
self-comparison, 281
semi-random walk, 233, 237
sequence alignment, 161
sequence repeat, 195
sequence space, 297
SET domain, 223, 226
Shakespeare sonnet, 302
shapes and sizes, 6
shapes of proteins, 32
shear and stagger, 120
shear number, 85, 120, 127, 138
sheet bend and curl, 111
sheet chirality, 93
shoelace, 227
significance of the RMSD, 175, 176
simplicial polyhedra, 73
simplified architecture, 63
simplified fold space, 187
simulated annealing, 161
skein move, 217
slip knot, 220
smoothed chain, 153, 187, 218, 233, 248
smoothed matrix, 282
smoothed protein structure, 134
smoothed traces, 20
smoothing algorithm, 218
snRNPs, 297
solenoid fold, 305
solvent accessible surface, 64, 91, 110, 252
sonnet structure, 302
spatial embedding, 212
spectrum, 286
spliceosome, 297
splines, 20
squirrel-cage, 71
SSAP program, 165, 182, 183, 185
stable marriage algorithm, 171
stacked β-sheet model, 113
stacked β-sheet packing, 114
stacked ring of β structure, 128
STAMP program, 164
starship ‘Enterprise’, 35
steric constraint, 251
stick model, 63, 70, 74, 169, 188, 230, 259, 267
stick packing, 63, 266
STICK program, 50, 61, 62
stick-figure, 63, 69, 170, 172, 187
stick-figure comparison, 170, 191
strand stagger, 120
string descriptor uniqueness, 87
string of beads, 227, 238
strophoid, 118
structural genomics, 208, 227
structural motif, 259
structural superposition, 164
structural symmetry, 290
structure classification, 180, 187
structure comparison methods, 159
structure comparison significance, 172
structure evolution, 161, 198, 293, 297
structure graph matching, 169
structure in sonnets, 302
structure layers, 188

structure prediction, 261
structure tree, 203
subalignments, 175
subgraph discovery, 170
subgraph isomorphism, 170, 212
subspace embedding, 40
subspace projection, 46
substructure matching, 201
summed area change, 64
super-folds, 183
super-secondary structure, 9, 93, 176, 290
superfamily, 183, 297
superposition, 161, 162, 164, 181, 259, 281
symmetric pattern, 215
symmetry, 241, 280
symmetry operations, 175

table of forms, 193, 200
tangle, 220, 248
TBC angle distribution, 136
TBC angles, 136
tertiary structure prediction, 261
tetrahedron, 75, 77, 79
threaded loop, 223, 224, 226
threading method, 265
TIM barrel, 12, 13, 85, 118, 120, 138, 140, 154, 184, 243, 244, 288, 289, 297, 300, 305
topological accessibility, 243
topological bond, 212
topological distance, 239
topological index, 238
topological invariant, 212
topological knot, 226, 238
topology, 195, 196, 211
topology cartoon, 86, 188
topology matching, 87
topology of fake proteins, 236
topology of open chains, 218
topology of proteins, 214
topology of weak links, 223
topology string, 87, 88, 195, 200, 201
tornado plot, 223, 240, 241
torsion angle, 22, 175
total power, 286
transform function, 258
transmembrane models, 71
trees of structures, 203
trefoil knot, 217, 221, 223
triangle inequalities, 40
triosephosphate isomerase, 12, 85, 118, 184
trypsin, 152, 154
twilight zone, 176
twist in sheets, 93
twist-bend-curl (TBC), 111, 136
twisted β-sheet, 70, 106
twisted lattice, 71
twisted layer model, 70
twisted ribbon, 71
twisted sheet atomic model, 109
twisted sheet model, 94
twisted sheet surface, 99

ubiquitin, 221
under-crossing, 217
uniqueness of folds, 195
unit sheet, 120
up-curling diagonal, 103
up-up radius, 105

valance shells, 188
VAST program, 173
vertex mobility, 80, 82
virtual bond angle, 23
virtual torsion angles, 23
von Willebrand factor, 289

wings of vertebrates, 294
wooden puzzle, 306

Z-value statistic, 173
Zimm–Bragg method, 144